Progress in Inflammation Research

Forthcoming titles:

Complement and Kidney Disease, P. F. Zipfel (Editor), 2005
Chemokine Biology – Basic Research and Clinical Application, Volume I: Immunobiology of Chemokines, B. Moser, G. L. Letts, K. Neote (Editors), 2005
Chemokine Biology – Basic Research and Clinical Application, Volume II: Pathophysiology of Chemokines, K. Neote, G. L. Letts, B. Moser (Editors), 2006
Toll-like Receptors in Inflammation, L. A. J. O'Neill, E. Brint (Editors), 2005
The Hereditary Basis of Rheumatic Diseases, R. Holmdahl (Editor), 2006

(Already published titles see last page.)

The NPY Family of Peptides in Immune Disorders, Inflammation, Angiogenesis and Cancer

Zofia Zukowska
Giora Z. Feuerstein

Editors

Birkhäuser Verlag
Basel · Boston · Berlin

Editors

Zofia Zukowska
Department of Physiology and Biophysics
Georgetown University Medical Center
3900 Reservoir Rd., NW,
Washington, DC 20057
USA

Giora Z. Feuerstein
Wyeth Research
Building A3153
500 Arcola Rd
Collegeville, PA 19426
USA

Library of Congress Cataloging-in-Publication Data

The NPY family of peptides in immune disorders, inflammation, angiogenesis and cancer /
Zofia Zukowska, Giora Z. Feuerstein, editors.
p. ; cm. -- (Progress in inflammation research)
Includes bibliographical references and index.
ISBN-13: 978-3-7643-7159-3 (alk. paper)
ISBN-10: 3-7643-7159-5 (alk. paper)
1. Neuropeptide Y--Pathophysiology. I. Zukowska, Zofia. II. Feuerstein, Giora Z., 1946-.
III. PIR (Series)

QP552.N38N4767 2005
612'.015756--dc22

20050053007

Bibliographic information published by Die Deutsche Bibliothek
Die Deutsche Bibliothek lists this publication in the Deutsche Nationalbibliografie;
detailed bibliographic data is available in the internet at http://dnb.ddb.de

ISBN-10: 3-7643-7159-5 Birkhäuser Verlag, Basel – Boston – Berlin
ISBN-13: 978-3-7643-7159-3 Birkhäuser Verlag, Basel – Boston – Berlin

Part of Springer Science+Business Media
Printed on acid-free paper produced from chlorine-free pulp. TCF ∞
Cover design: Markus Etterich, Basel
Cover illustration: Figure from Li L, Lee EW, Ji H, Zukowska Z (2003) Neuropeptide Y-induced acceleration of post-angioplasty occlusion of rat carotid artery. *ATVB* 23: 1204–1210; with kind permission of the publisher (LWW).
Printed in Germany
ISBN-10: 3-7643-7159-5
ISBN-13: 978-3-7643-7159-3

e-ISBN: 3-7643-7427-6

9 8 7 6 5 4 3 2 1

www.birkhauser.ch

Contents

List of contributors

Ken Abe, Department of Physiology & Biophysics, Georgetown University Medical Center, 3900 Reservoir Rd NW, Washington, DC 20057, USA

Sammy Bedoui, Hannover Medical School, Department of Functional and Applied Anatomy, OE4120, Carl-Neuberg Str. 1, 30625 Hannover, Germany; e-mail: Bedoui.Sammy@web.de

Magnus M. Berglund, Biovitrum AB, Department of Lead Discovery, Stockholm, Sweden

Meit Björndahl, Karolinska Institutet, Microbiology and Tumor Biology Center, 17177 Stockholm, Sweden; e-mail: meit.bjorndahl@mtc.ki.se

Noureddine Brakch, Lausanne University Hospital School, Division of Angiology, Lausanne, Switzerland

Renhai Cao, Karolinska Institutet, Microbiology and Tumor Biology Center, 17177 Stockholm, Sweden; e-mail: renhai.cao@mtc.ki.se

Yihai Cao, Karolinska Institutet, Microbiology and Tumor Biology Center, 17177 Stockholm, Sweden; e-mail: yihai.cao@mtc.ki.se

Mónica De la Fuente, Department of Physiology (Animal Physiology II), Faculty of Biological Sciences, Complutense University, 28040 Madrid, Spain; e-mail: mondelaf@bio.ucm.es

Elena Dozio, Università degli Studi di Milano, Istituto di Endocrinologia, via G. Balzaretti, 9, 20133 Milano, Italy

Giora Z. Feuerstein, Wyeth Research, Building A3153, 500 Arcola Rd, Collegeville, PA 19426, USA; e-mail: FEUERSG@wyeth.com

Donald R. Gehlert, Eli Lilly and Company, Lilly Research Laboratories, Lilly Corporate Center, Indianapolis, IN 46285, USA; e-mail: gehlert_donald_r@lilly.com

William P. Gray, Division of Clinical Neurosciences, University of Southampton, Room 6207, Biomedical Sciences Building, Bassett Crescent East, Southampton SO16 7PX, UK; e-mail: w.p.gray@soton.ac.uk

Eric Grouzmann, CHUV, Division of Clinical Pharmacology and Toxicology, 1011 Lausanne, Switzerland; e-mail: eric.grouzmann@chuv.hospvd.ch

Philip A. Hipskind, Eli Lilly and Company, Lilly Research Laboratories, Lilly Corporate Center, Indianapolis, IN 46285, USA

Joanna B. Kitlinska, Department of Physiology & Biophysics, Georgetown University Medical Center, 3900 Reservoir Rd NW, Basic Science Building, Washington, DC 20057, USA; e-mail: jbk4@georgetown.edu

Lydia E. Kuo, Department of Physiology & Biophysics, Georgetown University Medical Center, 3900 Reservoir Rd NW, Basic Science Building, Washington, DC 20057, USA

Jean-Silvain Lacroix, Unité de Rhinologie-Olfactologie, Service d'Oto-Rhino-Laryngologie, Hôpitaux Universitaires de Genève, Rue Micheli-du-Crest 24, 1211 Geneva, Switzerland; e-mail: silvain.lacroix@hcuge.ch

Basile N. Landis, Unité de Rhinologie-Olfactologie, Service d'Oto-Rhino-Laryngologie, Hôpitaux Universitaires de Genève, Rue Micheli-du-Crest 24, 1211 Geneva, Switzerland

Dan Larhammar, Department of Neuroscience, Unit of Pharmacology, Uppsala University, Box 593, 75124 Uppsala, Sweden; e-mail: Dan.Larhammar@neuro.uu.se

Edward W. Lee, Department of Physiology & Biophysics, Georgetown University Medical Center, 3900 Reservoir Rd NW, Washington, DC 20057, USA;
e-mail: leeew@georgetown.edu

Lijun Li, Department of Physiology & Biophysics, Georgetown University Medical Center, 3900 Reservoir Rd NW, Washington, DC 20057, USA

Paolo Magni, Università degli Studi di Milano, Istituto di Endocrinologia, via G. Balzaretti, 9, 20133 Milano, Italy; e-mail: paolo.magni@unimi.it

Sonia Medina, Department of Physiology (Animal Physiology II), Faculty of Biological Sciences, Complutense University, 28040 Madrid, Spain;
e-mail: smedina@terra.es

Rolf Mentlein, Department of Anatomy, University of Kiel, Olshausenstrasse 40, 24098 Kiel, Germany; e-mail: rment@anat.uni-kiel.de

Marcella Motta, Università degli Studi di Milano, Istituto di Endocrinologia, via G. Balzaretti, 9, 20133 Milano, Italy

Isabelle Plouin-Gaudon, Unité de Rhinologie-Olfactologie, Service d'Oto-Rhino-Laryngologie, Hôpitaux Universitaires de Genève, Rue Micheli-du-Crest 24, 1211 Geneva, Switzerland

Jennifer Pons, Department of Physiology & Biophysics, Georgetown University Medical Center, 3900 Reservoir Rd NW, Basic Science Building, Washington, DC 20057, USA

Massimiliano Ruscica, Università degli Studi di Milano, Instituto di Endocrinologia, via G. Balzaretti, 9, 20133 Milano, Italy

Erik Salanek, Department of Neuroscience, Unit of Pharmacology, Uppsala University, Box 593, 75124 Uppsala, Sweden

Helen E. Scharfman, Helen Hayes Hospital, New York State Department of Health, Columbia University, New York, NY, USA

Bradley K. Taylor, Department of Pharmacology, SL83, Tulane University University Health Sciences Center, New Orleans, LA 70112, USA;
e-mail: taylorb@tulane.edu

Jason U. Tilan, Department of Physiology & Biophysics, Georgetown University Medical Center, 3900 Reservoir Rd NW, Washington, DC 20057, USA

Stephan von Hörsten, Hannover Medical School, Department of Functional and Applied Anatomy, OE4120, Carl-Neuberg Str. 1, 30625 Hannover, Germany;
e-mail: hoersten.stephan.von@mh-hannover.de

Luxun Xue, Zhenzhou University Cancer Center, Zhenzhou, Henan, China

Zofia Zukowska, Department of Physiology & Biophysics, Georgetown University Medical Center, 3900 Reservoir Rd NW, Washington, DC 20057, USA;
e-mail: zzukow01@georgetown.edu

Introduction

The NPY family of peptides: from neurotransmitters and hormones to immune modulators, cytokines and growth factors

Zofia Zukowska[1] *and Giora Z. Feuerstein*[2]

[1]Zofia Zukowska, Department of Physiology and Biophysics Georgetown University Medical Center, 3900 Reservoir Rd., NW, Washington, DC 20057, USA; [2]Giora Z. Feuerstein, Wyeth Research, Building A3153, 500 Arcola Rd, Collegeville, PA 19426, USA

"Nothing in biology makes sense except in the light of evolution"

Theodosius Dobzhansky

In light of this motto by a famous evolutionist, NPY and its related peptides, peptide YY and pancreatic polypeptide, are a highly conserved family of genes and proteins. Evolutionarily ancient and structurally conserved, these peptides regulate fundamental cellular processes of growth, metabolism and cell-cell communication, and orchestrate the body's integrated responses to ensure survival and adaptation to the changing demands of the internal and external environment.

In this two-volume set* on NPY biology in 'Health and Disease' the editors' goal is to provide a comprehensive state-of-the-art review of the field – to discuss the frontiers of NPY peptides' research and drug development, and lay the groundwork for new directions and therapeutic opportunities in the next decade. This book focuses on the novel role of NPY peptides in inflammation, immune disorders, angiogenesis and proliferative diseases, which grew out of more 'traditional' fields of research such as neuroscience, cardiovascular and endocrine biology. The latter are covered in the EXS volume, which addresses the NPY family of peptides in the cardiovascular, metabolic and central nervous system diseases. For completeness and to grasp the full breath of the field, we strongly encourage the readers to familiarize themselves with the complementary information contained in this second volume. Additional information and in-depth analysis of particularly neuroendocrine, molecular and pharmacologic aspects of NPY research can also be found in such previous texts as *Neuropeptide Y and Related Peptides*, edited by M.C. Michel, and many excellent recently published reviews on NPY.

* Zofia Zukowska, Giora Z. Feuerstein (eds) *NPY Family of Peptides in Neurobiology, Cardiovascular and Metabolic Disorders: from Genes to Therapeutics*. EXS series, Birkhäuser Verlag, Basel, ISBN: 3-7643-7155-2

Originally described as an ancillary, sympathetic cotransmitter that modulates the release and action of the primary neurotransmitter, norepinephrine, NPY and its family members, have recently emerged as pleiotropic regulators of biological functions such as:

1. energy and metabolism
2. neural and endocrine modulation
3. immune surveillance and inflammatory reactions
4. tissue growth, regeneration and remodeling
5. integrated responses to stress

Such diverse functions may serve a multitude of survival and adaptive behaviors such as hibernation (surviving in hostile cold conditions), defense and aggression (integrated neuroendocrine, cardiovascular and skeletomuscular responses), reproduction and foraging resources. The authors present key elements of recent scientific explorations to support the notion that peptides from the NPY family act not only as neurotransmitters and neuroendocrine modulators but also as immune and inflammatory modulators, cytokines and growth factors.

Setting the tone of this volume, Larhammar and Salaneck present a comprehensive review of the evolution of the NPY, PYY and PP, and their receptors (*Section I*). The readers will learn how these genes evolved and how their functions changed, or, in some remarkable examples such as food intake, appeared to have stayed the same in non-vertebrates and vertebrates. The authors also address redundancies inherent in this family of peptides and receptors, discuss their functional implications, and point to possibilities of new activities, such as control of cell growth and inflammation, based on the evolutionary history and chromosomal location of these genes.

The following chapter (*Section II*) focuses on the molecular pharmacology of the NPY family of peptides' receptor systems. This section provides a foundation for understanding the diversity of receptor function, signaling and regulation, and their spatial distribution in the central and peripheral nervous system and gut. The comprehensive review of NPY/PYY/PP receptor biology is supplemented by a thorough discussion of present and future therapeutic perspectives.

The next two chapters in this section contain more discrete elaborations on NPY/PYY peptide processing and metabolism, illustrating the richness of functional diversification in this family of peptides. Unlike in the case of classical neurotransmitters, such as acetylcholine, which is enzymatically inactivated proteolytic cleavage of the NPY peptide family, both the precursors and the mature peptides, NPY1-36 and PYY 1-36, yield fragments that exercise novel biological functions. How this feature of the system offers possibilities of new therapeutics by manipulation of specific peptidases is discussed here.

Section III of the book highlights the newly discovered role of the NPY system in immune surveillance and inflammatory reactions. This section illuminates the potential for this family of peptides to link numerous stress responses to host defense mechanisms, which may lead to resolution of diseases or their exacerbation. In particular, the role of NPY and its receptors in neurogenic inflammation, pain, and stress-dependent immune diseases are reviewed, and innovative opportunities to treat host-defense pathological reactions via selective NPY receptor and enzyme processing drugs are discussed.

Section IV deals with the most recent information regarding the potential role of the NPY family of peptides in regulation of vascular growth and angiogenesis. This new area of NPY research offers great potential in ischemic re-vascularization, stem cell biology and organ remodeling and regeneration. Its therapeutic implications span from the treatment of ischemic diseases, wound healing and retinopathy to atherosclerosis and hypertension.

Section V extends this theme and completes the book with highlights on potent growth factor activities of the NPY family members in nerve regeneration and plasticity, and in tumors. Remarkable progress has been made in these fields and new therapeutics have emerged for the treatment of certain malignant tumors as well as neurodegenerative diseases.

Biology of the NPY family of peptides, their receptors and processing enzymes

Evolutionary perspective on the NPY-PYY-PP peptides and their receptors

Dan Larhammar and Erik Salaneck

Department of Neuroscience, Unit of Pharmacology, Uppsala University, Box 593
75124 Uppsala, Sweden

Introduction

The information available in the scientific literature on the neuropeptide Y (NPY) system is primarily focused on mammals for obvious reasons. Both NPY and the related peptide YY (PYY) were initially discovered in the pig in 1982 [1, 2] and purification of the peptides soon followed in many other mammals. In parallel, NPY and PYY were also purified from the brain and other tissues of several more distantly related vertebrates including many species of fish as well as frogs and birds. Studies in these non-mammals revealed not only that the peptides of the NPY family arose before the origin of gnathostomes (jawed vertebrates), but also that the peptides had remained highly conserved over more than 450 million years of evolution. The strong conservative selection pressure suggested that the peptides were involved in important functions [3–5].

Which functions had kept the peptides so well conserved? Physiological studies in mammals have provided an extremely long list of effects of NPY and PYY. NPY occurs predominantly in the central nervous system and in peripheral nerves, primarily in blood vessels. Among the most prominent effects of NPY in mammals are vasoconstriction, stimulation of appetite, and influence on circadian rhythm, pain, anxiolysis and the release of pituitary hormones [6]. PYY is primarily produced in gut endocrine cells and is released after meals contributing to inhibition of gut motility [7] and pancreatic secretion [8]. More recently PYY has been found to reach the hypothalamus and reduce appetite [9].

The third member of the family, pancreatic polypeptide (PP), is found in tetrapods (mammals, birds, reptiles and amphibians) and is much more variable across species, although reasonably well conserved among mammals. PP was initially discovered in the chicken in 1975 [10] and is now known to have arisen through duplication of the PYY gene [11] in an early tetrapod ancestor [5]. In mammals, PP is released from pancreatic F cells after meals and shares the effects of PYY by reducing pancreatic secretion and gut motility [12] as well as appetite [13]. Thus, within the same peptide family, there are two peptides that evolve very slowly (NPY

The NPY Family of Peptides in Immune Disorders, Inflammation, Angiogenesis and Cancer,
edited by Zofia Zukowska and Giora Z. Feuerstein

and PYY) whereas the third (PP) has had an unusually high rate of evolutionary change, and one peptide has a prominent appetite-stimulating effect (NPY), while the other two (PYY and PP) have the opposite effect. Such opposing effects among members of a family of evolutionarily related genes constitute intriguing challenges for evolutionary biologists to resolve. Comparative studies among different extant animal groups are probably the best way to deduce how such functional differences may have arisen during evolution.

In order to trace the origin and the primordial functions of the common ancestor of NPY, PYY and PP, our laboratory began several years ago to clone the peptide precursors in a variety of vertebrate species, not only to deduce the peptide sequences but also to generate tools for mapping of mRNA distribution. In addition, the DNA sequences would allow the synthesis of endogenous peptides for physiological experiments in various species. We started by cloning the peptide precursors in species which are common experimental animals and for which DNA libraries were available, namely chicken, goldfish and torpedo ray [14], then zebrafish and river lamprey [15, 16]. Subsequently, several other labs have added information for additional species (see [17]). We mapped the anatomical distribution of peptide mRNA by *in situ* hybridisation which showed that not only NPY but also PYY was expressed in the nervous system, in addition to the PYY expression observed in gut endocrine cells [15, 16]. This neuronal localisation of PYY had previously been equivocal using immunohistochemical techniques due to the cross-reactivity of many antisera. Our sequence comparisons across species explained why antisera tended to cross-react, namely due to the high sequence identity between NPY and PYY in many species [18, 17].

Our results indicated that the ancestral NPY/PYY peptide might have been expressed in both neurons and gut endocrine cells. This suggested that studies in the more distantly related lineages would perhaps not reveal one or a few ancestral functions, but rather an ancestral peptide with a broader functional repertoire. More recent comparisons of duplicated genes with species that still have a single gene have shown that new gene duplicates may not necessarily evolve new functions, but may instead subspecialise on some of the functions of mother gene's repertoire. This process has been termed subfunctionalisation [19].

Another major approach to investigate the evolution of NPY-family functions is to study the peptide receptors, the Y receptors. The first mammalian NPY receptor, corresponding to the pharmacologically defined Y1 subtype, was cloned in the rat [20, 21] and human [22, 23]. As expected, this Y receptor like all other more recently cloned Y receptors belong to the superfamily of rhodopsin-like G-protein-coupled receptors and seem to function primarily by inhibiting cAMP accumulation. The second pharmacologically defined Y receptor, Y2, was soon to be cloned as were the more unanticipated subtypes Y4, a PP-binding receptor, Y5 and Y6 (see [24] for references). Our lab proceeded to clone these and novel subtypes (Y7 and Y8) in a large number of non-mammalian vertebrates (see below).

The results have revealed a much more complicated evolutionary history than we had ever imagined. Most surprising was the conclusion that the receptor repertoire was already quite complex in the common ancestor of tetrapods (including humans), bony fishes and cartilaginous fishes some 450 million years ago with no less than seven distinct receptor subtypes. After this burst of additional gene copies, the trend has been to lose receptors, and the losses differ between evolutionary lineages. Humans have retained only four of the ancestral seven receptors.

In this review we will describe some of the major events in the evolution of the NPY system. After a brief overview of invertebrate NPY-like peptides, we will review the many gene duplications and gene losses that have taken place among vertebrates and discuss the implications for the functions of the NPY system during evolution.

Invertebrate NPY and receptors

A true orthologue of the vertebrate NPY system has been difficult to identify in invertebrates. This is because of the great difficulties to compare sequences for short peptides over great evolutionary distances [25], as similarity may also arise through convergent evolution. For instance, numerous peptides in distantly related phyla contain the carboxyterminal sequence Arg–Tyr-amide (or Arg–Phe-amide), or with the single-letter codes, RY-amide or RF-amide, like the vertebrate NPY-family peptides do. It would seem unlikely that all such peptides share a common ancestor. Rather, many of them have probably arisen independently and found G-protein-coupled receptors to which they can bind. However, if an invertebrate species has an NPY-like peptide that binds to a receptor that resembles the vertebrate Y receptors this strongly suggests that the peptide-receptor pair shares a common ancestry with the vertebrate NPY system.

A ligand-receptor pair with similarity to vertebrate NPY-like peptides and Y receptors was first described for the pond snail *Lymnaea stagnalis* [26]. The peptide is 39 amino acids long and shares 11 of the 36 amino acids (31%) with the proposed ancestral gnathostome NPY [3]. The *Lymnaea stagnalis* Y receptors is approximately 30% identical to the different mammalian Y receptor subtypes. Intriguingly, a pond snail parasite, the schistosome *Trichobilharzia ocellata*, can upregulate NPY production in the snail which may lead to increased host body growth and thereby benefit the parasite [27].

In insects the search for NPY and Y receptor orthologues has been more problematic. However, recently a *Drosophila melanogaster* receptor was described [28] that bound with high affinity to an endogenous peptide called NPF with sequence similarity to vertebrate NPY [29]. Accordingly, the receptor is called the NPF receptor (NPFR). It has been reported that larval NPF expression drops coincident with reduction of feeding, and that larvae engineered to overexpress NPF display prolonged foraging behaviour [30], thereby probably confirming orthology with the mammalian NPY system. The orthologous NPF peptide has been identified in the yellow fever mosquito, *Aedes aegypti* [31], the African malaria mosquito, *Anopheles gambiae* [32] and the silk worm *Bombyx mori* [33] and NPFR has been reported in *Anopheles gambiae* [34]. Other Drosophila melanogaster receptors with some sequence similarity to the vertebrate Y receptors have been reported, but their relationships to the vertebrate NPY system nevertheless appear questionable, namely PRF [35] and sNPFR [36, 37].

The nematode *Caenorhabditis elegans* has been reported to have a Y-like receptor called NPR-1 involved in feeding behaviour as shown by influence of feeding by allelic variation in the receptor sequence [38]. However, this receptor has subsequently been found to bind a peptide called AF9 with no similarity to NPY [39].

Vertebrate NPY-family peptides

The NPY-family peptides consist in vertebrates of 36 amino acids, with some rare exceptions [17]. All peptides have a carboxyterminal amide group. Due to the great differences in evolutionary rate between NPY, PYY and PP it was not possible to deduce their evolutionary relationships until sequences had been determined from a large number of species [3, 18]. The picture still remains somewhat unclear due to the presence of additional peptides in some species [5, 40, 41], but overall the analyses point to a single gene in the chordate ancestor followed by duplication at an early stage in a vertebrate ancestor, because what seems to be NPY and PYY exist as separate gene products even in lampreys which belong to the class cyclostomes [15]. The origin of the second lamprey PYY gene [16, 42–44] has not yet been resolved, except that it seems to have appeared early in the lamprey lineage.

The mechanism for the initial peptide gene duplication seems to be a chromosome duplication, because NPY and PYY (+PP) are located in gene regions that share members of a large number of gene families (Fig. 1). The two genes are located close to the HOXA and HOXB clusters on chromosomes 7 and 17, respectively, in the human genome [11, 45] as well as in zebrafish [16]. This suggests that this whole block, or the entire chromosome, was duplicated before the radiation of the vertebrate species [16, 18]. A detailed examination of these chromosomal regions in the human genome database provides further support for this hypothesis, as several additional genes have been discovered that belong to the same duplicated block (Fig. 1). However, numerous local rearrangements have subsequently changed the gene order and orientation.

The location of NPY and PYY close to two HOX clusters suggests that two additional peptide genes probably existed adjacent to HOXC and chromosome 12 and HOXD on chromosome 2, but that these postulated peptide genes may have been lost (Fig. 1). Such a set of paralogous chromosome regions with members from the same gene families has been named a paralogon [46]. The human genome contains

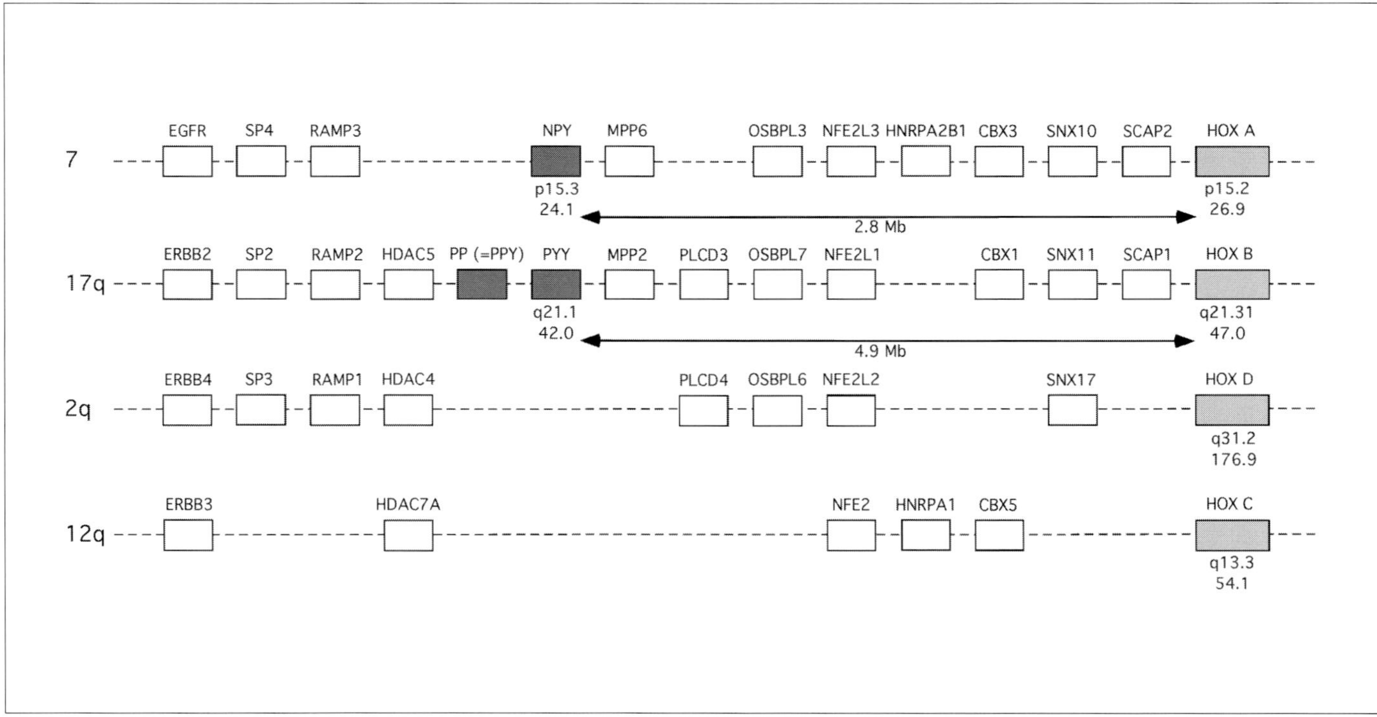

Figure 1
Schematic representation of the two chromosomal regions containing the NPY and PYY-PP genes and the homeobox gene clusters HOXA and HOXB, as well as the two related chromosomal regions harbouring the HOXC and HOXD clusters. The chromosomes are shown as dashed lines to indicate that they are not drawn to scale and that the physical gene order has been altered to highlight similarities between the four related chromosomes forming this paralogon. Abbreviations are those used in the Ensembl database.

several such paralogons [47, 48] which support the suggestion that the entire vertebrate genome underwent at least one and probably two tetraploidisations at an early stage [49–51].

The gene duplication that generated NPY and PYY apparently allowed specialisation or subfunctionalisation, as has been observed for other daughter genes after duplication [19, 52]. The specialisation of NPY on neuronal and PYY on endocrine functions may be most extreme in mammals, while some overlap still seems to exist in other vertebrates such as fishes, because zebrafish PYY was found to have extensive neuronal expression [16]. It remains virtually unexplored exactly which functions are carried out by NPY and which are performed by PYY (or both) in non-mammals due to the high sequence identity of the peptides.

Extensive genetic evidence now exists supporting an additional tetraploidisation in teleost fish evolution [53, 54]. Among the many gene families that provide evidence for this third tetraploidisation are again the HOX clusters which exist in 7–8 copies in the teleost fish that have been thoroughly investigated [55, 56]. This additional tetraploidisation and preservation of the extra HOX clusters would be expected to have generated additional copies of NPY and PYY. This is indeed the case: the pufferfish *Takifugu rubripes* and *Tetraodon nigriviridis* both possess two NPY genes and two PYY genes, whereas the zebrafish *Danio rerio* may have lost the NPYb gene [57]. All four genes in *Takifugu rubripes* are expressed as mRNA (B. Venkatesh and S. Brenner, personal communication) and appear to be functional. PYY is expressed in both endocrine cells and neurons [15, 58, 59], as is the NPY-related peptide NPF in fly and mosquito (see above), which supports the hypothesis that the ancestral vertebrate NPY/PYY-like peptide had broad expression.

The PP gene, as mentioned in the introduction, arose through tandem duplication of the PYY gene [11] early in the tetrapod lineage [5] before it split into amphibians and amniotes (reptiles, birds and mammals). The tetrapod origin is supported by the absence of PP in the fish genomes that have been sequenced. The duplication of PYY to generate PP may have led to functional specialisation with PP becoming the predominant peptide in the pancreas while PYY is expressed in endocrine in the intestine and colon [60]. It will be interesting to see whether the pattern is the same in all classes of tetrapods.

A few additional gene duplications have taken place for PYY and PP in various species and lineages, as reviewed previously [5, 40, 41]. Some of these duplicates have degenerated into pseudogenes whereas others are expressed. It still remains to be investigated whether the expressed PYY duplicates have acquired novel functions or become more specialised.

Vertebrate Y receptors

Five receptor subtypes exist in most mammals and are numbered from Y1 to Y6 in the chronological order of their discoveries. The Y family comprises one of the largest families of G-protein-coupled peptide receptors. During the early stages of Y receptor characterisation, pharmacological evidence was presented for a unique subtype and the postulated receptor was named Y3 [61]. Later it became evident that this receptor is not encoded by a separate gene, but by this time the nomenclature of the remaining subtypes had already been defined. The cloned receptor proposed to be a Y3 subtype [62] was found not to respond to NPY-family peptides [63, 64]. The Y3 binding and response profiles observed *in vivo* and *ex vivo* still remain to be accounted for in molecular terms. Also the Y6 gene is missing from the rat genome and is a pseudogene in several other mammals [65]. As will be discussed below, the redundancy of the Y6 subtype appears to be confined to the mammalian lineage.

The cloning of the various mammalian receptor subtypes has been described in several reviews [24]. Phylogenetic analyses of receptor sequences (Fig. 2) show that the Y1, Y4 and Y6 sequences are about 50% identical to each other and form a subfamily within the Y receptor family, called the Y1 subfamily. Y2 and Y5 exhibit about 30% identity to each other and to all members of the Y1 subfamily which indicates that the duplications that generated Y1–Y2–Y5 must be more ancient than the duplications that produced the members of the Y1 subfamily [41, 66, 67] (see Figs 2 and 3).

The Y1, Y2 and Y5 genes are located within a relatively short segment of chromosome 4 in the human genome, while Y4 and Y6 are located on two separate chromosomes, chromosome 10 and 5, respectively [67, 68] (Fig. 3). Based on these two types of information, i.e., phylogenetic analyses and chromosomal location, a likely scenario for the expansion of the gene family would be that two local duplications of an ancestral Y receptor gene resulted in a chromosomal segment harbouring three Y genes, the ancestors of Y1, Y2 and Y5 (Fig. 4). Later, this entire segment would have been duplicated twice resulting in the more closely related members of the Y1 subfamily. As will be discussed below, gene losses have occurred for both the Y1-subfamily and the novel subtype Y7 and the losses differ among the vertebrate lineages. This scenario is in agreement with the so-called 2R hypothesis that proposes that two large-scale or total genome duplications (tetraploidisations) occurred during early vertebrate evolution, some time prior to the appearance of jawed vertebrates some 450–500 million year ago [47, 48, 50, 69]. In the case of the Y family, the fourth chromosomal segment harbouring Y receptor genes seems to have been completely lost in mammals (cf the loss of the presumed extra duplicates of NPY and PYY discussed above), but not in other vertebrate classes.

All mammals have the genes for the Y1, Y2, Y4 and Y5 subtypes [41, 67], while the Y6 gene has been lost or has been rendered non-functional due to mutation in certain mammalian species [65]. The species in which a functional Y6 subtype is missing are spread throughout the mammalian clade. Surprisingly, the loss of function seems to be due to distinct mutations in primates [70–72], pig [68], and guinea

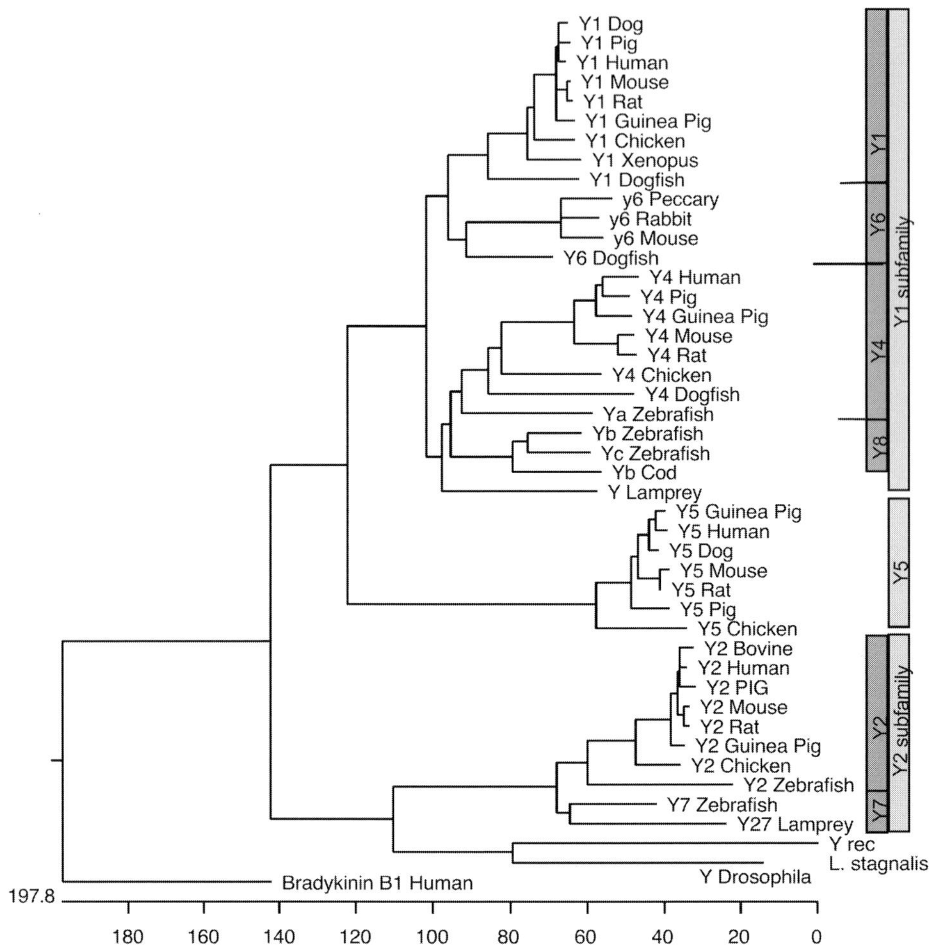

Figure 2
*A phylogenetic tree showing the relationships of Y receptor sequences generated with the Neighbor–Joining method. Three major branches are formed by the Y1 subfamily, the Y2 subfamily, and the Y5 sequences, respectively. The subfamilies and subtypes are indicated on the right. The distantly related snail NPY receptor (*Lymnaea stagnalis*) and fly NPFR (*Drosophila melanogaster*) sequences appear in this tree together with the Y2 subfamily, but their position depends on the phylogenetic method and the outgroup used to root the tree. In this case, the human bradykinin B1 sequence was used as outgroup.*

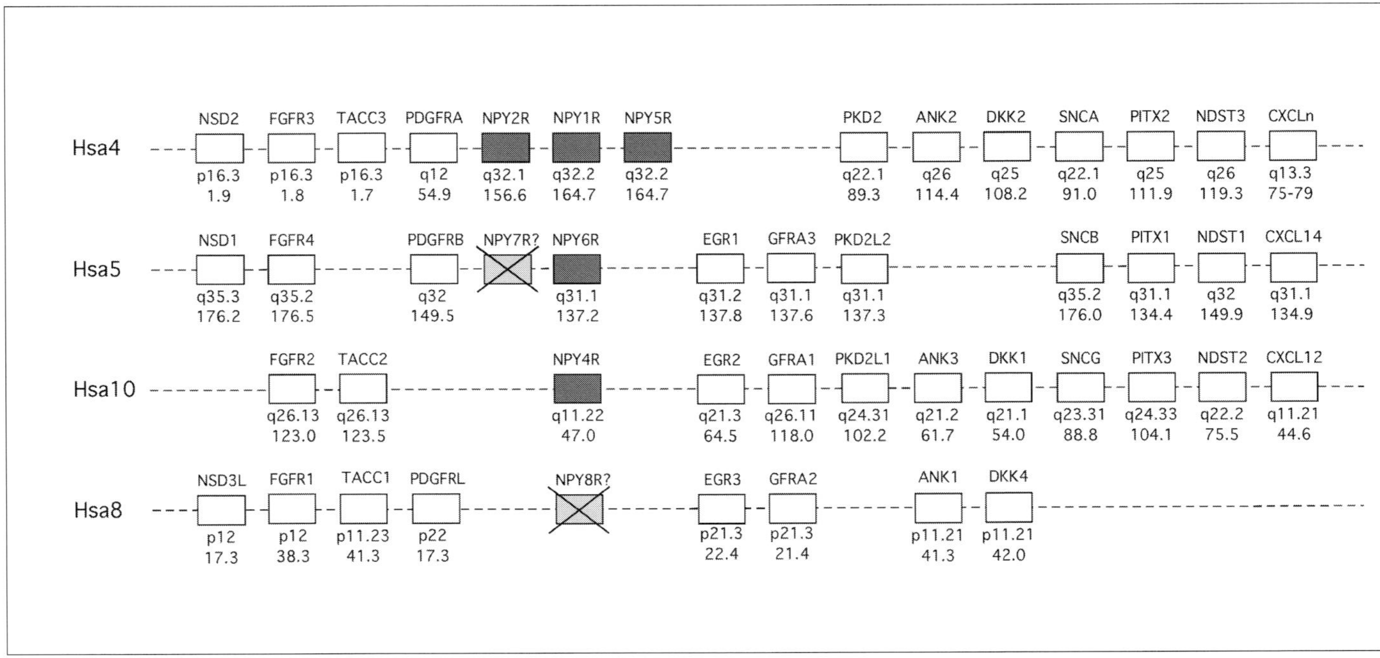

Figure 3
Schematic representation of the three chromosomal regions containing the Y genes as well as a fourth chromosome (Hsa8) that also belongs to this paralogon. The Y1 receptor gene is called NPY1R, Y2 is NPY2R, etc. The hypothetical position for the Y7 gene thought to once have existed is shown and crossed-out on Hsa5. The chromosomes are shown as dashed lines to indicate that they are not drawn to scale and that the physical gene order has been altered to highlight similarities. Abbreviations are those used in the Ensembl database.

pig [65] while the rat lacks the gene altogether [73]. This suggests that the gene most likely lost importance in the mammalian ancestor and because of this redundancy became prone to mutation. In rabbit and mouse the Y6 gene appears to be functional [74, 75]. In the collared peccary, a relative of the domestic pig (separated by approximately 30 million years of evolution), the Y6 gene also appears to code for a functional transcript [76], although its possible function is not known.

The functions of each of the receptor subtypes has been mostly studied in mammals (primarily rodents) and have been recently reviewed [77]. In the periphery, Y1 mediates vasoconstriction [78] and in the hypothalamus, along with Y5, it stimulates food intake. Y1 also participates in many other contexts such as regulation of circadian rhythm, hippocampal activity, anxiolysis, pituitary function, pain, etc [6, 77]. The distribution of the Y1 and Y5 subtypes differs considerably, with Y5 being expressed only in the central nervous system [79, 80], and Y1 exhibiting a much more widespread distribution [81, 82]. Interestingly, Y5 seems to be expressed only in cells that also express Y5 [83], suggesting coordinated regulation of expression through their overlapping promoter regions [84]. In contrast to Y1 and Y5, the subtypes Y2 and Y4 inhibit food intake [9, 13, 85]. Subtype Y2 is also involved in many other functions, primarily as a presynaptic receptor mediating auto-feedback and thereby having opposite effects of Y1, for instance on vasoconstriction and anxiety. Y2 also stimulates angiogenesis [86] and influences circadian rhythm [87, 88].

The repertoire of Y receptors in the avian lineage is the same as that of mammals, with one interesting difference. Orthologues of all of the mammalian subtypes have been identified in the domestic chicken [89–91] (Fredriksson and Larhammar, unpublished). The exciting difference is that chicken has retained the ancient duplicate of Y2, namely the Y7 gene (Bromée, Sjödin and Larhammar, unpublished), which has been lost in mammals. Its properties are presently being investigated. The chicken orthologues of the functional mammalian subtypes are quite well conserved and have similar pharmacological binding profiles. The major difference is that the Y4 receptor is not as PP-selective in mammals, but responds to both PYY and PP. As the Y4 receptor existed before PP was formed, the chicken Y4 receptor presumably maintains more of the ancestral PYY-binding properties. In contrast to the observation that the Y6 subtype is non-functional in several mammals, it appears to be functional in chicken which is supported by a slower rate of evolution than its mammalian orthologue (Fredriksson and Larhammar, unpublished). The chicken Y6 mRNA is expressed in the hypothalamus but its function and pharmacological properties are still unknown.

Amphibians have all of the mammalian subtypes and in addition Y7, as in chicken, and also the Y1-subfamily receptor previously described in fish, namely Y8, previously called Yb [92]. Only the Y1, Y2 and Y7 sequences have yet been published [93, 94] and work is in progress to characterise all receptors. The large amphibian Y receptor repertoire obviously strengthens the 2R hypothesis and supports the early origin of the Y genes lost in mammals.

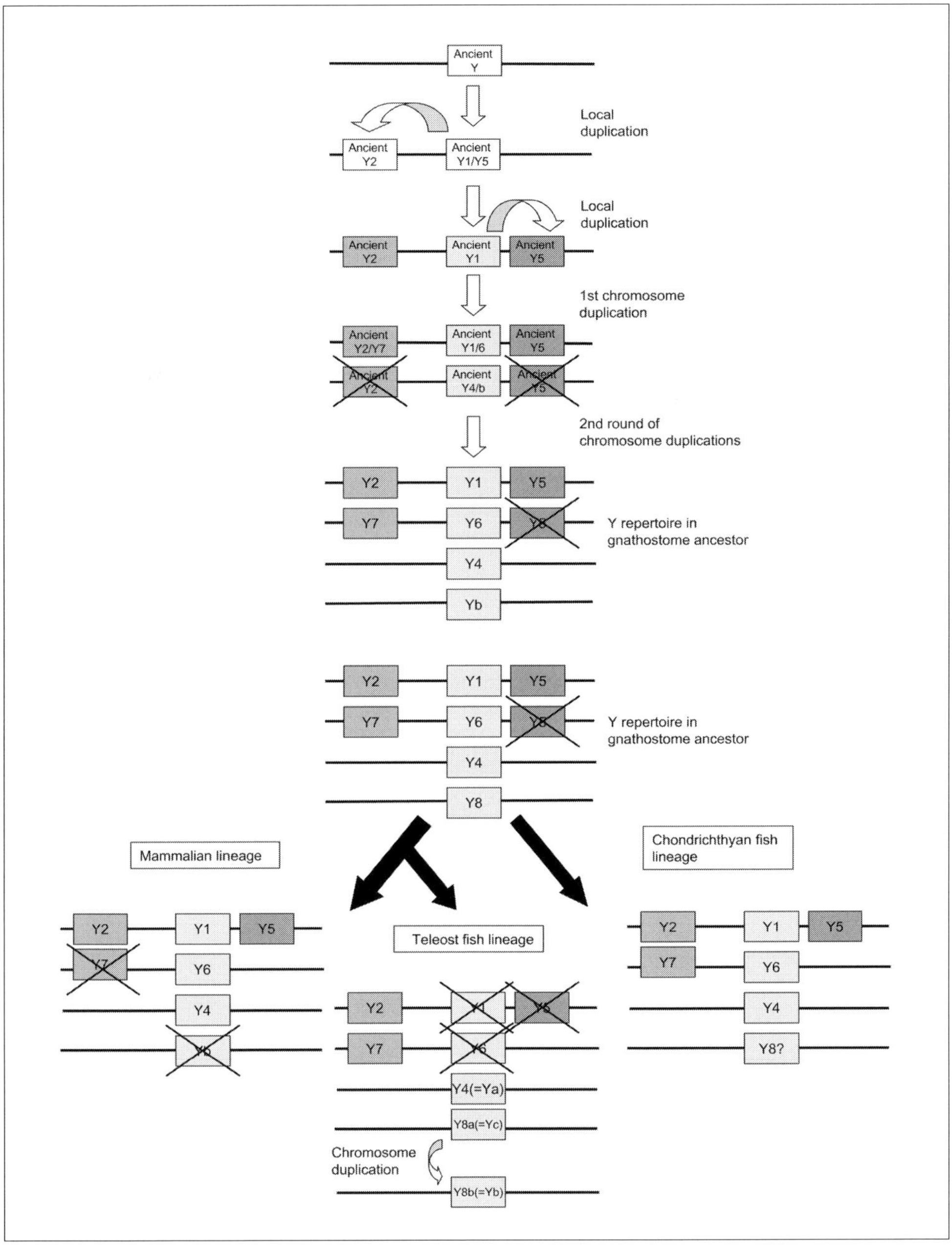

Figure 4
Proposed gene duplication scheme for the Y receptor family. The crossed-out genes were probably lost after each duplication event

In addition to the two proposed early vertebrate tetraploidisations, a subsequent third tetraploidisation has occurred in a teleost, or ray-finned fish, ancestor. Genome sequencing projects of several teleost fishes such as zebrafish, fugu and medaka have confirmed this theory [53, 54]. In accordance with this, mammalian (tetrapod) genes often have two orthologues in fish. Prior to the initiation of these genome sequencing projects three Y1-like receptors had been cloned from zebrafish, but did not seem to represent orthologues to any of the mammalian Y1, Y4 or Y6 genes. These fish sequences were therefore tentatively regarded as novel genes and given distinct gene names, namely Ya, Yb and Yc [92, 95, 96]. As additional Y receptor sequences were reported, the relationships of these genes have become clearer. First, it seems that Ya actually is a rapidly evolving orthologue of Y4 [97]. As described above, PP, which is the preferred ligand for mammalian Y4, had not appeared at the time of the teleost split from the tetrapod lineage, but arose due to a local duplication of the PYY gene in early tetrapod evolution. Subsequent co-evolution with PP in tetrapods has endowed different evolutionary constraints upon tetrapod Y4 compared to teleost Ya, which must obviously have quite different pharmacological and physiological characteristics. Furthermore, the genes surrounding Y4 on human chromosome 10 and Ya in zebrafish seem to be orthologous.

The relationships of zebrafish Yb and Yc also became clearer with the sequence information from teleost genomes. According to the 2R theory, the early vertebrate genome was duplicated twice within a relatively short time span. This should have then resulted in four chromosome segments harbouring Y receptors. However, the fourth candidate chromosome region in the human genome, chromosome 8, harbours no Y genes [41, 67], but other genes in this chromosome segment have orthologues adjacent to or near both Yb and Yc in zebrafish (Larsson, Sundström, Olsson and Larhammar, unpublished). This suggests that Yb and Yc are descendants of the fourth Y1 subfamily member, Y8, lost in mammals. As mentioned above, an orthologue of Y8 is indeed found in the frog. The fish Y8 gene was likely duplicated in the teleost tetraploidisation, resulting in the two daughter genes Yb and Yc. Preliminary results from some basal teleost species that diverged prior to the teleost radiation (such as sturgeon, gar and bowfin) show that a single Yb/Yc gene may exist in these (Salaneck et al., unpublished). Given the ancient age of Yb, the name Y8 for this gene is now justified, and the two zebrafish receptors have been renamed Y8a (formerly Yc) and Y8b (formerly Yb).

As in amphibians and chicken, both Y2 and the related Y7 sequence have been identified in teleosts. The zebrafish receptors have been characterised pharmacologically and display some differences [94] (Fredriksson et al., submitted). The Y5 gene seems to be missing in teleosts, like Y1 [98], but has been identified in basal ray-finned fish [99]. Thus, both Y1 and Y5 seem to have been lost in teleosts.

The first branching in the gnathostome tree was the split of cartilaginous fishes (sharks, rays, and chimaeras) from the lineage leading to bony fishes and tetrapods.

One of the most studied species among the cartilaginous fishes is a shark, the spiny dogfish. The Y2 and Y7 genes, as well as Y5 and the three Y1-subfamily genes Y1, Y4, Y6, have been cloned from this species [97] (Larson, Salaneck and Larhammar, unpublished). So far no Y8 gene has been found in the dogfish. If indeed this gene is missing, it must have been lost twice; once in the shark lineage and once in the lineage leading to amniotes (birds, reptiles and mammals). The shark sequences have proven extremely valuable for phylogenetic analyses. However, no full-length clones exist so no functional data have been obtained regarding their ligand preferences. Some information exists regarding the anatomical distribution of mRNA for the three shark Y1-subfamily genes, which seem to differ considerably from the mammalian orthologues (see below).

Lampreys and hagfish comprise the two living groups of jawless fishes or agnathans. These two lineages are thought to have branched off from the chordate tree between the two pre-gnathostome tetraploidisations, thereby being affected by only one of two major duplication events. Two Y receptor genes have been cloned from the European river lamprey, *Lampetra fluviatilis*. One gene shows high sequence identity to Y4 and Yb [100] and the other shows identity to Y2 and Y7 sequences (unpublished). The phylogenetic analyses support the duplication scenario (Fig. 4), suggesting that the lamprey genes are descendants of a common ancestor of two separate genes in gnathostomes. Further cloning will be necessary to see if an expected Y1/Y6 orthologue and a Y5 orthologue exist in agnathans.

The closest vertebrate relatives are the lancelet *Amphioxus* and the sea squirt *Ciona intestinalis*. As orthologues of NPY and Y receptors exist in snails and insects (see above), it was expected to find orthologues also in the recently published *Ciona* genome. The absence suggests gene loss. Cloning efforts in *Amphioxus* have not yet revealed any orthologues and the complete genome sequence is eagerly awaited. The animal groups represented by *Amphioxus* and *Ciona* intestinalis are particularly important as they diverged prior to the vertebrate tetraploidisations and thus have a simpler gene repertoire, for instance for the Hox cluster [49].

Discussion

Taken together, the Y receptor sequences described above for a wide range of species show that a basic repertoire of Y receptors with at least six and presumably seven (including Y8) genes was already established prior to the appearance of jawed vertebrates, the gnathostomes, somewhere between 450 and 500 million years ago (Fig. 4). The duplication scenario with an ancestral triplet that was quadrupled by chromosome duplications leaves open the possibility that additional Y receptor genes may have been part of the ancestral repertoire and may have survived in some vertebrate lineages, but the increasing number of complete or almost complete genome sequences make it unlikely that additional Y receptors exist. Separate duplications may have taken place in some lineages, as discussed below for fish, but it seems improbable that any Y receptors remain to be discovered in mammals. The reported deviating pharmacological profiles seen in some mammals *in vivo* or in *ex vivo* preparations might be due to species differences, combinations of receptor subtypes (heterodimers) or accessory proteins [77]. On the other hand, given the large number of species it is not unlikely that additional duplications may have occurred in some.

After the burst of new receptor genes in the vertebrate ancestor, the trend has been to lose genes, with Y8 disappearing in the amniote lineage and Y7 having been lost in mammals. In the mammalian class, Y6 has been lost in some species, e.g., *Homo sapiens* and other primates, but still lingers in other (mouse and rabbit), albeit with unknown function. Intriguingly, the two appetite-stimulating receptors Y1 and Y5 have been lost in teleost fishes [98, 99]. This was unexpected for two reasons. First, teleost fishes have undergone an additional genome duplication, why one should expect additional gene copies, not fewer. It seems like only one duplicate has survived the teleost tetraploidisation, namely the duplicate of Y8 (Fig. 4). Secondly, there is evidence that teleost fish respond to NPY with increased feeding [101], and that starvation upregulates NPY [102] and Y receptors (E. T. Larson et al., unpublished). If both Y1 and Y5 are missing, the appetite-stimulating effect of NPY must be mediated by one of the other Y subtypes, perhaps Y8.

Another interesting line of research will be to explore the role of Y2 and Y4 (Ya) in fishes, as these receptors inhibit appetite in mammals. Important aspects to consider are that PP (the preferred Y4 ligand in mammals) does not exist in fish, and whether PYY can be processed to PYY3-36 as in mammals to become more selective for Y2. Sequence analyses of fish PYY suggest that it may perhaps not be processed. If so, Y2 and Y4 may have other roles in fish that remain to be identified. The importance of PYY and PP in appetite inhibition will also be interesting to explore in the chicken where the Y4 receptor binds to both of these ligands [91]. It is possible that appetite inhibition by PYY and PP through Y2 and Y4, respectively, is a more recent function, perhaps evolved in mammals.

The mechanism of duplication for NPY and PYY from their common ancestor, as well as the Y receptors from the original receptor triplet, was through duplication of a large chromosome block (or entire chromosomes through tetraploidisation). This means that the gene duplicates initially had identical promotor regions and thus also identical gene regulation, before differences began to accumulate that caused differential expression. An interesting question concerning the block duplications is whether the different gene families that were duplicated simultaneously have some type of functional interaction. For instance, do NPY and PYY have functional connections to the Hox genes or the EGF (epidermal growth factor) receptor genes that are located in the same chromosome regions? Functional constraints that conserve linkage seem unlikely because the distances between genes of the same two families often differ from one chromosome to another in a paralogon, and also between species as shown for the Y receptor genes in human, pig and mouse [68]. However, a few functional connections do seem possible, but remain to be demonstrated, for instance concerning NPY/PYY or Y receptors and other genes that are involved in growth, such as EGFR and PDGF. Also it is intriguing that CRH receptors and NPY/PYY are on the same chromosomes [5, 40, 41] as these genes have opposing effects on both metabolism and anxiety. Another connection with the CRH–ACTH system is that the genes for the glucocorticoid receptor and the mineralocorticoid receptor are on the same chromosomes as the Y6 and the Y1–Y2–Y5 receptors, respectively. But as pointed out above, functional connections and selective constraints remain to be demonstrated. It is possible that these linkages are simply due to common ancestry for the chromosomal regions and lack of gross rearrangements such as translocations.

Stimulation of appetite as an ancestral function of the NPY system is supported by the observations in the snail *Lymnaea stagnalis* as well as in *Drosophila melanogaster* described above. Also, NPY seems to stimulate food intake in almost all vertebrates that have been investigated in addition to the predominating rat and mouse: rhesus monkey [103], pig [104], sheep [105], rabbit [106], golden hamster [107], guinea pig [108, 109], golden-mantled ground squirrel *Spermophilus saturatus* [110], chicken [111], white-crowned sparrow [112], red-sided garter snake *Thamnophis sirtalis parietalis* [113], South African clawed frog *Xenopus laevis* [114], and goldfish [115]. Indeed, NPY has been reported to be the most potent orexigenic factor in goldfish [116], as in rat. Furthermore, NPY is upregulated in fasting salmon of two species, chinook salmon *Oncorhynchus tshawytscha* and coho salmon *Oncorhynchus kisutch* [102]. One species where NPY did not induce feeding is the dog (mongrel dog) [117]. Thus, overall, there is massive evidence for an appetite-stimulating role of NPY in both tetrapods and fish suggesting that this is an ancestral function.

With regard to functions of NPY and PYY in other contexts than feeding, the information from non-mammalian vertebrates is still quite limited. Vascular and cardiac effects of NPY have only been studied in a few species. In the longnose skate, *Raja rhina*, NPY was found to potentiate the contractile response of the coronary artery to norepinephrine [118]. NPY contracts the afferent branchial artery in the common dogfish *Scyliorhinus canicula* [119]. In dogfish hearts, NPY increases cardiac output by increasing heart rate and, in combination with norepinephrine, there was increased pressure [120]. Both NPY and PYY contracted isolated gut arteries in three species of elasmobranchs [121]. In eel, NPY enhanced the contractile force of the isolated atrium [122, 123]. In contrast, NPY produced vasorelaxation of celiac arteries of the Atlantic cod, *Gadus morhua* [123]. Thus, the most common effect of NPY on blood vessels is contraction, as in mammals. However, it is still unknown which Y receptor subtypes mediate these effects (and which Y subtypes exist in these species).

In frogs, NPY reduces pigmentation and was in fact discovered as a melanotropin(α-MSH)-release-inhibiting factor in the European green frog, *Rana ridibunda* [124]. Therefore, NPY was initially named melanostatin. The same response could be produced with PYY [125]. This effect has subsequently been proposed to be mediated by the Y5 receptor whereas the TRH-induced release of α-MSH is suppressed via Y1 [126].

A largely unexplored area in the evolutionary perspective is expression of NPY/PYY and Y receptors in the immune system. Several reports have been published for mammals [127, 128], but as for so many other immune system components, there are extensive differences even among mammals such as human, rat and mouse [129]. The immune system is one of the most rapidly evolving systems why evolutionary studies over larger distances will be particularly challenging. Now that molecular tools are available for many species, it should at least be possible to search for the presence of mRNA in immune system cells.

In the absence of extensive comparative functional studies *in vivo* or *ex vivo*, and as a first step to determine functional specialisation of the various Y receptor subtypes, it is interesting to consider their anatomical distribution. We have investigated the presence of Y receptor mRNA in different organs by reverse transcriptase-PCR in chicken, the shark spiny dogfish (*Squalus acanthias*), and the river lamprey (*Lampetra fluviatilis*). The overall impression is that extensive differences exist across species for individual receptor subtypes. Among the most conspicuous differences to mammals it is worth noting that chicken Y4 mRNA [91] is expressed in liver and spleen, in addition to adipose tissue like in mammals which otherwise have expression mostly in the gastrointestinal tract [130, 131]. The shark Y4 receptor mRNA is expressed in brain, retina, kidney and muscle [97] (see Fig. 5). Surprisingly, the shark Y1 receptor mRNA is not detectable in brain or heart, but is prominent in liver and kidney, while the Y6 receptor mRNA is present in retina, kidney and gastrointestinal tract. The lamprey Y1-subfamily receptor, which is thought to be a Y1/Y6 pro-orthologue [100], is expressed in CNS, liver and gonads. The detection method is not quantitative and conclusions should be cautious, but there is little doubt that prominent differences in mRNA distribution have evolved for orthologous receptors across vertebrate classes.

In summary, the repertoire of vertebrate Y receptors expanded prior to the gnathostome radiation, at least some 400 million years ago, whereupon differential gene loss has occurred in various vertebrate lineages. Despite these events, the NPY system seems to have retained its appetite-stimulating properties in the vertebrates that have been studied. More detailed studies are greatly needed in non-mammalian vertebrates regarding the roles of the NPY system in feeding, cardiovascular regulation, circadian rhythm and many other physiological functions. With the clarified picture of receptor and peptide evolution, as well as access to molecular tools in many species, such studies should now be possible.

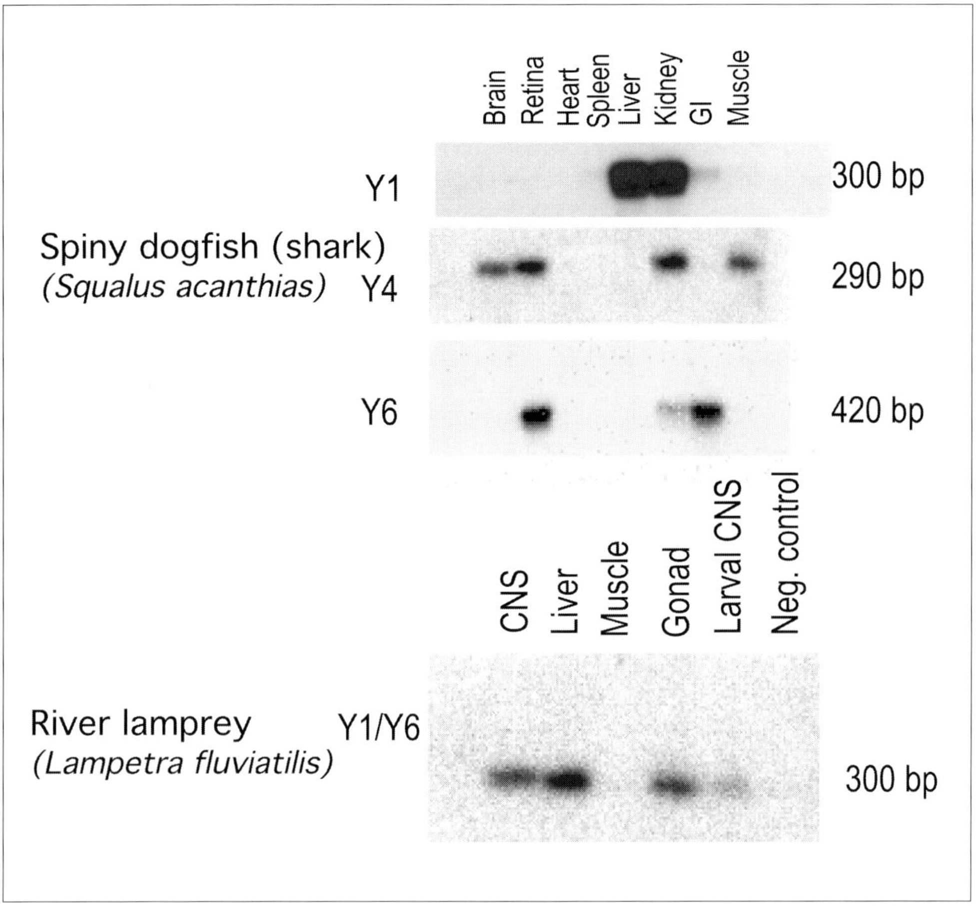

Figure 5
*Reverse transcriptase-PCR autoradiograms showing tissue distribution of the three Y1-subfamily receptor genes in a shark, the spiny dogfish (*Squalus acanthias*), and the single Y1-subfamily gene (proposed Y1/Y6 pro-orthologue) so far found in an agnathan, the river lamprey (*Lampetra fluviatilis*). GI stands for gastrointestinal tract. The expression patterns of the shark genes differ considerably from their mammalian orthologues. Note that the assay is not quantitative.*

Acknowledgements

This work was supported by grants from the Swedish Research Council, Formas (The Swedish Research Council for Environment, Agricultural Sciences and Spatial Planning), and Carl Trygger's Foundation. We thank our colleagues whose unpublished observations we cite.

References

1 Tatemoto K (1982) Neuropeptide Y: Complete amino acid sequence of the brain peptide. *Proc Natl Acad Sci USA* 79: 5485–5489

2 Tatemoto K (1982) Isolation and characterization of peptide YY (PYY), a candidate gut hormone that inhibits pancreatic exocrine secretion. *Proc Natl Acad Sci USA* 79: 2514–2518

3 Larhammar D, Blomqvist AG, Söderberg C (1993) Evolution of neuropeptide Y and its related peptides. *Comp Biochem Physiol* 106C: 743–752

4 Larhammar D (1996) Structural diversity of receptors for neuropeptide Y, peptide YY and pancreatic polypeptide. *Regul Pept* 65: 165–174

5 Cerdá-Reverter JM, Larhammar D (2000) Neuropeptide Y family of peptides: structure, anatomical expression, function, and molecular evolution. *Biochem Cell Biol* 78: 371–392

6 Pedrazzini T, Pralong F, Grouzmann E (2003) Neuropeptide Y: the universal soldier. *Cell Mol Life Sci* 60: 350–377

7 Fujimiya M, Inui A (2000) Peptidergic regulation of gastrointestinal motility in rodents. *Peptides* 21: 1565–1582

8 Naruse S, Kitagawa M, Ishiguro H, Hayakawa T (2002) Feedback regulation of pancreatic secretion by peptide YY. *Peptides* 23: 359–365

9 Batterham RL, Cowley MA, Small CJ, Herzog H, Cohen MA, Dakin CL, Wren AM, Brynes AE, Low MJ, Ghatei MA et al (2002) Gut hormone PYY(3-36) physiologically inhibits food intake. *Nature* 418: 650–654

10 Kimmel JR, Hayden LJ, Pollock HG (1975) Isolation and characterization of a new pancreatic polypeptide hormone. *J Biol Chem* 250: 9369–9374

11 Hort Y, Baker E, Sutherland GR, Shine J, Herzog H (1995) Gene duplication of the human peptide YY gene (PYY) generated the pancreatic polypeptide gene (PPY) on chromosome 17q21.1. *Genomics* 26: 77–83

12 Hazelwood RL (1993) The pancreatic polypeptide (PP-fold) family: Gastrointestinal, vascular, and feeding behavioral implications. *Proc Soc Expl Biol Med* 202: 44–63

13 Batterham RL, Le Roux CW, Cohen MA, Park AJ, Ellis SM, Patterson M, Frost GS, Ghatei MA, Bloom SR (2003) Pancreatic polypeptide reduces appetite and food intake in humans. *J Clin Endocrinol Metab* 88: 3989–3992

14 Blomqvist AG, Söderberg C, Lundell I, Milner RJ, Larhammar D (1992) Strong evolutionary conservation of neuropeptide Y: Sequences of chicken, goldfish, and *Torpedo marmorata* DNA clones. *Proc Natl Acad Sci USA* 89: 2350–2354

15 Söderberg C, Pieribone VA, Dahlstrand J, Brodin L, Larhammar D (1994) Neuropeptide role of both peptide YY and neuropeptide Y in vertebrates suggested by abundant expression of their mRNAs in a cyclostome brain. *J Neurosci Res* 37: 633–640

16 Söderberg C, Wraith A, Ringvall M, Yan Y-L, Postlethwait J, Brodin L, Larhammar D (2000) Zebrafish genes for neuropeptide Y and peptide YY reveal origin by chromo-

some duplication from an ancestral gene linked to the homeobox cluster. *J Neurochem* 75: 908–918

17 Cerdá-Reverter JM, Martínez-Rodríguez G, Zanuy S, Carrillo M, Larhammar D (2000) Molecular evolution of the neuropeptide Y (NPY) family of peptides: cloning of NPY-related peptides from the sea bass (*Dicentrarchus labrax*). *Regul Peptides* 95: 25–34

18 Larhammar D (1996) Evolution of neuropeptide Y, peptide YY, and pancreatic polypeptide. *Regul Peptides* 62: 1–11

19 Force A, Lynch M, Pickett FB, Amores A, Yan Y, Postlethwait J (1999) Preservation of duplicate genes by complementary, degenerative mutations. *Genetics* 151: 1531–1545

20 Eva C, Keinänen K, Monyer H, Seeburg P, Sprengel R (1990) Molecular cloning of a novel G protein-couple receptor that may belong to the neuropeptide receptor family. *FEBS Lett* 271: 81–84

21 Krause J, Eva C, Seeburg PH, Sprengel R (1992) Neuropeptide Y1 subtype pharmacology of a recombinantly expressed neuropeptide receptor. *Mol Pharmacol* 41: 817–821

22 Herzog H, Hort YJ, Ball HJ, Hayes G, Shine J, Selbie LA (1992) Cloned human neuropeptide Y receptor couples to two different second messenger systems. *Proc Natl Acad Sci USA* 89: 5794–5798

23 Larhammar D, Blomqvist AG, Yee F, Jazin E, Yoo H, Wahlestedt C (1992) Cloning and functional expression of a human neuropeptide Y/peptide YY receptor of the Y1 type. *J Biol Chem* 267: 10935–10938

24 Larhammar D, Wraith A, Berglund MM, Holmberg SKS, Lundell I (2001) Origins of the multiple NPY-family receptors in mammals. *Peptides* 22: 295–307

25 Dores RM, Rubin DA, Quinn TW (1996) Is it possible to construct phylogenetic trees using polypeptide hormone sequences? *Gen Comp Endocrinol* 103: 1–12

26 Tensen CP, Cox KJA, Burke JF, Leurs R, van der Schors RC, Geraerts WPM, Vreugdenhil E, van Heerikhuizen H (1998) Molecular cloning and characterization of an invertebrate homologue of a neuropeptide Y receptor. *Eur J Neurosci* 10: 3409–3416

27 Hoek RM, van Kesteren RE, Smit AB, de Jong-Brink M, Geraerts WP (1997) Altered gene expression in the host brain caused by a trematode parasite: neuropeptide genes are preferentially affected during parasitosis. *Proc Natl Acad Sci USA* 94: 14072–14076

28 Garczynski SF, Brown MR, Shen P, Murray TF, Crim JW (2002) Characterization of a functional neuropeptide F receptor from *Drosophila melanogaster*. *Peptides* 23: 773–780

29 Brown MR, Crim JW, Arata RC, Cai HN, Chun C, Shen P (1999) Identification of a *Drosophila* brain-gut peptide related to the neuropeptide Y family. *Peptides* 20: 1035–1042

30 Wu Q, Wen T, Lee G, Park JH, Cai HN, Shen P (2003) Developmental control of foraging and social behavior by the *Drosophila* neuropeptide Y-like system. Neuron 39: 147–161

31 Stanek DM, Pohl J, Crim JW, Brown MR (2002) Neuropeptide F and its expression in the yellow fever mosquito, *Aedes aegypti*. *Peptides* 23: 1367–1378

32 Riehle MA, Garczynski SF, Crim JW, Hill CA, Brown MR (2002) Neuropeptides and peptide hormones in Anopheles gambiae. *Science* 298: 172–175

33 Xia Q, Zhou Z, Lu C, Cheng D, Dai F, Li B, Zhao P, Zha X, Cheng T, Chai C et al (2004) A draft sequence for the genome of the domesticated silkworm (*Bombyx mori*). *Science* 306: 1937–1940

34 Garczynski SF, Crim JW, Brown MR (2005) Characterization of neuropeptide F and its receptor from the African malaria mosquite, *Anopheles gambiae. Peptides* 26: 99–107

35 Li X-J, Wu Y-N, North A, Forte M (1992) Cloning, functional expression, and developmental regulation of a neuropeptide Y receptor from *Drosophila melanogaster. J Biol Chem* 267: 9–12

36 Mertens I, Meeusen T, Huybrechts R, De Loof A, Schoofs L (2002) Characterization of the short neuropeptide F receptor from Drosophila melanogaster. *Biochem Biophys Res Commun* 297: 1140–1148

37 Feng G, Reale V, Chatwin H, Kennedy K, Venard R, Ericsson C, Yu K, Evans P D, Hall L M (2003) Functional characterization of a neuropeptide F-like receptor from *Drosophila melanogaster. Eur J Neurosci* 18: 227–238

38 de Bono M, Bargmann CI (1998) Natural variation in a neuropeptide Y receptor homolog modifies social behavior and food response in *C. elegans. Cell* 94: 679–689

39 Kubiak TM, Larsen MJ, Nulf SC, Zantello MR, Burton KJ, Bowman JW, Modric T, Lowery DE (2003) Differential activation of “social” and “solitary” variants of the *C. elegans* GPCR NPR-1 by its cognate ligand AF9. *J Biol Chem* 278: 33724–33729

40 Conlon JM (2002) The origin and evolution of peptide YY (PYY) and pancreatic polypeptide (PP). *Peptides* 23: 269–278

41 Larhammar D, Fredriksson R, Larson ET, Salaneck E (2004) In: Michel MC (ed): *Neuropeptide Y and Related Peptides*, Springer-Verlag, Berlin-Heidelberg pp 75–100

42 Conlon JM, Bjørnholm B, Jørgensen FS, Youson JH, Schwartz TW (1991) Primary structure and conformational analysis of peptide methionine-tyrosine, a peptide related to neuropeptide Y and peptide YY isolated from lamprey intestine. *Eur J Biochem* 199: 293–298

43 Wang Y, Nielsen PF, Youson JH, Potter IC, Lance VA, Conlon JM (1999) Molecular evolution of peptide tyrosine-tyrosine: primary structure of PYY from the lampreys *Geotria australis* and *Lampetra fluviatilis*, bichir, python and desert tortoise. *Regulatory Peptides* 79: 103–108

44 Montpetit CJ, Chatalov V, Yuk J, Rasaratnam I, Youson JH (2005) Expression of neuropeptide Y family peptides in the brain and gut during stages of the life cycle of a parasitic lamprey (*Petromyzon marinus*) and a nonparasitic lamprey (*Ichthyomyzon gagei*). In: Vaudry H, Roubos E, Schoofs L, Flik G, Larhammar D (eds): *Trends in Comparative Endocrinology and Neurobiology. Ann NY Acad Sci* 1040: 140–149

45 Baker E, Hort YJ, Ball H, Sutherland GR, Shine J, Herzog H (1995) Assignment of the human neuropeptide Y gene to chromosome 7p15.1 by nonisotopic *in situ* hybridization. *Genomics* 26: 163–164

46 Coulier F, Popovici C, Villet R, Birnbaum D (2000) Meta*HOX* gene clusters. *J Exp Zool* 288: 345–351

47 Popovici C, Leveugle M, Birnbaum D, Coulier F (2001) Homeobox gene clusters and the human paralogy map. *FEBS Lett* 491: 237–242

48 Lundin LG, Hallböök F, Larhammar D (2003) Numerous groups of chromosomal regional paralogies strongly indicate two genome doublings at the root of the vertebrates. *J Struct Funct Genomics* 3: 53–63

49 Garcia-Fernàndez J, Holland PWH (1994) Archetypal organization of the amphioxus *Hox* gene cluster. *Nature* 370: 563–566

50 Holland PWH, Garcia-Fernàndez J, Williams NA, Sidow A (1994) Gene duplications and the origins of vertebrate development. *Dev Suppl* 125–133

51 Shimeld MS, Holland PWH (2000) Vertebrate innovations. *Proc Natl Acad Sci USA* 97: 4449–4452

52 Postlethwait J, Amores A, Cresko W, Singer A, Yan YL (2004) Subfunction partitioning, the teleost radiation and the annotation of the human genome. *Trends Genet* 20: 481–490

53 Taylor JS, Braasch I, Frickey T, Meyer A, Van de Peer Y (2003) Genome duplication, a trait shared by 22,000 species of ray-finned fish. *Genome Res* 13: 382–390

54 Jaillon O, Aury JM, Brunet F, Petit JL, Stange-Thomann N, Mauceli E, Bouneau L, Fischer C, Ozouf-Costaz C, Bernot A et al (2004) Genome duplication in the teleost fish *Tetraodon nigroviridis* reveals the early vertebrate proto-karyotype. *Nature* 431: 946–957

55 Powers TP, Amemiya CT (2004) Evolutionary plasticity of vertebrate Hox genes. *Curr Genomics* 5: 459–472

56 Hoegg S, Meyer A (2005) Hox clusters as models for vertebrate genome evolution. *Trends in Genetics* 21: 421–424

57 Sundström G, Larsson TA, Brenner S, Venkatesh B, Larhammar D (2005) Ray-fin fish tetraploidization gave rise to pufferfish duplicates of NPY and PYY, but zebrafish NPY duplicate was lost. In: Vaudry H, Roubos E, Schoofs L, Flik G, Larhammar D (eds): *Trends in Comparative Endocrinology and Neurobiology. Ann NY Acad Sci* 1040: 476–478

58 Pieribone VA, Brodin L, Friberg K, Dahlstrand J, Söderberg C, Larhammar D, Hökfelt T (1992) Differential expression of mRNAs for neuropeptide Y-related peptides in rat nervous tissues: possible evolutionary conservation. *J Neuroscience* 12: 3361–3371

59 Söderberg C (1996) Molecular evolution of the neuropeptide Y family peptides. Uppsala University, Uppsala, Sweden

60 Ekblad E, Sundler F (2002) Distribution of pancreatic polypeptide and peptide YY. *Peptides* 23: 251–261

61 Michel MC, Beck-Sickinger A, Cox H, Doods HN, Herzog H, Larhammar D, Quirion R, Schwartz T, Westfall T (1998) XVI. International Union of Pharmacology recommendations for the nomenclature of neuropeptide Y, peptide YY and pancreatic polypeptide receptors. *Pharmacol Rev* 50: 143–150

62 Rimland J, Xin W, Sweetnam P, Saijoh K, Nestler EJ, Duman RS (1991) Sequence and expression of a neuropeptide Y receptor cDNA. *Mol Pharmacol* 40: 869–875

63 Herzog H, Hort YJ, Shine J, Selbie LA (1993) Molecular cloning, characterization and localization of the human homolog to the reported bovine NPY Y3 receptor: Lack of NPY binding and activation. *DNA Cell Biol* 12: 465–471

64 Jazin EE, Yoo H, Blomqvist AG, Yee F, Weng G, Walker MW, Salon J, Larhammar D, Wahlestedt C (1993) A proposed bovine neuropeptide Y (NPY) receptor, or its human homologue, confers neither NPY binding sites nor NPY responsiveness on transfected cells. *Regul Peptides* 47: 247–258

65 Starbäck P, Wraith A, Eriksson H, Larhammar D (2000) Neuropeptide Y receptor gene y6: multiple deaths or resurrection? *Biochem Biophys Res Comm* 277: 264–269

66 Larhammar D, Fredriksson R, Salaneck E, Larsson TA, Holmberg SKS, Lundell I, Larson ET (2002) Evolution of vertrebrate neuropeptide Y receptors. In: Keller R, Dircksen H, Sedlmeier D, Vaudry H (eds): *Proceedings of the 21st Conference of European Comparative Endocrinologists*. Monduzzi Editore, Bologna 17–22

67 Larhammar D, Salaneck E (2004) Molecular evolution of NPY receptor subtypes. *Neuropeptides* 38: 141–151

68 Wraith A, Törnsten A, Chardon P, Harbitz I, Chowdhary BP, Andersson L, Lundin L-G, Larhammar D (2000) Evolution of the neuropeptide Y receptor family: gene and chromosome duplications deduced from the cloning of the five receptor subtype genes in pig. *Genome Res* 10: 302–310

69 Sidow A (1996) Gen(om)e duplications in the evolution of early vertebrates. *Curr Opin Genetics Devel* 6: 715–722

70 Gregor P, Feng Y, DeCarr LB, Cornfield LJ, McCaleb ML (1996) Molecular characterization of a second mouse pancreatic polypeptide receptor and its inactivated human homologue. *J Biol Chem* 271: 27776–27781

71 Matsumoto M, Nomura T, Momoses K, Ikeda Y, Kondou Y, Akiho H, Togami J, Kimura Y, Okada M, Yamaguchi T (1996) Inactivation of a novel neuropeptide Y/peptide YY receptor gene in primate species. *J Biol Chem* 271: 27217–27220

72 Rose PM, Lynch JS, Frazier ST, Fisher SM, Chung W, Battaglino P, Fathi Z, Leibel R, Prabhavathi F (1997) Molecular genetic analysis of a human neuropeptide Y receptor. The human homolog of the murine "Y5" receptor may be a pseudogene. *J Biol Chem* 272: 3622–3627

73 Burkhoff A, Linemeyer DL, Salon JA (1998) Distribution of a novel hypothalamic neuropeptide Y receptor gene and its absence in rat. *Mol Brain Res* 53: 311–316

74 Nakamura M, Yokoyama M, Watanabe H, Matsumoto T (1997) Molecular cloning, organization and localization of the gene for the mouse neuropeptide Y-Y5 receptor. *Biochim Biophys Acta* 1328: 83–89

75 Borowsky B, Walker MW, Bard J, Weinshank RL, Laz TM, Vaysse P, Branchek TA, Gerald C (1998) Molecular biology and pharmacology of multiple NPY Y5 receptor species homologs. *Regul Pept* 75–76: 45–53

76 Wraith A (1999) Molecular evolution of the neuropeptide Y receptor family. Insights from mammals and fish. Uppsala University, Uppsala, Sweden
77 Berglund MM, Hipskind PA, Gehlert DR (2003) Recent developments in our understanding of the physiological role of PP-fold peptide receptor subtypes. *Exp Biol Med (Maywood)* 228: 217–244
78 Pedrazzini T, Seydoux J, Kunstner P, Aubert J-F, Grouzmann E, Beermann F, Brunner H-R (1998) Cardiovascular response, feeding behavior and locomotor activity in mice lacking the NPY Y1 receptor. *Nature Med* 4: 722–726
79 Campbell RE, Ffrench-Mullen JM, Cowley MA, Smith MS, Grove KL (2001) Hypothalamic circuitry of neuropeptide Y regulation of neuroendocrine function and food intake via the Y5 receptor subtype. *Neuroendocrinology* 74: 106–119
80 Wolak ML, DeJoseph MR, Cator AD, Mokashi AS, Brownfield MS, Urban JH (2003) Comparative distribution of neuropeptide Y Y1 and Y5 receptors in the rat brain by using immunohistochemistry. *J Comp Neurol* 464: 285–311
81 Kopp J, Xu ZQ, Zhang X, Pedrazzini T, Herzog H, Kresse A, Wong H, Walsh JH, Hokfelt T (2002) Expression of the neuropeptide Y Y1 receptor in the CNS of rat and of wild-type and Y1 receptor knock-out mice. Focus on immunohistochemical localization. *Neuroscience* 111: 443–532
82 Matsuda H, Brumovsky PR, Kopp J, Pedrazzini T, Hokfelt T (2002) Distribution of neuropeptide Y Y1 receptors in rodent peripheral tissues. *J Comp Neurol* 449: 390–404
83 Parker RM, Herzog H (1999) Regional distribution of Y-receptor subtype mRNAs in rat brain. *Eur J Neurosci* 11: 1431–1448
84 Herzog H, Darby K, Ball H, Hort Y, Beck-Sickinger A, Shine J (1997) Overlapping gene structure of the human neuropeptide Y receptor subtypes Y1 and Y5 suggests coordinate transcriptional regulation. *Genomics* 41: 315–319
85 Batterham RL, Cohen MA, Ellis SM, Le Roux CW, Withers DJ, Frost GS, Ghatei MA, Bloom SR (2003) Inhibition of food intake in obese subjects by peptide YY3-36. *N Engl J Med* 349: 941–948
86 Zukowska-Grojec Z, Karwatowska-Prokopczuk E, Fisher TA, Ji H (1998) Mechanisms of vascular growth-promoting effects of neuropeptide Y: role of its inducible receptors. *Regul Pept* 75–76: 231–238
87 Golombek DA, Biello SM, Rendon RA, Harrington ME (1996) Neuropeptide Y phase shifts the circadian clock *in vitro* via a Y2 receptor. *Neuroreport* 7: 1315–1319
88 Huhman KL, Gillespie CF, Marvel CL, Albers HE (1996) Neuropeptide Y phase shifts circadian rhythms *in vivo* via a Y2 receptor. *Neuroreport* 7: 1249–1252
89 Salaneck E, Holmberg SK, Berglund MM, Boswell T, Larhammar D (2000) Chicken neuropeptide Y receptor Y2: structural and pharmacological differences to mammalian Y2. *FEBS Lett* 484: 229–234
90 Holmberg SK, Mikko S, Boswell T, Zoorob R, Larhammar D (2002) Pharmacological characterization of cloned chicken neuropeptide Y receptors Y1 and Y5. *J Neurochem* 81: 462–471
91 Lundell I, Boswell T, Larhammar D (2002) Chicken neuropeptide Y-family receptor Y4:

a receptor with equal affinity for pancreatic polypeptide, neuropeptide Y and peptide YY. *J Mol Endocrinol* 28: 225–235
92 Lundell I, Berglund MM, Starbäck P, Salaneck E, Gehlert DR, Larhammar D (1997) Cloning and characterization of a novel neuropeptide Y receptor subtype in the zebrafish. DNA Cell Biol 16: 1357–1363
93 Blomqvist AG, Roubos EW, Larhammar D, Martens GJ (1995) Cloning and sequence analysis of a neuropeptide Y/peptide YY receptor Y1 cDNA from *Xenopus laevis*. *Biochim Biophys Acta* 1261: 439–441
94 Fredriksson R, Larson ET, Yan Y-L, Postlethwait JH, Larhammar D (2003) Novel neuropeptide Y Y2-like receptor subtype in zebrafish and frogs supports early vertebrate chromosome duplications. *J Mol Evol* 58: 106–114
95 Ringvall M, Berglund MM, Larhammar D (1997) Multiplicity of neuropeptide Y receptors: cloning of a third distinct subtype in the zebrafish. *Biochem Biophys Res Comm* 241: 749–755
96 Starbäck P, Lundell I, Fredriksson R, Berglund MM, Yan Y-L, Wraith A, Söderberg C, Postlethwait JH, Larhammar D (1999) Neuropeptide Y receptor subtype with unique properties cloned in the zebrafish; the zYa receptor. *Mol Brain Res* 70: 242–252
97 Salaneck E, Ardell D, Larson ET, Larhammar D (2003) Three neuropeptide Y receptors in the spiny dogfish, *Squalus acanthias*, support chromosome doublings in early vertebrate evolution. *Mol Biol Evol* 20: 1271–1280
98 Larsson TA, Olsson F, Sundström G, Brenner S, Venkatesh B, Larhammar D (2005) Pufferfish and zebrafish have five distinct NPY receptor subtypes, but have lost appetite receptors Y1 and Y5. *Ann NY Acad Sci* 1040: 375–377
99 Salaneck E, Larson ET, Larsson TA, Larhammar D (2005) Effects of a teleost tetraploidization on neuropeptide Y receptor gene repertoire in ray-finned fishes. *Ann NY Acad Sci* 1040: 457–459
100 Salaneck E, Fredriksson R, Larson ET, Conlon JM, Larhammar D (2001) A neuropeptide Y receptor Y1-subfamily gene from an agnathan, the European river lamprey. A potential ancestral gene. *Eur J Biochem* 268: 6146–6154
101 Narnaware YK, Peter RE (2001) Neuropeptide Y stimulates food consumption through multiple receptors in goldfish. *Physiol Behav* 74: 185–190
102 Silverstein JT, Breininger J, Baskin DG, Plisetskaya EM (1998) Neuropeptide Y-like gene expression in the salmon brain increases with fasting. *Gen Comp Endocrinol* 110: 157–165
103 Larsen P, Tang-Christensen M, Stidsen C, Madsen K, Smith M, Cameron J (1999) Activation of central neuropeptide Y Y1 receptors potently stimulates food intake in male rhesus monkeys. *J Clin Endocrinol Metab* 84: 3781–3791
104 Parrott RF, Heavens RP, Baldwin BA (1986) Stimulation of feeding in the satiated pig by intracerebroventricular injection of neuropeptide Y. *Physiol Behav* 36: 523–525
105 Miner JL, Della-Fera MA, Paterson JA, Baile CA (1989) Lateral cerebroventricular injection of neuropeptide Y stimulates feeding in sheep. *Am J Physiol* 257: R383–R387

106 Pau MY, Pau KY, Spies HG (1988) Characterization of central actions of neuropeptide Y on food and water intake in rabbits. *Physiol Behav* 44: 797–802

107 Kulkosky PJ, Glazner GW, Moore HD, Low CA, Woods SC (1988) Neuropeptide Y: behavioral effects in the golden hamster. *Peptides* 9: 1389–1393

108 Lecklin A, Lundell I, Paananen L, Wikberg JE, Männistö PT, Larhammar D (2002) Receptor subtypes Y1 and Y5 mediate neuropeptide Y induced feeding in the guinea-pig. *Br J Pharmacol* 135: 2029–2037

109 Lecklin A, Lundell I, Salmela S, Beck-Sickinger AG, Larhammar D (2003) Agonists for neuropeptide Y receptors Y1 and Y5 stimulate different phases of feeding in guinea pigs. *Br J Pharmacol* 139: 1433–1440

110 Boswell T, Richardson RD, Schwartz MW, D'Alessio DA, Woods SC, Sipols AJ, Baskin D G, Kenagy GJ (1993) NPY and galanin in a hibernator: hypothalamic gene expression and effects on feeding. *Brain Res Bull* 32: 379–384

111 Kuenzel WJ, Douglass LW, Davison BA (1987) Robust feeding following central administration of neuropeptide Y or peptide YY in chicks, *Gallus domesticus*. *Peptides* 8: 823–828

112 Richardson RD, Boswell T, Raffety BD, Seeley RJ, Wingfield JC, Woods SC (1995) NPY increases food intake in white-crowned sparrows: effect in short and long photoperiods. *Am J Physiol* 268: R1418–R1422

113 Morris YA, Crews D (1990) The effects of exogenous neuropeptide Y on feeding and sexual behavior in the red-sided garter snake (*Thamnophis sirtalis parietalis*). *Brain Res* 530: 339–341

114 Crespi EJ, Vaudry H, Denver RJ (2004) Roles of corticotropin-releasing factor, neuropeptide Y and corticosterone in the regulation of food intake in *Xenopus laevis*. *J Neuroendocrinol* 16: 279–288

115 Volkoff H, Peter RE (2001) Interactions between orexin A, NPY and galanin in the control of food intake of the goldfish, *Carassius auratus*. *Regul Pept* 101: 59–72

116 Volkoff H, Canosa LF, Unniappan S, Cerda-Reverter JM, Bernier NJ, Kelly SP, Peter RE (2005) Neuropeptides and the control of food intake in fish. *Gen Comp Endocrinol* 142: 3–19

117 Inui A, Okita M, Nakajima M, Inoue T, Sakatani N, Oya M, Morioka H, Okimura Y, Chihara K, Baba S (1991) Neuropeptide regulation of feeding in dogs. *Am J Physiol* 261: R588–R594

118 Bjenning C, Holmgren S, Farrell AP (1993) Neuropeptide Y potentiates contractile response to norepinephrine in skate coronary artery. *Am J Physiol* 265: H661–H665

119 Bjenning C, Hazon N, Balasubramaniam A, Holmgren S, Conlon JM (1993) Distribution and activity of dogfish NPY and peptide YY in the cardiovascular system of the common dogfish. *Am J Physiol* 264: R1119–R1124

120 Xiang H (1994) Comparative aspects of the role of neuropeptide Y in the regulation of the vertebrate heart. Cardiosci 5: 209–213

121 Preston E, Jonsson AC, McManus CD, Conlon JM, Courtice GP (1998) Comparative

vascular responses in elasmobranchs to different structures of neuropeptide Y and peptide YY. *Regul Pept* 78: 57–67

122 Uesaka T (1996) Synergistic action of neuropeptide Y and adrenaline in the eel atrium. *J Exp Biol* 199: 1873–1880

123 Shahbazi F, Holmgren S, Larhammar D, Jensen J (2002) Neuropeptide Y effects on vasorelaxation and intestinal contraction in the Atlantic cod *Gadus morhua*. *Am J Physiol Regul Integr Comp Physiol* 282: R1414–R1421

124 Chartrel N, Conlon JM, Danger J-M, Fournier A, Tonon N-C, Vaudry H (1991) Characterization of melanotropin-release-inhibiting factor (melanostatin) from frog brain: Homology with human neuropeptide Y. *Proc Natl Acad Sci USA* 88: 3862–3866

125 Mor A, Chartrel N, Vaudry H, Nicolas P (1994) Skin peptide tyrosine-tyrosine, a member of the pancreatic polypeptide family: Isolation, structure, synthesis, and endocrine activity. *Proc Natl Acad Sci USA* 91: 10295–10299

126 Galas L, Tonon MC, Beaujean D, Fredriksson R, Larhammar D, Lihrmann I, Jegou S, Fournier A, Chartrel N, Vaudry H (2002) Neuropeptide Y inhibits spontaneous alpha-melanocyte-stimulating hormone (alpha-MSH) release via a Y(5) receptor and suppresses thyrotropin-releasing hormone-induced alpha-MSH secretion via a Y(1) receptor in frog melanotrope cells. *Endocrinology* 143: 1686–1694

127 Zukowska Z, Pons J, Lee EW, Li L (2003) Neuropeptide Y: a new mediator linking sympathetic nerves, blood vessels and immune system? *Can J Physiol Pharmacol* 81: 89–94

128 Bedoui S, Miyake S, Straub RH, von Horsten S, Yamamura T (2004) More sympathy for autoimmunity with neuropeptide Y? *Trends Immunol* 25: 508–512

129 Ericsson A, Larhammar D, McIntyre KR, Persson H (1987) A molecular genetic approach to the identification of genes expressed predominantly in the neuroendocrine and immune systems. *Immunol Rev* 100: 261–277

130 Lundell I, Blomqvist AG, Berglund MM, Schober DA, Johnson D, Statnick MA, Gadski RA, Gehlert DR, Larhammar D (1995) Cloning of a human receptor of the NPY receptor family with high affinity for pancreatic polypeptide and peptide YY. *J Biol Chem* 270: 29123–29128

131 Lundell I, Statnick MA, Johnson D, Schober DA, Starbäck P, Gehlert DR, Larhammar D (1996) The cloned rat pancreatic polypeptide receptor exhibits profound differences to the orthologous human receptor. *Proc Natl Acad Sci USA* 93: 5111–5115

Function, distribution and molecular pharmacology of NPY-family receptors

Magnus M. Berglund[1], *Philip A. Hipskind*[2] *and Donald R. Gehlert*[2]

[1]Biovitrum AB, Department of Lead Discovery, Stockholm, Sweden; [2]Eli Lilly and Company, Lilly Research Laboratories, Lilly Corporate Center, Indianapolis, IN 46285, USA

Introduction

Pancreatic polypeptide (PP) was the first member of the Neuropeptide Y (NPY) peptide family to be discovered. It was originally isolated as a contaminant observed during the preparation of insulin [1, 2]. NPY was first isolated by Tatemoto and Mutt using a method to identify peptides from pig brain with an amidated carboxy-terminus [3, 4]. In a similar manner, the final member of the peptide family was isolated from preparations of pig intestine and called peptide YY (PYY). While NPY is primarily localized to neurons and PP is found mainly in endocrine pancreas, PYY is found in both endocrine cells and neurons and, therefore, may exhibit neurotransmitter functions as well. The general structure of these peptides was first established using X-ray crystallography of avian PP [5]. In this structural conformation, amino acid residues 1–8 form a type II proline helix followed by a loop. Residues 15–32 form an a-helix, and the four most carboxy-terminal residues are in a flexible loop conformation. This hairpin-like three-dimensional structure has been called the PP-fold [6] and is present in all three peptides. There is remarkable evolutionary divergence in the peptide sequence conservation within this family. When comparing the amino acid sequences from a variety of mammals, only two of the 36 amino acids of NPY are variable. As such, NPY is one of the most evolutionary conserved peptides known. On the other hand, the PYY sequence exhibits eight variable amino acids between different orders of mammals. PP is one of the least conserved peptides known and appears to have undergone rapid and recent evolution. Despite the limited conservation observed in the PP and PYY amino acid sequences, the general three-dimensional structure is maintained across a wide variety of species indicating its potential importance in function.

The NPY gene is composed of four exons and results in the synthesis of a 97 amino acid pre-pro NPY and is located on human chromosome 7 at the locus 7p15.1 [7]. PreproNPY is proteolytically processed into the C-terminal peptide of NPY (CPON) and NPY. The C-terminal amidation of NPY is essential for its bio-

The NPY Family of Peptides in Immune Disorders, Inflammation, Angiogenesis and Cancer,
edited by Zofia Zukowska and Giora Z. Feuerstein

logical activity. NPY is further processed to $NPY_{3\text{-}36}$ and $NPY_{2\text{-}36}$ by two enzymes, dipeptidyl peptidase IV and aminopeptidase P, respectively. While the functional significance of CPON is unknown, the C-terminal fragments of NPY have some selectivity for the Y2 receptor subtype (for review see [8]). In the rat brain, NPY is found in numerous brain regions including the hypothalamus, amygdala, hippocampus, nucleus of the solitary tract, locus coeruleus, nucleus accumbens and cerebral cortex [9]. In the rat brain, NPY is colocalized with norepinephrine, GABA, somatostatin in agouti-related protein (AGRP) containing neurons (for review see [10]). The adrenal medulla is the primary source of circulating NPY [11] though it is also expressed in liver, heart, spleen and in endothelial cells of blood vessels [12]. NPY is abundant in peripheral neurons including sympathetic neurons where it is co-stored and co-released with norepinephrine [13]. It is also expressed in a subpopulation of parasympathetic neurons.

The human PYY and PP genes are found 10 kb apart on chromosome 17q21.1 [14]. Like NPY, the PYY and PP genes are organized with four exons and three introns. The rat preproPYY sequence is 98 amino acids consisting of a preceding signal peptide sequence and followed by a 31 amino acid C-terminal extension peptide called CPOP [15]. PYY is found primarily in the L cells of the gastrointestinal tract with the highest concentrations found in the rectum followed by the ileum and the colon [16, 17]. Like NPY, PYY is cleaved by dipeptidyl peptidase to the 3–36 fragment that is thought to be the principal active circulating peptide. PYY and $PYY_{3\text{-}36}$ are stored in the gut mucosal endocrine cells and both are found in the circulation [18]. Rat preproPP has a signal peptide sequence and a 30 residue C-terminal extension peptide [19]. PP is found in the pancreas and is localized to the islets of Langerhans [20].

NPY function

NPY is one of the most potent orexigenic peptides known and this aspect of its biological activity has been studied extensively. NPY has been found to induce food intake with a preference for carbohydrate-rich food when injected ICV in many species ranging from fish [21, 22] to rhesus monkey [23]. When administered centrally to rats, antisense oligonucleotides that reduce NPY translation decrease feeding [24, 25]. Centrally administered NPY also regulates metabolism by decreasing energy expenditure [26]. NPY levels are changed in all conditions involving a disturbed energy balance such as anorexia, bulimia nervosa, and diabetes [27].

Other centrally mediated effects of NPY include decreased thermogenesis [28], anti-convulsant activity [29, 30], modulation of arousal [31], anxiolysis [32] and modulation of cognition [33]. NPY deficient mice were found to have an impaired development of neurons in the olfactory epithelium [34] suggesting a role in neuronal development. Furthermore, central NPY is involved in regulation of reproduction [35] and the regulation of circadian rhythms [36]. NPY is frequently colocalized with noradrenaline (NA) in sympathetic nerves and enhances NA-mediated vasoconstriction, especially upon strong stimulation (see [37] for review on vascular effects). In addition, NPY appears to stimulate smooth muscle proliferation [38]. Contrary to this, centrally administered NPY reduces arterial blood pressure and heart tone in both rats and dogs [39]. In addition, NPY can act as an antinociceptive peptide [40], probably through inhibition of substance P release in the dorsal horn of the spinal cord.

Several cardiovascular dysfunctions as well as some tumor diseases [41] are associated with increased plasma levels of NPY. One of the most important recent developments in NPY research is the finding that NPY regulates ethanol consumption. In transgenic mice [42], an inverse correlation between NPY levels and drinking has been shown. Furthermore, a point mutation (leucine^7 to proline) in the pre-pro NPY gene has been associated with higher alcohol consumption in humans [43] and linkage analysis suggesting that dysfunction in the NPY gene locus contributes to ethanol preference in rats [44]. NPY appears to be involved in regulation of neuronal excitability as mice lacking NPY are more susceptible to seizures [29] and people with temporal lobe epilepsy have increased NPY expression in CA3 regions as well as prominent rearrangements in receptor distribution [45].

PYY and PYY3-36 function

PYY is released in the gastrointestinal (GI) tract in response to meals. Many of the GI effects described for PP including inhibition of gall bladder secretion, gut motility and pancreatic secretion can also be produced by PYY [46]. Other peripheral effects produced by PYY include the inhibition of fluid and electrolyte secretion in the intestinal tract [47] and vasoconstriction [48]. Similar to NPY, PYY when administered centrally, results in increased food intake. Through dipeptidyl peptidase IV, PYY is post-translationally processed to PYY_{3-36}, a peptide selective for the Y2 receptor. Selective Y2 agonism has been found to decrease food intake and peripherally administered PYY_{3-36} produces a potent and decrease in food intake in both human [49, 50] and rodents [51–55].

PP function

PP is almost exclusively expressed in endocrine pancreas and is released in response to meals [56]. PP produces its biological effects mainly in the gastrointestinal tract where it inhibits pancreatic secretion, gall bladder activity, intestinal motility [46], ileum contractions [57], gastric emptying [58] and stimulates colon contractions

[59]. In addition, PP affects metabolic functions including glycogenolysis and decreases fatty acid levels [27]. However, binding sites for PP have been found in several rat brain regions including the interpeduncular nucleus, hypothalamus, and brainstem [60–62] suggesting that PP may also have direct effects on brain function. This may explain why intracerebroventricular (ICV) injection of PP has been found to stimulate feeding in several different species of experimental animals [63–65]. In contrast, peripherally administered PP inhibits food intake in a similar fashion to $PYY_{3\text{-}36}$ in humans [66, 67] and reduces food intake and increases energy expenditure in rodents [58].

Receptors for the NPY family of peptides

The human NPY-family peptides bind to a family of four receptors, namely Y1, Y2, Y4 and Y5, all belonging to the G-protein coupled receptor (GPCR) superfamily. Some species including mouse have an additional receptor (y6) that is non-functional in humans and absent in rats. Y1, Y2 and Y5 bind NPY and PYY with similar high affinities. PP binds to Y4 with higher affinity than NPY and PYY and has moderate affinity for Y5. Pharmacologically, the NPY-family receptors can be distinguished by various synthetic peptides such as Pro^{34} substituted versions of NPY and PYY that have decreased Y2 potency. On the other hand, Y2 binds truncated versions of NPY and PYY such as $PYY_{3\text{-}36}$ and $NPY_{13\text{-}36}$ with similar affinity as the native peptides (see Tab. 1).

The Y1 receptor

The Y1 receptor was the first NPY receptor to be cloned [68, 69]. Full length NPY and PYY are potent agonists at Y1 whereas PP and C-terminally truncated NPY/PYY fragments display lower affinity. The gene for Y1 is located in a cluster together with Y2 and Y5 on human chromosome 4q31. Based on several lines of evidence, the Y1 receptor is involved in the feeding response of NPY [23, 70–75]. In addition, hypothalamic Y1 mRNA levels decrease during fasting [76]. While ICV injection of Y1 selective agonists increase feeding [77], Y1 antagonists can inhibit NPY-induced feeding [23, 78, 79]. The Y1 receptor is also involved in the cross-talk between NPY and another orexigenic peptide in the hypothalamus as the feeding response produced by central administration of melanin concentrating hormone (MCH) is attenuated by administration of Y1 receptor antagonists [80].

Most of the vascular effects of NPY and PYY are transduced via the Y1 receptor [38, 81–83] as shown by the inhibitory affects of various Y1-selective antagonists [83–86] on NPY induced vasoconstriction. Interestingly, the centrally regu-

Table 1 - Cloned NPY receptors with ligand binding profiles

Receptor	Ligand binding profile
Y1	$NPY \approx PYY \approx [Leu^{31},Pro^{34}]NPY > NPY_{2\text{-}36} > NPY_{3\text{-}36} \geq PP > NPY_{13\text{-}36}$
Y2	$NPY \geq NPY_{2\text{-}36} \approx NPY_{3\text{-}36} \approx NPY_{13\text{-}36} >> [Leu^{31},Pro^{34}]NPY$
Y4	$PP > PYY \geq NPY > NPY_{2\text{-}36}$
Y5	$NPY \approx PYY \approx NPY_{2\text{-}36} > hPP > [D\text{-}Trp^{32}]NPY > NPY_{13\text{-}36} > rPP$

lated effects of NPY, i.e., reduction of blood pressure and heart rate are also signaled mainly through Y1 [39] as are many of the behavioral effects of NPY such as anxiolysis and antidepressant activity [87–90]. Recent reports described an emerging role for NPY in the regulation of ethanol consumption (for review see [91]). Injections of NPY into the paraventricular nucleus increase ethanol intake in rats [92]. The importance of Y1 in the regulation of ethanol consumption has been confirmed using both Y1 specific antagonists [92, 93] and knockout animals [94]. In addition, knockout studies of the Y1 receptor have shown that the ability of NPY to regulate arousal [31, 95], neurogenic inflammation and antinociception through inhibition of substance P release in the dorsal horn [96] are mediated by Y1.

The Y2 receptor

Originally, the Y2 receptor was identified using vascular preparations and defined by the activity of amino-terminally truncated fragments of NPY and PYY such as $NPY_{3\text{-}36}$ and $NPY_{13\text{-}36}$ that are full agonists with similar potency as the native peptides [97] (Tab. 1). When position 34 of NPY or PYY (Gln) is replaced by the corresponding amino acid in PP (Pro), the resulting peptides do not bind to Y2 [98, 99]. When homology screening failed to identify Y2 clones, several research groups turned to various expression cloning approaches and found cDNA clones coding for proteins with PYY-binding abilities [100–102]. The Y2 receptor gene codes for a 381 amino acid protein that is located close to the Y1 gene on chromosome 4 [103]. Like Y1, the Y2 receptor is highly conserved between species with more than 90 % identity between orders of mammals [82, 100–102, 104, 105]. Surprisingly, the Y2 receptor was only about 30% identical to the Y1 and Y4 receptors explaining the failure of homology screening approaches.

The Y2 receptor is mainly located presynaptically where it acts as an autoreceptor inhibiting further release of neurotransmitter [97, 106, 107]. This may explain why agonists specific for the Y1 receptor are anxiolytic [87, 88] while Y2 agonists like NPY_{13-36} and C2-NPY appear to be anxiogenic [87, 108]. The same opposing relationship between Y1 and Y2 is evident for the central effects of NPY on blood pressure as Y2 specific agonists increase blood pressure while activation of central Y1 receptors decrease it [39]. The Y2 receptor is also directly involved in some of the vascular effects of NPY [109]. For instance, in pig spleen, a Y2 specific agonist evoked potent vasoconstriction [82] that could be inhibited by a Y2-selective antagonist, BIIE0246, [48]. In addition, Y2 is involved in NPY-induced angiogenesis [38, 110] and NPY-induced effects on circadian rhythms [36, 111–113]. Using Y2 receptor deficient mice, the importance of Y2 in the feeding response to NPY [114] as well as in bone formation [115] were demonstrated.

Whereas centrally administered NPY and PYY stimulate food intake, the Y2 selective agonist PYY_{3-36} administered peripherally to humans, rhesus monkey and rodents as well as centrally to rodents inhibits food intake and weight gain [49, 50, 116, 117]. Mice lacking the Y2 receptor are insensitive to the anorexic effects of PYY_{3-36} [49]. Based on these findings, clinical trials were initiated at Amylin and Nastech to develop PYY_{3-36} as an anti-obesity medication.

The Y4 receptor

The third Y receptor to be identified by cloning had high affinity for PP. The most interesting feature of the Y4 receptor may be the low degree of sequence identity between species. When comparing the rat and human sequences, only 75% of the amino acids are identical, making Y4 one of the most rapidly evolving GPCRs known [118, 119]. Considerable differences in peptide pharmacology between the rat and human receptors have been noted. For instance, when position 34 of NPY or PYY is replaced by proline [119–121] there is an increase in rat affinity while the human receptor is unaffected by this change [118, 122].

Since PP is the preferred endogenous ligand for the Y4 receptor, it is likely that this receptor mediates many of the GI effects produced by PP like rabbit ileum contractions [57]. Centrally located Y4 receptors may be involved in the regulation of reproduction as the Y1 antagonist/Y4 agonist 1229U91 [123, 124] induced release of luteinizing hormone when injected intracerebroventricularly [125].

Similar to the Y2 receptor, Y4 agonism may serve as a potential inhibitory signal on food intake. PP injected i.v. to human subjects reduced food intake [67] confirming several earlier studies suggesting PPs role as an anorexic peptide.

The Y5 receptor

NPY and NPY_{2-36} are equally potent in producing large increases in feeding after ICV administration [126] suggesting the receptor mediating the feeding response to

NPY differed from Y1, Y2, and Y4. In addition, NPY with position 32 replaced with D-tryptophan ([D-Trp^{32}]NPY) selectively inhibited NPY-induced feeding [127] though it had relatively low affinity for Y1 and Y2. Thus, a feeding receptor was proposed with the pharmacological profile NPY=PYY=NPY$_{2\text{-}36}$>NPY$_{3\text{-}36}$≥[D-Trp^{32}]NPY (Tab. 1). Expression cloning from a hypothalamic rat cDNA library resulted in a gene encoding a 446 amino acid protein [128, 129]. [D-Trp^{32}]NPY was found to be a modestly selective Y5 agonist that inhibited cAMP synthesis but with a lower potency than NPY, PYY and NPY$_{2\text{-}36}$ [128]. Although the Y5 receptor is highly conserved (88–90% overall amino acid identity and 95–98% when the third intracellular loop is not accounted for) [130, 131] between orders of mammals, there may be species differences in the endogenous ligand for the Y5 receptor. For instance, rat PP binds with very low affinity to the Y5 receptor from various species whereas PP from human and other species has much higher affinity [130]. Thus, it is possible that PP is involved in Y5 signaling in humans but not in rats. This may indicate a difference of potential importance when extrapolating effects produced by PP-fold peptides in rodents to physiological and behavioral effects in humans.

The role of the Y5 receptor in NPY-induced feeding has been confirmed by studies involving antisense knockdown [132–135], knockout animals [136], and Y5-selective agonists [26, 128, 137–139]. The selective Y5 antagonist CGP71683A [140] was reported to antagonize NPY induced feeding; however, recent findings suggest that it inhibited feeding by a non-Y5 mechanism [141, 142]. Other effects of NPY that are mediated by the Y5 receptor are reproduction through inhibition of luteinizing hormone release [143] and regulation of brain excitability and seizures [144, 145]. Furthermore, it has been shown that the Y5 selective agonist [D-Trp^{32}]NPY inhibited neuronal activity in the suprachiasmatic nucleus without generating a phase-shift [36] indicating that Y5 may also be indirectly involved in regulation of circadian rhythms [36, 146].

Signaling pathways of NPY family receptors

Cloned Y1, Y2, Y4, and Y5 receptors have all been shown to couple to Gi and thus mediate inhibition of cAMP synthesis [68, 69, 118, 128]. Signaling involves pertussis toxin-sensitive phosphorylation of extracellular signal regulated protein kinase (ERK) 1 and 2 in CHO cells confirming that these receptors couple to Gi/Go. The Y1 receptor has also been shown to activate mitogen-activated protein kinase in gut epithelial cells (IEC-6) [147]. Furthermore, whereas protein kinase C (PKC) mediates the phosphorylation of the Y5 receptor, a PKC independent pathway may also be involved in Y1, Y2, and Y4 signaling [148]. Y1, Y2, Y4, and Y5 receptors can also couple to phospholipase C to provoke release of Ca^{2+} from intracellular stores [68, 69, 122, 128].

Internalization and desensitization of NPY-family receptors

Agonist driven internalization has been reported for all NPY receptors but the kinetics and the degree of internalization varies. Internalization was found using fluorescent ligands allowing detection of sub-cellular distribution of receptors by confocal microscopy [149]. About 20–30% of Y1 receptors expressed in the neuroblastoma cell line SK-N-MC were internalized after stimulation [149]. Internalization of the Y1 receptor expressed in CHO cells has been found to follow the endosome pathway followed by rapid recycling to the cell surface [150–152]. Agonist driven internalization was also found for the Y4 receptor whereas the Y2 receptor did not display detectable internalization [150–152]. An elegant method to study receptor fate in real time is to link the receptor to a fluorescent marker such as green fluorescent protein (GFP). When expressed in HEK293 cells, the Y1 receptor fused to GFP was found to rapidly internalize after agonist (PYY) stimulation into endosomes and recycled to the surface within 60 min [151].

Bioluminescence energy transfer (BRET) is a method that uses the transfer of light from one luminescent molecule to a fluorophore enabling the measurement of distance between two closely located proteins. When all four NPY-receptors in rhesus monkey (Y1, Y2, Y4 and Y5) was fused to Renilla luciferase (Rluc) and cotransfected, one at a time, into HEK293 cells stably expressing the β-arrestin 2 fused to GFP, it was found that Y1 displayed a strong and rapid agonist driven interaction with β-arrestin 2 [153]. Y5 also displayed a rapid recruitment of β-arrestin 2 but with a lower maximal signal. Y2 showed very weak and slow increase in BRET signal suggesting a limited interaction with β-arrestin 2. As β-arrestin recruitment is a key part in the internalization, these data suggest that the Y1 and probably Y5 receptors internalize rapidly upon agonist stimulation whereas the Y2 receptor does not. The Y4 receptor may display an intermediate behavior. One obvious drawback of fusing luciferase or GFP to a receptor is the massive addition to the receptor protein as GFP and Rluc are only about 100 amino acids smaller than the entire receptor protein. Addition of the large GFP or Rluc molecules to the C-terminus of a receptor appears to interfere surprisingly little with function as binding and internalization seem to be intact [154]. However, in that study, G-protein coupling was impaired when GFP was coupled to the Y4 receptor [154]. Thus, it is possible that the binding pocket and the internalization machinery are less sensitive for carboxy-terminal modifications than the G-protein interaction is. Addition of GFP at the very amino-terminus of the Y1 receptor did produce a receptor that was able to functionally couple to G-proteins in HEK293 cells [151] upon agonist binding. Also, the Y1 receptor has been successfully expressed carrying GFP in the carboxy-terminus [155]. Very recently, using various C-terminally truncated Y1 receptors, the C-terminal region between Ser^{352} and Thr^{361} was identified as an important site for desensitization and β-arrestin 2 interaction of the Y1 receptor [156] as truncation after Ser352 abolished desensitization and β-arrestin 2 recruitment whereas truncation after Thr^{361} behaved as the WT receptor. Curiously however, both truncations resulted in receptors with a high level of ligand-independent internalization [156].

Dimerization among NPY-family receptors

An emerging concept for GPCRs is that the receptors form homo- and heterodimers and these complexes have functional implications [157]. These dimers can be covalently linked as described for the metabotropic glutamate receptors or held together by non-covalent forces as for many of the rhodopsin like receptors (family A). All four Y receptors have been found to homodimerize when expressed in HEK293 cells [154, 158]. This was shown using biochemical methods (i.e., Western blot) and biophysical methods such as photo-bleach FRET together with confocal microscopy or BRET. Y1, Y2 and Y5 homodimers were not affected by agonist treatment [158] whereas Y4 appears to be assembled as a dimer that is dissociated upon agonist stimulation. Interestingly, in a BRET study, the Y1 and Y5 receptors were found to heterodimerize when co-expressed in HEK293 cells [159]. Similar to the Y4 homodimer, Y1-Y5 heterodimerization was affected by ligand occupancy. Contrary to what was observed with Y4 homodimerization, Y1-Y5 heterodimerization was increased selectively by the agonists $NPY_{2\text{-}36}$ and NPY(19-23)-PP [139]. Non-selective agonists such as NPY failed to affect the BRET ratio. Moreover, co-expression of Y1 and Y5 increased the efficacy of another Y5 selective agonist, PP(1-17)-NPY(18-36) [139], to inhibit forskolin stimulated cAMP synthesis whereas the efficacy of non-selective peptides was unaffected by co-expression of the two receptors. It was also found that antagonists selective for either the Y1 or Y5 receptor decreased the BRET ratio suggesting that the heterodimer is either dissociated by antagonists or that a conformational-shift further separated the BRET counterparts (GFP and *Renilla luciferase*).

Distribution of NPY-family receptors

A number of approaches have been taken to localize NPY receptor subtypes. In general, Northern blot or PCR analysis of mRNA encoding the receptors has been used to evaluate general organ distribution while *in situ* hybridization was used to look at distributions within organs, especially within the brain. Distribution of the receptor protein has been evaluated using quantitative receptor autoradiography and, when specific antisera have been available, immunohistochemistry has also been utilized. Immunohistochemistry has the advantage of cellular resolution but provides only a qualitative representation of the receptor distribution.

Y1 receptors are primarily localized to the central nervous system and in peripheral blood vessels. Within the central nervous system, high levels of Y1 receptors were localized to regions such as the anterior thalamus, cerebral cortex, medial geniculate and the amygdala [160–163]. This distribution generally correlates with the distribution of Y1 mRNA using *in situ* hybridization [164]. In the peripheral nervous system, Y1 mRNA was found in the superior cervical and dorsal root ganglion [165]. Y1 mRNA has been detected in a number of peripheral tissues including the colon, kidney, adrenal gland, heart and placenta [166]. Within the kidney, Y1 mRNA was detected in the renal collecting ducts, loop of Henle and the juxtaglomerular apparatus. Within the other organs, Y1 mRNA was localized to intramyocardial, colonic and renal blood vessels with the expression being localized to the intima as well as the media.

Using a variety of techniques, the Y2 receptor was found in the central and peripheral nervous system, the intestine and certain blood vessels. Using receptor autoradiography, brain Y2 receptors are found in the hypothalamus, lateral septum and hippocampus and amygdala [160–163]. In the peripheral nervous system, Y2 receptors are found in the superior cervical ganglion and dorsal root ganglia [165]. Using PCR detection, mRNA encoding the Y2 receptor has been found in a variety of regions of the gastrointestinal tract including the crypt, villus, colonic epithelium, and non-epithelial jejunum [167].

In contrast to the Y1 and Y2 receptors, the Y4 receptor exhibits a rather restricted distribution within the brain. While mRNA encoding this receptor exhibits a rather broad distribution [164], studies using radiolabeled PP indicate a high level expression in the interpeduncular nucleus, locus coeruleus in the nucleus of the solitary tract [62]. A lower-level of expression is seen within the hypothalamus, thalamus as well as a number of other brain regions. Interestingly, the binding site labeled by ^{125}I-PP in the interpeduncular nucleus does not pharmacologically match the Y4 receptor indicating this may be an additional PP binding site [168]. Using Northern blot analysis, Y4 mRNA was found to be abundant in the human colon with somewhat lower levels in the prostate and pancreas [118]. In addition, mRNA encoding Y4 has been found in skeletal muscle, thyroid gland, heart, stomach, small intestine, adrenal medulla, cerebral cortex and the nasal mucosa [122].

The Y5 receptor was originally isolated from hypothalamus where it was believed to mediate orexigenic effects of NPY. The mRNA encoding the Y5 receptor been found have a broad distribution in the central nervous system [129, 164, 169], however, only low levels of Y5 receptor protein have been detected in the brain using receptor autoradiography [170, 171] and immunohistochemistry [172]. In the periphery, Y5 mRNA is been reported to be present in the intestine, ovary, testes, prostate, spleen, pancreas, kidney, skeletal muscle, liver, placenta and heart [128].

Agonists and antagonists for NPY-family receptors

Peptide ligands

Over the past four years there's been considerable progress in the development of novel selective analogs of NPY. A number of these analogs have been directed toward discovering tools to understand the role of the Y1 and Y5 receptors in the feeding produced by central administration of NPY. For instance, several novel analogs of NPY have been found to be highly selective for the Y1 receptor [77]. Three of these analogs are included in Table 2. These all produce an increase in feeding after intracerebroventricular administration. Over 4 h, [D-Arg^{25}]-NPY produced a similar increase in feeding compared to NPY. This increase in feeding was blocked by the peptidic Y1 antagonist 1229U91. However, it was found that 1229U91, besides being a Y1 antagonist, also is a potent agonist at the Y4 receptor [123, 124]. Soll and co-workers [173] have reported a very potent and Y1 selective peptide [Phe^7, Pro^{34}]-NPY, though no work has been reported on its effects *in vivo*. In addition to novel Y1 receptor agonists, an analog of the Y1 selective peptide antagonist 1229U91 was recently reported [174]. This analog consisted of an OMe replacement for the amide in 1229U91 (GR231118). This peptide retained high affinity for the Y1 receptor but had much reduced affinity for the Y4 receptor when compared to 1229U91. In functional assays, this peptide analog potently antagonized the NPY inhibition of forskolin-stimulated adenylate cyclase in Y1 receptor containing cells while having no effect (agonist or antagonist) in Y4 receptor containing cells. When administered intrahypothalamically, this Y1 antagonist inhibited both NPY stimulated feeding as well as feeding in schedule fed rats.

More limited work has been performed to obtain Y2 selective peptide analogs. One of the more promising peptide analogs was recently published by Soll et al. [173]. This peptide (Cyclo S-S [Cys^{20}, Cys^{24}]pNPY) exhibits subnanomolar affinity for the Y2 receptor with low micromolar affinity at Y1 and Y5. No additional information is available in the scientific literature that further describes the properties of this compound. On the other hand, considerable progress has been made for developing Y5 selective agonists. However, several of these peptide analogs have considerable affinity for both the Y4 and Y5 receptors (Tab. 2). One of these analogs, 2-36[K4, $RYYSA^{19\text{-}23}$]-PP, produced a dose-dependent increase in food consumption after intracerebroventricular administration [138]. The magnitude of this increase in food intake exceeded that produced by identical doses of NPY. More selective Y5 agonist peptides have also been reported that have little affinity for the Y4 receptor. One of these, [Ala^{31}, Aib^{32}]-NPY was found to produce a small increase in food intake when compared to control animals [139]. Another analog, [$cPP^{1\text{-}7}$, $NPY^{19\text{-}23}$, Ala^{31}, Aib^{32}, Gln^{34}]-hPP, which exhibited an approximately 20-fold higher affinity for the Y5 receptor when compared to analog [Ala^{31}, Aib^{32}]-NPY, produced increases in food intake that exceeded NPY at similar doses [139]. Finally, the relatively

Table 2 - Summary of in vivo *feeding effects for compounds 9-14*

Compound	Y5-driven feeding	NPY-driven feeding	Fasted or ad lib feeding	Refs.
9	100 (3, PO)	NA (3, PO)	NA (30, PO)	[192]
10	–	–	NA (40, PO)	[193]
11	53 (30, PO)	–	NA (30, PO)	[195]
12	active	–	NA	[198]
13	100 (30, PO)	–	NA (30, PO)	[196]
14	65 (100, ug icv)	24 (100, ug icv)	–	[197]

(xx (mg/Kg); xx = % inhib; NA, not active; MA, minimally active)

low affinity analog, p[D-Trp^{34}]-NPY, was reported to be a selective Y5 agonist [175]. This peptide produced a similar inhibition of forskolin-stimulated adenylate cyclase to that seen for NPY and was found to increase food intake after intracerebroventricular administration. However, the increase was not as substantial is that observed with similar doses of NPY. The putative Y5 selective antagonist, CGP71683A, antagonized the increase in food intake produced by this peptide, however, the specificity of CGP71683A has recently been called into question [141].

Small molecule NPY medicinal chemistry

This portion of the review will concentrate on the primary, small molecule NPY medicinal chemistry literature appearing over the last 2 years. Although an effort was made to catalog all ligands that appeared in journal publications, a particular emphasis has been placed on those series that have reported *in vivo* biological data. The structures of all compounds discussed below are presented in Figures 1–3.

NPY Y1 ligands

Bristol-Myers-Squibb (BMS) continues to be active in the Y1 arena with molecules that improve upon the properties of their first entry into the field, BMS 193885 (compound **1**). In 2002 they had reported a 52% reduction in spontaneous nocturnal food intake by the intraperitoneal route with **1** [176]. As this route of administration can be fraught with complicated interpretations in food consumption assays, they sought an orally available agent. Unfortunately poor bioavailability (<0.1%) was observed with compound **1**, and, as the team rationalized that an improvement

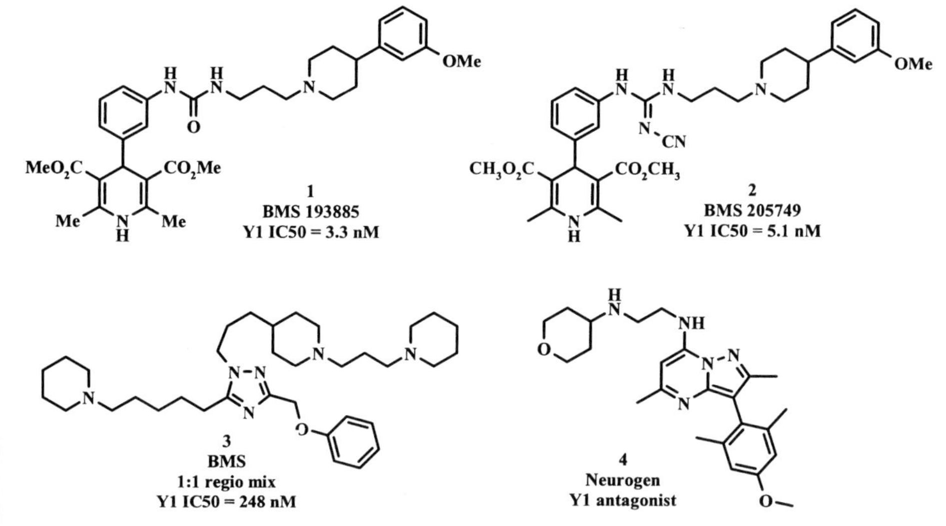

Figure 1
Novel non-peptide analogs with selectivity for the Y1 receptor

in permeability was required, the urea functionality was replaced. Cyanoquanidine **2** was discovered as a selective and potent (Y1 IC_{50} = 5.1 nM), analog with improved CACO-2 permeability [177]. Unfortunately no feeding data with compound **2** was disclosed.

BMS also reported on the preparation of a combinatorial library based on NPY Y1 ligands disclosed first in 1997 by Lilly [178–181]. Unfortunately the highest potency achieved with their 1,2,4-triazole analogs was 248 nM (compound **3**) [182].

Neurogen recently disclosed a series of bicyclic NPY Y1 ligands likely derived from CRF1 active molecules [183–185]. A multitude of various core structures adorned with amine substituents were claimed in this series of three extensive patents. Compound **4** was a prominent example.

NPY Y2 ligands

Over the last two years, there has been only two additional reports (beyond the original disclosures of **5** by Karl Thomae [186, 187] describing selective NPY Y2

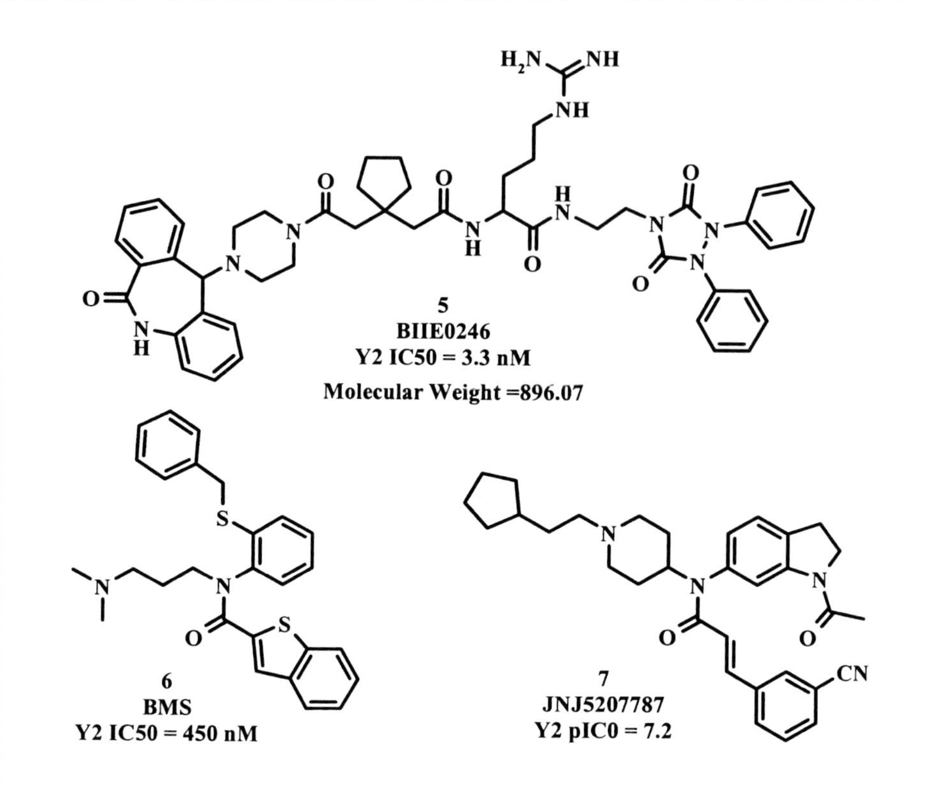

Figure 2
Novel non-peptide analogs with selectivity for the Y2 receptor

receptor ligands in the primary medicinal chemistry literature. Beginning with an HTS hit and again using a combinatorial approach, workers at BMS were able to find a modestly potent 450 nM ligand (compound **6**) [188]. Unfortunately, they were unable to further optimize this series as no further work is evident in the literature. More recently workers at Johnson & Johnson reported on a series of cinnamide analogs exemplified by JNJ5207787 (compound **7**). This analog exhibited an IC_{50} of approximately 100 nM in a radioligand binding assay and inhibited PYY stimulated [^{35}S]GTPgS binding with a pIC_{50} of 7.2. Additional details including *in vivo* activity were recently reported [189]. NPY Y2 binding sites ([^{125}I]PYY) sensitive to 7 were found in rat brain regions known to express Y2 receptors (*versus* the converse with respect to Y1 sites), and in addition, 7 was shown to occupy Y2 receptor binding sites as revealed by *ex vivo* receptor occupancy (45% at 1 h, 30 mg/Kg IP). It should be noted that while **6** and **7** are of

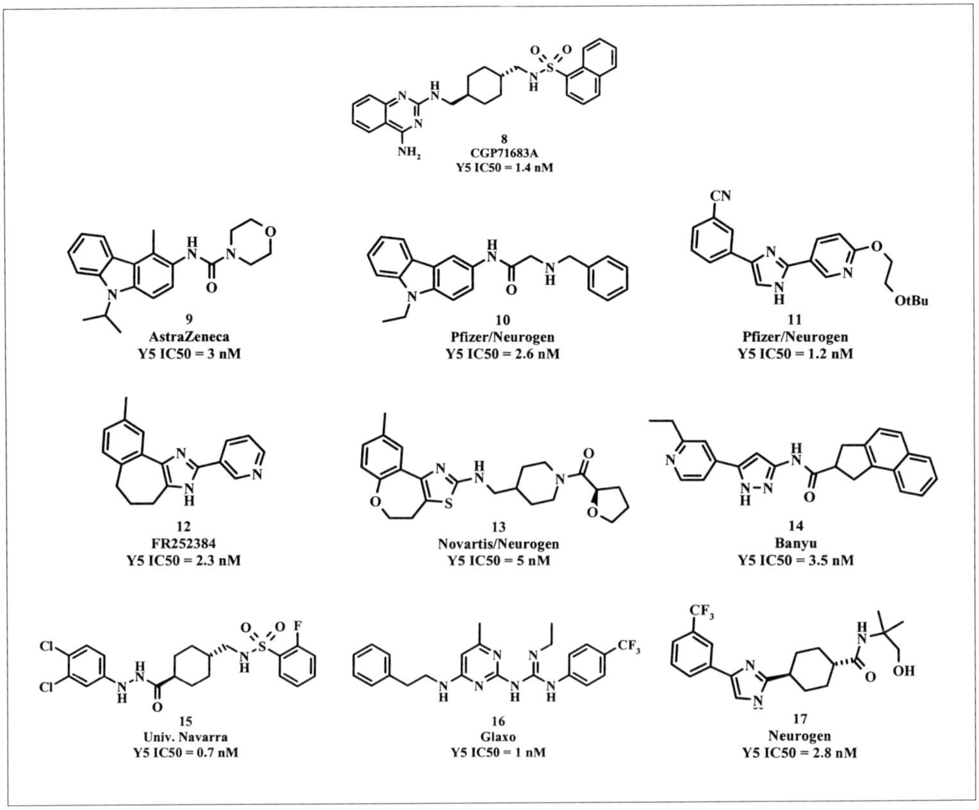

Figure 3
Novel non-peptide analogs with selectivity for the Y5 receptor

modest potency, they represent considerable advancements *versus* compound 5 with respect to molecular weight.

NPY Y5 ligands

Over the last two years, several groups have been quite active in the field of small molecule NPY Y5 ligands. Yielding one of the more diverse arrays of ligand structure types, most pharmaceutical companies found it straightforward to at least identify a starting NPY Y5 small molecule structure for *in vitro* potency optimization. For example, compound 8 (CGP71683A) was reported to reduce food consumption in a variety of NPY and non-NPY driven models [140, 190]. Unfortunately these results were clouded by similar activity observed in NPY Y5 knockout animals [141,

Table 3 - Novel peptide analogs with selectivity for PP-fold peptide receptor subtypes

Peptides	K_i (nM) or IC_{50} (nM) Y1	Y2	Y4	Y5	Ref.
Y1 agonist					
[D-Arg^{25}]-NPY*	0.9	11.6	74.6	43.4	[77]
[D-His^{26}]-NPY*	2.0	29.0	20.1	34.6	[77]
Des-AA$^{11\text{-}18}$[Cys^{7,21}, D-Lys^{9}(Ac), D-His^{26}, Pro^{34}]-NPY*	1.2	801	31.4	2363	[77]
[Phe^{7}, Pro^{34}]-pNPY*	0.009	32.1	ND	34	[173]
Y1 antagonist					
1229U91 (GR231118) –OMe substituted*	0.46	624	65.5	7890	[174]
Y2 agonist					
Cyclo S-S [Cys^{20}, Cys^{24}]pNPY*	1525	0.38	ND	6296	[173]
Y4/Y5 agonist					
[cPP$^{1\text{-}7}$, pNPY$_{19\text{-}23}$, His^{34}]-hPP	5.7	22.2	0.06	0.04	[213]
2-36[K^{4}, RYYSA$_{19\text{-}23}$]-PP*	0.87	1.95	0.004	0.029	[138]
Y5 agonist					
[Ala^{31}, Aib^{32}]-NPY	>700	>500	>1000	6.0	[139]
[cPP$^{1\text{-}7}$, NPY$_{19\text{-}23}$, Ala^{31}, Aib^{32}, Gln^{34}]-hPP	530	>500	51	0.24	[139]
[hPP$^{1\text{-}17}$,Ala^{31},Aib^{32}]hNPY	>1000	>500	170	0.92	[139]
p[D-Trp^{34}]-NPY**	>10,000	631	1905	41.7	[175]

*K_i values; **converted to K_i values from pK_i values; ND, not done*

191]. This has spurred a number of other groups to investigate alternative structure types. For example, compounds **9–14** are representative of the various platforms that have some type of *in vivo* food consumption data reported in the literature. AstraZeneca developed compound **9**, and it was reported to completely block NPY Y5 agonist induced feeding at 3 mg/Kg. Unfortunately **9** did not block NPY-driven feeding (3 mg/Kg) and also had no effect on free or fast-induced feeding (30 mg/Kg) in rats (see Tab. 3) [192]. Starting from a similar hit structure, the Pfizer group developed compound 10 [193]. A highly potent NPY Y5 antagonist *in vitro*, compound **10** did not block fasted feeding in rats at 40 mg/Kg, PO.

A group of companies identified 1,3-disubstituted, five-membered ring heterocycles as NPY Y5 screening hits. Compounds **11–14** and **17** are examples [194–198].

In vivo data for the first group 11–14 is also summarized in Table 3. As one can see, a number of researchers have been able to develop potent compounds capable of blocking NPY Y5-driven feeding *in vivo*, however it has been a struggle to observe a robust block of an endogenous or a general NPY driven food effect. This has lead many of these authors to conclude that NPY Y5 is likely not a major mediator of feeding in rodents.

Three other investigators have reported additional structures without additional *in vivo* feeding data. The University of Navarra reported on compound **15**, a simplification of **8** (CGP71683A) [199]. Glaxo disclosed a series of metabolically stable, brain permeable guanidine analogs represented by **16** [200]. Finally, starting from a potent, low molecular weight hit with minimal oral bioavailability, compound 17 was engineered with markedly improved bioavailability and low hERG potential [194].

Mutagenesis studies of NPY-family receptors

Several mutagenesis studies have been conducted to identify the key amino acids in receptor-ligand interaction at the Y1 receptor [201–207]. All in all, 68 positions out of the 384 of the Y1 receptor have been mutated and the some key findings are summarized in Table 4. Initially, the significance of acidic amino acids in the extracellular regions of the Y1 receptor [201] were proposed to interact with basic amino acids of NPY and PYY. Furthermore, the amino terminal amide of the ligands was found to interact with amino acids in the TM regions forming a hydrophobic pocket between TM4, TM6 and TM7 [202]. There are numerous positions that have been mutated by different labs with contradictory results (see [8] for a more extensive discussion on Y1 mutagenesis). In some cases, this can be explained by the differences in expression systems used. These include a vaccinia virus vector to express the human Y1 receptor in HeLa cells [201, 202], *E. coli* [203] and mammalian cell lines [205–207]. The wild type Y1 receptor has also successfully been expressed in insect (Sf21) cells [208]. Considering the differences between studies, it is apparent that the initial model of the Y1 receptor with the acidic amino acids in the loops needs to be re-evaluated. As far as we know, only one study has been published regarding mutagenesis of a non-Y1 NPY-family receptor. This study of the human and chicken Y2 receptors employed species differences in the protein sequences and correlated it with pharmacological differences with respect to antagonist BIIE0246 binding [209]. Three amino acids in TM regions 3 (Gln^{135}), 5 (Leu^{227}) and 6 (Leu^{284}) were replaced for His, Gln and Phe in the chicken Y2 receptor preventing BIIE0246 binding. Reciprocal mutations of these amino acids could swap the functions (human–chicken and *vice versa*). Several additional mutagenesis studies of the Y1, Y2 and Y5 receptors have been presented as posters at meetings but remain to be published in manuscript form [210].

Table 4 Summary of mutagenesis and other structural studies of the Y1 receptor

Mutated amino acid(s)	Major findings	Refs.
Δ2-20, [N2Q, N11Q, N17Q], Y100R	No detection of surface expression	[201, 206]
N-term FLAG,[L35R, D200A], D104G, W106A, E110G, C113S, Q120Y, F221A, H306G	No binding of ^{3}H-NPY or ^{125}I-PYY	[205, 206]
D86A, D86N, D86E, Q120E, N283D	No binding of ^{3}H-BIBP3226	[204]
W163A, N283A, D287A	No binding of agonists and antagonists	[201, 203, 204, 206, 207]
[N2Q, N17Q], E10A, N17Q E20A, E29A, [D31A,D32A],L35R, P36H, L43A, F96A, Δ333-384, His-FLAG in C-term	Bind ^{3}H-NPY with the same affinity as WT	[201, 202, 206, 207]
L174A, Y176A, V178A, H207A, S210A, T212A, L279A, T280A, N289A, Q291A	Bind NPY and BIBP3226 with the same affinity as WT	[203–205]
F41A, V89A, E110A, S123A, L171A, Y176A, Q177A, D181A, E182A, L216A, N297A, T308A	Bind NPY and antagonists 1229U91 and J-104870 as WT	[202, 207]
Y100A, D104A,	Binds antagonists but not peptidic agonists	[201, 202, 204, 206, 207]
H105A, S170A, D205A	Binds BIBP3226 as WT but decreased affinity for peptide agonists	[201, 202, 207]
Y47A, Y211A, W276A, H298G	Bind peptide agonists as WT but decreased affinity for BIBP3226	[204, 205]
C-terminal GFP	Binds NPY as WT	[151, 155, 158]
C-terminal deletion at S^{352}	No desensitization and no β-arrestin2 recruitment. Binding of agonists is un-altered	[156]

Conclusions and future perspectives

The more we study the NPY family of receptors, the more functions are assigned to them. Developments of knockout animals and receptor selective antagonists have revealed several novel functions of the NPY-family receptors. However, the complete picture of what receptor is responsible for a particular function is still illusive, perhaps due to the large degree of redundancy in the NPY peptides and receptors. Double or conditional knockout animals will no doubt become valuable tools here [211, 212]. In addition, developmental compensation to the loss of NPY or Y receptors is poorly understood and this may account for some of the phenotypes or lack thereof observed in behavioral studies.

Successful attempts to develop clinically useful Y1 or Y5 selective antagonists have failed due to lack of efficacy in preclinical models of food intake. The vast co-expression of the Y1 and Y5 receptors in the brain along with the evidence for a Y1/Y5 hetrodimer suggests that the alleged "feeding receptor" responsible for the centrally regulated orexigenic effects of NPY and PYY may be via a heterodimer. However, the recently re-discovered anorectic effects of peripheral agonism at the Y2 and Y4 receptors may, in fact, constitute more appealing treatment strategies for obesity and the metabolic syndrome. Other interesting areas where drugs acting at NPY receptor have prospects as novel therapeutics include alcoholism and various mood disorders. Finally, several findings suggest that the NPY family peptides and receptors are involved in nociception and inflammation, aspects that have been somewhat neglected compared to the predominant interest in the metabolic functions of this fascinating peptide family.

References

1 Kimmel JR, Hayden LJ, Pollock HG (1975) Isolation and characterization of a new pancreatic polypeptide hormone. *J Biol Chem* 250: 9369–9376

2 Chance RE, Moon NE, Johnson MG (1979) Human pancreatic-polypeptide (HPP) and bovine pancreatic-polypeptide (BPP). In: BM Jaffe, HR Behlman (eds): *Methods in hormone radioimmunoassay*. Academic Press, New York

3 Tatemoto K (1982) Neuropeptide Y: complete amino acid sequence of the brain peptide. *Proc Natl Acad Sci USA* 79: 5485–5489

4 Tatemoto K, Carlquist M, Mutt V (1982) Neuropeptide Y-a novel brain peptide with structural similarities to peptide YY and pancreatic polypeptide. *Nature* 296: 659–660

5 Blundell TL, Pitts JE, Tickle IJ, Wood SP, Wu C-W (1981) X-ray analysis (1.4 Å resolution) of avian pancreatic polypeptide: small globular protein hormone. *Proc Natl Acad Sci* 78: 4175–4179

6 Fuhlendorff J, Johansen NL, Melberg SG, Thogersen H, Schwartz TW (1990) The antiparallel pancreatic polypeptide fold in the binding of neuropeptide Y to Y1 and Y2 receptors. *J Biol Chem* 265: 11706–11712

7 Cerda-Reverter JM, Martinez-Rodriguez G, Zanuy S, Carrillo M, Larhammar D (2000) Molecular evolution of the neuropeptide Y (NPY) family of peptides: cloning of three NPY-related peptides from the sea bass (Dicentrarchus labrax). *Regul Pept* 95: 25–34

8 Berglund MM, Hipskind PA, Gehlert DR (2003) Recent developments in our understanding of the physiological role of PP-fold peptide receptor subtypes. *Exp Biol Med (Maywood)* 228: 217–244

9 Chronwall BM, DiMaggio DA, Massari VJ, Pickel VM, Ruggiero DA, O'Donohue TL (1985) The anatomy of neuropeptide-Y-containing neurons in rat brain. *Neuroscience* 15: 1159–1181

10 Kask A, Harro J, von Horsten S, Redrobe JP, Dumont Y, Quirion R (2002) The neurocircuitry and receptor subtypes mediating anxiolytic-like effects of neuropeptide Y. *Neurosci Biobehav Rev* 26: 259–283

11 Cavadas C, Ribeiro CA, Cotrim M, Mosimann F, Brunner HR, Grouzmann E (2002) Catecholamine and neuropeptide Y secretion from human adrenal chromaffin cells: effect of nicotine and KCl. *Ann NY Acad Sci* 971: 332–334

12 Strand FL (1999) *Neuropeptides:regulators of physiological processes.* Charles F. Stevens, London

13 Sundler F, Böttcher G, Ekblad E, Håkanson R (1993) PP, PYY and NPY – Occurence and distribution in the periphery. In: WF Colmers, C Wahlested (eds): *The biology of neuropeptide Y and related peptides.* Humana Press, Totowa, NJ, 157–183

14 Hort Y, Baker E, Sutherland GR, Shine J, Herzog H (1995) Gene duplication of the human peptide YY gene (PYY) generated the pancreatic polypeptide gene (PPY) on chromosome 17q21.1. *Genomics* 26: 77–83

15 Krasinski SD, Wheeler MB, Leiter AB (1991) Isolation, characterization, and developmental expression of the rat peptide-YY gene. *Mol Endocrinol* 5: 433–440

16 Adrian TE, Ferri GL, Bacarese-Hamilton AJ, Fuessl HS, Polak JM, Bloom SR (1985) Human distribution and release of a putative new gut hormone, peptide YY. *Gastroenterology* 89: 1070–1077

17 Ekblad E, Sundler F (2002) Distribution of pancreatic polypeptide and peptide YY. *Peptides* 23: 251–261

18 Grandt D, Schimiczek M, Beglinger C, Layer P, Goebell H, Eysselein VE, Reeve JR, Jr. (1994) Two molecular forms of peptide YY (PYY) are abundant in human blood: characterization of a radioimmunoassay recognizing PYY 1-36 and PYY 3-36. *Regul Pept* 51: 151–159

19 Yamamoto H, Nata K, Okamoto H (1986) Mosaic evolution of prepropancreatic polypeptide. *J Biol Chem* 261: 6156–6159

20 Kimmel JR, Pollock HG, Chance RE, Johnson MG, Reeve JR Jr, Taylor IL, Miller C, Shively JE (1984) Pancreatic polypeptide from rat pancreas. *Endocrinology* 114: 1725–1731

21 Lopez-Patino MA, Guijarro AI, Isorna E, Delgado MJ, Alonso-Bedate M, de Pedro N (1999) Neuropeptide Y has a stimulatory action on feeding behavior in goldfish (*Carassius auratus*). *Eur J Pharmacol* 377: 147–153

22 Narnaware YK, Peter RE (2002) Influence of diet composition on food intake and neuropeptide Y (NPY) gene expression in goldfish brain. *Regul Pept* 103: 75–83
23 Larsen PJ, Tang-Christensen M, Stidsen CE, Madsen K, Smith MS, Cameron JL (1999) Activation of central neuropeptide Y Y1 receptors potently stimulates food intake in male rhesus monkeys. *J Clin Endocrinol Metab* 84: 3781–3791
24 Akabayashi A, Wahlestedt C, Alexander JT, Leibowitz SF (1994) Specific inhibition of endogenous neuropeptide Y synthesis in arcuate nucleus by antisense oligonucleotides suppresses feeding behavior and insulin secretion. *Brain Res Mol Brain Res* 21: 55–61
25 Hulsey MG, Pless CM, White BD, Martin RJ (1995) ICV administration of anti-NPY antisense oligonucleotide: effects on feeding behavior, body weight, peptide content and peptide release. *Regul Pept* 59: 207–214
26 Hwa JJ, Witten MB, Williams P, Ghibaudi L, Gao J, Salisbury BG, Mullins D, Hamud F, Strader CD, Parker EM (1999) Activation of the NPY Y5 receptor regulates both feeding and energy expenditure. *Am J Physiol* 277: R1428–R1434
27 Gehlert DR (1998) Multiple receptors for the pancreatic polypeptide (PP-fold) family: physiological implications. *Proc Soc Exp Biol Med* 218: 7–22
28 Lopez-Valpuesta FJ, Nyce JW, Griffin-Biggs TA, Ice JC, Myers RD (1996) Antisense to NPY-Y1 demonstrates that Y1 receptors in the hypothalamus underlie NPY hypothermia and feeding in rats. *Proc R Soc Lond B Biol Sci* 263: 881–886
29 Erickson JC, Clegg KE, Palmiter RD (1996) Sensitivity to leptin and susceptibility to seizures of mice lacking neuropeptide Y. *Nature* 381: 415–421
30 Kopp J, Nanobashvili A, Kokaia Z, Lindvall O, Hokfelt T (1999) Differential regulation of mRNAs for neuropeptide Y and its receptor subtypes in widespread areas of the rat limbic system during kindling epileptogenesis. *Brain Res Mol Brain Res* 72: 17–29
31 Naveilhan P, Canals JM, Valjakka A, Vartiainen J, Arenas E, Ernfors P (2001) Neuropeptide Y alters sedation through a hypothalamic Y1-mediated mechanism. *Eur J Neurosci* 13: 2241–2246
32 Heilig M, Widerlov E (1995) Neurobiology and clinical aspects of neuropeptide Y. *Crit Rev Neurobiol* 9: 115–136
33 Flood JF, Hernandez EN, Morley JE (1987) Modulation of memory processing by neuropeptide Y. *Brain Res* 421: 280–290
34 Hansel DE, Eipper BA, Ronnett GV (2001) Neuropeptide Y functions as a neuroproliferative factor. *Nature* 410: 940–944
35 Kalra SP, Xu B, Dube MG, Moldawer LL, Martin D, Kalra PS (1998) Leptin and ciliary neurotropic factor (CNTF) inhibit fasting-induced suppression of luteinizing hormone release in rats: role of neuropeptide Y. *Neurosci Lett* 240: 45–49
36 Gribkoff VK, Pieschl RL, Wisialowski TA, van den Pol AN, Yocca FD (1998) Phase shifting of circadian rhythms and depression of neuronal activity in the rat suprachiasmatic nucleus by neuropeptide Y: mediation by different receptor subtypes. *J Neurosci* 18: 3014–3022
37 Franco-Cereceda A, Liska J (1998) Neuropeptide Y Y1 receptors in vascular pharmacology. *Eur J Pharmacol* 349: 1–14

38 Zukowska-Grojec Z, Karwatowska-Prokopczuk E, Rose W, Rone J, Movafagh S, Ji H, Yeh Y, Chen WT, Kleinman HK, Grouzmann E et al (1998) Neuropeptide Y: a novel angiogenic factor from the sympathetic nerves and endothelium. *J Neurosci* 83: 187–195

39 Morton KD, McCloskey MJ, Potter EK (1999) Cardiorespiratory responses to intracerebroventricular injection of neuropeptide Y in anaesthetised dogs. *Regul Pept* 81: 81–88

40 Broqua P, Wettstein JG, Rocher MN, Gauthier-Martin B, Riviere PJ, Junien JL, Dahl SG (1996) Antinociceptive effects of neuropeptide Y and related peptides in mice. *Brain Res* 724: 25–32

41 Dotsch J, Christiansen H, Hanze J, Lampert F, Rascher W (1998) Plasma neuropeptide Y of children with neuroblastoma in relation to stage, age and prognosis, and tissue neuropeptide Y. *Regul Pept* 75–76: 185–190

42 Thiele TE, Marsh DJ, Ste. Marie L, Bernstein IL, Palmiter RD (1998) Ethanol consumption and resistance are inversely related to neuropeptide Y levels. *Nature* 396: 366–369

43 Ilveskoski E, Kajander OA, Lehtimaki T, Kunnas T, Karhunen PJ, Heinala P, Virkkunen M, Alho H (2001) Association of neuropeptide y polymorphism with the occurrence of type 1 and type 2 alcoholism. *Alcohol Clin Exp Res* 25: 1420–1422

44 Carr LG, Foroud T, Bice P, Gobbett T, Ivashina J, Edenberg H, Lumeng L, Li TK (1998) A quantitative trait locus for alcohol consumption in selectively bred rat lines. *Alcohol Clin Exp Res* 22: 884–887

45 Furtinger S, Pirker S, Czech T, Baumgartner C, Ransmayr G, Sperk G (2001) Plasticity of Y1 and Y2 receptors and neuropeptide Y fibers in patients with temporal lobe epilepsy. *J Neurosci* 21: 5804–5812

46 Hazelwood RL (1993) The pancreatic polypeptide (PP-fold) family: gastrointestinal, vascular, and feeding behavioral implications. *Proc Soc Exp Biol Med* 202: 44–63

47 Eto B, Boisset M, Anini Y, Voisin T, Desjeux JF (1997) Comparison of the antisecretory effect of endogenous forms of peptide YY on fed and fasted rat jejunum. *Peptides* 18: 1249–1255

48 Malmstrom RE (2001) Vascular pharmacology of BIIE0246, the first selective non-peptide neuropeptide Y Y(2) receptor antagonist, *in vivo*. *Br J Pharmacol* 133: 1073–1080

49 Batterham RL, Cowley MA, Small CJ, Herzog H, Cohen MA, Dakin CL, Wren AM, Brynes AE, Low MJ, Ghatei MA et al (2002) Gut hormone PYY(3-36) physiologically inhibits food intake. *Nature* 418: 650–654

50 Batterham RL, Cohen MA, Ellis SM, Le Roux CW, Withers DJ, Frost GS, Ghatei MA, Bloom SR (2003) Inhibition of food intake in obese subjects by peptide YY3-36. *N Engl J Med* 349: 941–948

51 Challis BG, Pinnock SB, Coll AP, Carter RN, Dickson SL, O'Rahilly S (2003) Acute effects of PYY3-36 on food intake and hypothalamic neuropeptide expression in the mouse. *Biochem Biophys Res Commun* 311: 915–919

52 van den Hoek AM, Heijboer AC, Corssmit EP, Voshol PJ, Romijn JA, Havekes LM, Pijl

H (2004) PYY3-36 reinforces insulin action on glucose disposal in mice fed a high-fat diet. *Diabetes* 53: 1949–1952

53 Adams SH, Won WB, Schonhoff SE, Leiter AB, Paterniti JR Jr (2004) Effects of peptide YY[3-36] on short-term food intake in mice are not affected by prevailing plasma ghrelin levels. *Endocrinology* 145: 4967–4975

54 Pittner RA, Moore CX, Bhavsar SP, Gedulin BR, Smith PA, Jodka CM, Parkes DG, Paterniti JR, Srivastava VP, Young AA (2004) Effects of PYY[3-36] in rodent models of diabetes and obesity. *Int J Obes Relat Metab Disord* 28: 963–971

55 Chelikani PK, Haver AC, Reidelberger RD (2005) Intravenous infusion of PYY(3-36) potently inhibits food intake in rats. *Endocrinology* 146: 879–888

56 Schwartz TW, Rehfeld JF, Stadil F, Larson LI, Chance RE, Moon N (1976) Pancreatic-polypeptide response to food in duodenal-ulcer patients before and after vagotomy. *Lancet* 1: 1102–1105

57 Feletou M, Nicolas JP, Rodriguez M, Beauverger P, Galizzi JP, Boutin JA, Duhault J (1999) NPY receptor subtype in the rabbit isolated ileum. *Br J Pharmacol* 127: 795–801

58 Asakawa A, Inui A, Yuzuriha H, Ueno N, Katsuura G, Fujimiya M, Fujino MA, Niijima A, Meguid MM, Kasuga M (2003) Characterization of the effects of pancreatic polypeptide in the regulation of energy balance. *Gastroenterology* 124: 1325–1336

59 Pheng LH, Perron A, Quirion R, Cadieux A, Fauchere JL, Dumont Y, Regoli D (1999) Neuropeptide Y-induced contraction is mediated by neuropeptide Y Y2 and Y4 receptors in the rat colon. *Eur J Pharmacol* 374: 85–91

60 Trinh T, van Dumont Y, Quirion R (1996) High levels of specific neuropeptide Y/pancreatic polypeptide receptors in the rat hypothalamus and brainstem. *Eur J Pharmacol* 318: R1–3

61 Gehlert DR, Schober DA, Gackenheimer SL, Beavers L, Gadski R, Lundell I, Larhammar D (1997) [125I]Leu31, Pro34-PYY is a high affinity radioligand for rat PP1/Y4 and Y1 receptors: evidence for heterogeneity in pancreatic polypeptide receptors. *Peptides* 18: 397–401

62 Whitcomb DC, Puccio AM, Vigna SR, Taylor IL, Hoffman GE (1997) Distribution of pancreatic polypeptide receptors in the rat brain. *Brain Res* 760: 137–149

63 Clark JT, Kalra PS, Crowley WR, Kalra SP (1984) Neuropeptide Y and human pancreatic polypeptide stimulate feeding behavior in rats. Endocrinology 115: 427–429

64 Inui A, Okita M, Nakajima M, Inoue T, Sakatani N, Oya M, Morioka H, Okimura Y, Chihara K, Baba S (1991) Neuropeptide regulation of feeding in dogs. *Am J Physiol* 261: R588–R594

65 Asakawa A, Inui A, Ueno N, Fujimiya M, Fujino MA, Kasuga M (1999) Mouse pancreatic polypeptide modulates food intake, while not influencing anxiety in mice. *Peptides* 20: 1445–1448

66 Berntson GG, Zipf WB, O'Dorisio TM, Hoffman JA, Chance RE (1993) Pancreatic polypeptide infusions reduce food intake in Prader-Willi syndrome. *Peptides* 14: 497–503

67 Batterham RL, Le Roux CW, Cohen MA, Park AJ, Ellis SM, Patterson M, Frost GS,

Ghatei MA, Bloom SR (2003) Pancreatic polypeptide reduces appetite and food intake in humans. *J Clin Endocrinol Metab* 88: 3989–3992

68 Larhammar D, Blomqvist AG, Yee F, Jazin E, Yoo H, Wahlested C (1992) Cloning and functional expression of a human neuropeptide Y/peptide YY receptor of the Y1 type. *J Biol Chem* 267: 10935–10938

69 Herzog H, Hort YJ, Ball HJ, Hayes G, Shine J, Selbie LA (1992) Cloned human neuropeptide Y receptor couples to two different second messenger systems. *Proc Natl Acad Sci* USA 89: 5794–5798

70 Kask A, Rago L, Harro J (1998) Evidence for involvement of neuropeptide Y receptors in the regulation of food intake: studies with Y1-selective antagonist BIBP3226. *Br J Pharmacol* 124: 1507–1515

71 Wieland HA, Engel W, Eberlein W, Rudolf K, Doods HN (1998) Subtype selectivity of the novel nonpeptide neuropeptide Y Y1 receptor antagonist BIBO 3304 and its effect on feeding in rodents. *Br J Pharmacol* 125: 549–555

72 Kanatani A, Ito J, Ishihara A, Iwaasa H, Fukuroda T, Fukami T, MacNeil DJ, Van der Ploeg LH, Ihara M (1998) NPY-induced feeding involves the action of a Y1-like receptor in rodents. *Regul Pept* 75–76: 409–415

73 Pedrazzini T, Seydoux J, Kunstner P, Aubert JF, Grouzmann E, Beermann F, Brunner HR (1998) Cardiovascular response, feeding behavior and locomotor activity in mice lacking the NPY Y1 receptor. *Nat Med* 4: 722–726

74 Kushi A, Sasai H, Koizumi H, Takeda N, Yokoyama M, Nakamura M (1998) Obesity and mild hyperinsulinemia found in neuropeptide Y-Y1 receptor- deficient mice. *Proc Natl Acad Sci* USA 95: 15659–15664

75 Corpa ES, McQuade J, Krasnicki S, Conze DB (2001) Feeding after fourth ventricular administration of neuropeptide Y receptor agonists in rats. *Peptides* 22: 493–499

76 Cheng X, Broberger C, Tong Y, Yongtao X, Ju G, Zhang X, Hokfelt T (1998) Regulation of expression of neuropeptide Y Y1 and Y2 receptors in the arcuate nucleus of fasted rats. *Brain Res* 792: 89–96

77 Mullins D, Kirby D, Hwa J, Guzzi M, Rivier J, Parker E (2001) Identification of potent and selective neuropeptide Y Y(1) receptor agonists with orexigenic activity *in vivo*. Mol Pharmacol 60: 534–540

78 Iyengar S, Li DL, Simmons RM (1999) Characterization of neuropeptide Y-induced feeding in mice: do Y1-Y6 receptor subtypes mediate feeding? *J Pharmacol Exp Ther* 289: 1031–1040

79 Kanatani A, Hata M, Mashiko S, Ishihara A, Okamoto O, Haga Y, Ohe T, Kanno T, Murai N, Ishii Y et al (2001) A typical Y1 receptor regulates feeding behaviors: effects of a potent and selective Y1 antagonist, J-115814. *Mol Pharmacol* 59: 501–505

80 Chaffer CL, Morris MJ (2002) The feeding response to melanin-concentrating hormone is attenuated by antagonism of the NPY Y(1)-receptor in the rat. *Endocrinology* 143: 191–197

81 Grundemar L, Ekelund M (1996) Effects of the neuropeptide Y (NPY)-receptor antagonist BIBP3226 on vascular NPY-receptors with different ligand requirements. *Pharmacol Toxicol* 79: 266–269

82 Malmstrom RE, Hokfelt T, Bjorkman JA, Nihlen C, Bystrom M, Ekstrand AJ, Lundberg JM (1998) Characterization and molecular cloning of vascular neuropeptide Y receptor subtypes in pig and dog. *Regul Pept* 75–76: 55–70

83 Capurro D, Huidobro-Toro JP (1999) The involvement of neuropeptide Y Y1 receptors in the blood pressure baroreflex: studies with BIBP 3226 and BIBO 3304. *Eur J Pharmacol* 376: 251–255

84 Serradeil-Le Gal C, Valette G, Rouby PE, Pellet A, Oury-Donat F, Brossard G, Lespy L, Marty E, Neliat G, de Cointet P et al (1995) SR 120819A, an orally-active and selective neuropeptide Y Y1 receptor antagonist. *FEBS Lett* 362: 192–196

85 Abrahamsson C (2000) Neuropeptide Y1- and Y2-receptor-mediated cardiovascular effects in the anesthetized guinea pig, rat, and rabbit. *J Cardiovasc Pharmacol* 36: 451–458

86 Malmstrom RE, Balmer KC, Weilitz J, Nordlander M, Sjolander M (2001) Pharmacology of H 394/84, a dihydropyridine neuropeptide Y Y(1) receptor antagonist, *in vivo*. *Eur J Pharmacol* 418: 95–104

87 Nakajima M, Inui A, Asakawa A, Momose K, Ueno N, Teranishi A, Baba S, Kasuga M (1998) Neuropeptide Y produces anxiety via Y2-type receptors. *Peptides* 19: 359–363

88 Sajdyk TJ, Vandergriff MG, Gehlert DR (1999) Amygdalar neuropeptide Y Y1 receptors mediate the anxiolytic-like actions of neuropeptide Y in the social interaction test. *Eur J Pharmacol* 368: 143–147

89 Kask A, Nguyen HP, Pabst R, Von Horsten S (2001) Neuropeptide Y Y1 receptor-mediated anxiolysis in the dorsocaudal lateral septum: functional antagonism of corticotropin-releasing hormone-induced anxiety. *Neuroscience* 104: 799–806

90 Redrobe JP, Dumont Y, Fournier A, Quirion R (2002) The neuropeptide Y (NPY) Y1 receptor subtype mediates NPY-induced antidepressant-like activity in the mouse forced swimming test. *Neuropsychopharmacology* 26: 615–624

91 Thiele TE, Sparta DR, Hayes DM, Fee JR (2004) A role for neuropeptide Y in neurobiological responses to ethanol and drugs of abuse. *Neuropeptides* 38: 235–244

92 Kelley SP, Nannini MA, Bratt AM, Hodge CW (2001) Neuropeptide-Y in the paraventricular nucleus increases ethanol self-administration. *Peptides* 22: 515–522

93 Sparta DR, Fee JR, Hayes DM, Knapp DJ, MacNeil DJ, Thiele TE (2004) Peripheral and central administration of a selective neuropeptide Y Y1 receptor antagonist suppresses ethanol intake by C57BL/6J mice. *Alcohol Clin Exp Res* 28: 1324–1330

94 Thiele TE, Koh MT, Pedrazzini T (2002) Voluntary alcohol consumption is controlled via the neuropeptide Y Y1 receptor. *J Neurosci* 22: RC208

95 Naveilhan P, Canals JM, Arenas E, Ernfors P (2001) Distinct roles of the Y1 and Y2 receptors on neuropeptide Y-induced sensitization to sedation. *J Neurochem* 78: 1201–1207

96 Naveilhan P, Hassani H, Lucas G, Blakeman KH, Hao JX, Xu XJ, Wiesenfeld-Hallin Z, Thoren P, Ernfors P (2001) Reduced antinociception and plasma extravasation in mice lacking a neuropeptide Y receptor. *Nature* 409: 513–517

97 Wahlestedt C, Yanaihara N, Hakanson R (1986) Evidence for different pre-and post-junctional receptors for neuropeptide Y and related peptides. *Regul Pept* 13: 307–318

98 Fuhlendorff J, Gether U, Aakerlund L, Langeland-Johansen N, Thogersen H, Melberg SG, Olsen UB, Thastrup O, Schwartz TW (1990) [Leu31, Pro34]neuropeptide Y: a specific Y1 receptor agonist. *Proc Natl Acad Sci* USA 87: 182–186

99 Potter EK, Fuhlendorff J, Schwartz TW (1991) [Pro34]neuropeptide Y selectively identifies postjunctional-mediated actions of neuropeptide Y *in vivo* in rats and dogs. *Eur J Pharmacol* 193: 15–19

100 Gerald C, Walker MW, Vaysse PJ, He C, Branchek TA, Weinshank RL (1995) Expression cloning and pharmacological characterization of a human hippocampal neuropeptide Y/peptide YY Y2 receptor subtype. *J Biol Chem* 270: 26758–26761

101 Rimland JM, Seward EP, Humbert Y, Ratti E, Trist DG, North RA (1996) Coexpression with potassium channel subunits used to clone the Y2 receptor for neuropeptide Y. *Mol Pharmacol* 49: 387–390

102 Gehlert DR, Beavers LS, Johnson D, Gackenheimer SL, Schober DA, Gadski RA (1996) Expression cloning of a human brain neuropeptide Y Y2 receptor. *Mol Pharmacol* 49: 224–228

103 Ammar DA, Eadie DM, Wong DJ, Ma YY, Kolakowski LF Jr, Yang-Feng TL, Thompson DA (1996) Characterization of the human type 2 neuropeptide Y receptor gene (NPY2R) and localization to the chromosome 4q region containing the type 1 neuropeptide Y receptor gene. *Genomics* 38: 392–398

104 Sharma P, Holmberg SK, Eriksson H, Beck-Sickinger AG, Grundemar L, Larhammar D (1998) Cloning and functional expression of the guinea pig neuropeptide Y Y2 receptor. *Regul Pept* 75–76: 23–28

105 Wraith A, Tornsten A, Chardon P, Harbitz I, Chowdhary BP, Andersson L, Lundin LG, Larhammar D (2000) Evolution of the neuropeptide Y receptor family: gene and chromosome duplications deduced from the cloning and mapping of the five receptor subtype genes in Pig. *Genome Res* 10: 302–310

106 King PJ, Williams G, Doods H, Widdowson PS (2000) Effect of a selective neuropeptide Y Y(2) receptor antagonist, BIIE0246 on neuropeptide Y release. *Eur J Pharmacol* 396: R1–3

107 Smith-White MA, Hardy TA, Brock JA, Potter EK (2001) Effects of a selective neuropeptide Y Y2 receptor antagonist, BIIE0246, on Y2 receptors at peripheral neuroeffector junctions. *Br J Pharmacol* 132: 861–868

108 Sajdyk TJ, Schober DA, Smiley DL, Gehlert DR (2002) Neuropeptide Y-Y(2) receptors mediate anxiety in the amygdala. *Pharmacol Biochem Behav* 71: 419–423

109 Smith-White MA, Herzog H, Potter EK (2002) Role of neuropeptide Y Y(2) receptors in modulation of cardiac parasympathetic neurotransmission. *Regul Pept* 103: 105–111

110 Ekstrand AJ, Cao R, Bjorndahl M, Nystrom S, Jonsson-Rylander AC, Hassani H, Hallberg B, Nordlander M, Cao Y (2003) Deletion of neuropeptide Y (NPY) 2 receptor in mice results in blockage of NPY-induced angiogenesis and delayed wound healing. *Proc Natl Acad Sci* USA 100: 6033–6038

111 Golombek DA, Biello SM, Rendon RA, Harrington ME (1996) Neuropeptide Y phase shifts the circadian clock *in vitro* via a Y2 receptor. *Neuroreport* 7: 1315–1319

112 Huhman KL, Gillespie CF, Marvel CL, Albers HE (1996) Neuropeptide Y phase shifts circadian rhythms *in vivo* via a Y2 receptor. *Neuroreport* 7: 1249–1252

113 Soscia SJ, Harrington ME (2005) Neuropeptide Y does not reset the circadian clock in NPY Y2–/– mice. *Neurosci Lett* 373: 175–178

114 Naveilhan P, Hassani H, Canals JM, Ekstrand AJ, Larefalk A, Chhajlani V, Arenas E, Gedda K, Svensson L, Thoren P et al (1999) Normal feeding behavior, body weight and leptin response require the neuropeptide Y Y2 receptor. *Nat Med* 5: 1188–1193

115 Baldock PA, Sainsbury A, Couzens M, Enriquez RF, Thomas GP, Gardiner EM, Herzog H (2002) Hypothalamic Y2 receptors regulate bone formation. *J Clin Invest* 109: 915–921

116 Halatchev IG, Ellacott KL, Fan W, Cone RD (2004) Peptide YY3-36 inhibits food intake in mice through a melanocortin-4 receptor-independent mechanism. *Endocrinology* 145: 2585–2590

117 Moran TH, Smedh U, Kinzig KP, Scott KA, Knipp S, Ladenheim EE (2005) Peptide YY (3-36) inhibits gastric emptying and produces acute reductions in food intake in rhesus monkeys. *Am J Physiol Regul Integr Comp Physiol* 288: R384–388

118 Lundell I, Blomqvist AG, Berglund MM, Schober DA, Johnson D, Statnick MA, Gadski RA, Gehlert DR, Larhammar D (1995) Cloning of a human receptor of the NPY receptor family with high affinity for pancreatic polypeptide and peptide YY. *J Biol Chem* 270: 29123–29128

119 Lundell I, Statnick MA, Johnson D, Schober DA, Starback P, Gehlert DR, Larhammar D (1996) The cloned rat pancreatic polypeptide receptor exhibits profound differences to the orthologous receptor. *Proc Natl Acad Sci USA* 93: 5111–5115

120 Yan H, Yang J, Marasco J, Yamaguchi K, Brenner S, Collins F, Karbon W (1996) Cloning and functional expression of cDNAs encoding human and rat pancreatic polypeptide receptors. *Proc Natl Acad Sci* USA 93: 4661–4665

121 Eriksson H, Berglund MM, Holmberg SKS, Kahl U, Gehlert DR, Larhammar D (1998) The cloned guinea pig pancreatic polypeptide receptor Y4 resembles more the human Y4 than does the rat Y4. *Regul Pept* 75–76: 29–37

122 Bard JA, Walker MW, Branchek TA, Weinshank RL (1995) Cloning and functional expression of a human Y4 subtype receptor for pancreatic polypeptide, neuropeptide Y, and peptide YY. *J Biol Chem* 270: 26762–26765

123 Parker EM, Babij CK, Balasubramaniam A, Burrier RE, Guzzi M, Hamud F, Mukhopadhyay G, Rudinski MS, Tao Z, Tice M et al (1998) GR231118 (1229U91) and other analogues of the C-terminus of neuropeptide Y are potent neuropeptide Y Y1 receptor antagonists and neuropeptide Y Y4 receptor agonists. *Eur J Pharmacol* 349: 97–105

124 Schober DA, Van Abbema AM, Smiley DL, Bruns RF, Gehlert DR (1998) The neuropeptide Y Y1 antagonist, 1229U91, a potent agonist for the human pancreatic polypeptide-preferring (NPY Y4) receptor. *Peptides* 19: 537–542

125 Raposinho PD, Broqua P, Hayward A, Akinsanya K, Galyean R, Schteingart C, Junien

J, Aubert ML (2000) Stimulation of the gonadotropic axis by the neuropeptide Y receptor Y1 Antagonist/ Y4 agonist 1229U91 in the male rat. *Neuroendocrinology* 71: 2–7

126 Stanley BG, Magdalin W, Seirafi A, Nguyen MM, Leibowitz SF (1992) Evidence for neuropeptide Y mediation of eating produced by food deprivation and for a variant of the Y1 receptor mediating this peptide's effect. *Peptides* 13: 581–587

127 Balasubramaniam A, Sheriff S, Johnson ME, Prabhakaran M, Huang Y, Fischer JE, Chance WT (1994) [D-TRP32]neuropeptide Y: a competitive antagonist of NPY in rat hypothalamus. J Med Chem 37: 811–815

128 Gerald C, Walker MW, Criscione L, Gustafson EL, Batzl-Hartmann C, Smith KE, Vaysse P, Durkin MM, Laz TM, Linemeyer DL et al (1996) A receptor subtype involved in neuropeptide-Y-induced food intake. *Nature* 382: 168–171

129 Hu Y, Bloomquist BT, Cornfield LJ, DeCarr LB, Flores-Riveros JR, Friedman L, Jiang P, Lewis-Higgins L, Sadlowski Y, Schaefer J et al (1996) Identification of a novel hypothalamic neuropeptide Y receptor associated with feeding behavior. *J Biol Chem* 271: 26315–26319

130 Borowsky B, Walker MW, Bard J, Weinshank RL, Laz TM, Vaysse P, Branchek TA, Gerald C (1998) Molecular biology and pharmacology of multiple NPY Y5 receptor species homologs. *Regul Pept* 75–76: 45–53

131 Lundell I, Eriksson H, Marklund U, Larhammar D (2001) Cloning and characterization of the guinea pig neuropeptide Y receptor Y5. *Peptides* 22: 357–363

132 Schaffhauser AO, Stricker-Krongrad A, Brunner L, Cumin F, Gerald C, Whitebread S, Criscione L, Hofbauer KG (1997) Inhibition of food intake by neuropeptide Y Y5 receptor antisense oligodeoxynucleotides. *Diabetes* 46: 1792–1798

133 Tang-Christensen M, Kristensen P, Stidsen CE, Brand CL, Larsen PJ (1998) Central administration of Y5 receptor antisense decreases spontaneous food intake and attenuates feeding in response to exogenous neuropeptide Y. *J Endocrinol* 159: 307–312

134 Flynn MC, Turrin NP, Plata-Salaman CR, Ffrench-Mullen JM (1999) Feeding response to neuropeptide Y-related compounds in rats treated with Y5 receptor antisense or sense phosphothio-oligodeoxynucleotide. *Physiol Behav* 66: 881–884

135 Campbell RE, Ffrench-Mullen JM, Cowley MA, Smith MS, Grove KL (2001) Hypothalamic circuitry of neuropeptide Y regulation of neuroendocrine function and food intake via the Y5 receptor subtype. *Neuroendocrinology* 74: 106–119

136 Marsh DJ, Hollopeter G, Kafer KE, Palmiter RD (1998) Role of the Y5 neuropeptide Y receptor in feeding and obesity. *Nat Med* 4: 718–721

137 Wyss P, Stricker-Krongrad A, Brunner L, Miller J, Crossthwaite A, Whitebread S, Criscione L (1998) The pharmacology of neuropeptide Y (NPY) receptor-mediated feeding in rats characterizes better Y5 than Y1, but not Y2 or Y4 subtypes. *Regul Pept* 75–76: 363–371

138 McCrea K, Wisialowski T, Cabrele C, Church B, Beck-Sickinger A, Kraegen E, Herzog H (2000) 2-36[K4,RYYSA(19-23)]PP a novel Y5-receptor preferring ligand with strong stimulatory effect on food intake. *Regul Pept* 87: 47–58

139 Cabrele C, Langer M, Bader R, Wieland HA, Doods HN, Zerbe O, Beck-Sickinger AG

(2000) The first selective agonist for the neuropeptide YY5 receptor increases food intake in rats. *J Biol Chem* 275: 36043–36048

140 Criscione L, Rigollier P, Batzl-Hartmann C, Rueger H, Stricker-Krongrad A, Wyss P, Brunner L, Whitebread S, Yamaguchi Y, Gerald C et al (1998) Food intake in free-feeding and energy-deprived lean rats is mediated by the neuropeptide Y5 receptor. *J Clin Invest* 102: 2136–2145

141 Della-Zuana O, Sadlo M, Germain M, Feletou M, Chamorro S, Tisserand F, de Montrion C, Boivin JF, Duhault J, Boutin JA et al (2001) Reduced food intake in response to CGP 71683A may be due to mechanisms other than NPY Y5 receptor blockade. *Int J Obes Relat Metab Disord* 25: 84–94

142 Levens NR, Della-Zuana O (2003) Neuropeptide Y Y5 receptor antagonists as anti-obesity drugs. *Curr Opin Investig Drugs* 4: 1198–1204

143 Raposinho PD, Pierroz DD, Broqua P, White RB, Pedrazzini T, Aubert ML (2001) Chronic administration of neuropeptide Y into the lateral ventricle of C57BL/6J male mice produces an obesity syndrome including hyperphagia, hyperleptinemia, insulin resistance, and hypogonadism. *Mol Cell Endocrinol* 185: 195–204

144 Marsh DJ, Baraban SC, Hollopeter G, Palmiter RD (1999) Role of the Y5 neuropeptide Y receptor in limbic seizures. *Proc Natl Acad Sci* USA 96: 13518–13523

145 Guo H, Castro PA, Palmiter RD, Baraban SC (2002) Y5 receptors mediate neuropeptide Y actions at excitatory synapses in area CA3 of the mouse hippocampus. *J Neurophysiol* 87: 558–566

146 Yannielli PC, Harrington ME (2001) The neuropeptide Y Y5 receptor mediates the blockade of “photic-like” NMDA-induced phase shifts in the golden hamster. *J Neurosci* 21: 5367–5373

147 Mannon PJ, Mele JM (2000) Peptide YY Y1 receptor activates mitogen-activated protein kinase and proliferation in gut epithelial cells via the epidermal growth factor receptor. *Biochem J* 350: 655–661

148 Mullins DE, Zhang X, Hawes BE (2002) Activation of extracellular signal regulated protein kinase by neuropeptide Y and pancreatic polypeptide in CHO cells expressing the NPY Y1, Y2, Y4 and Y5 receptor subtypes. *Regul Pept* 105: 65–73

149 Fabry M, Langer M, Rothen-Rutishauser B, Wunderli-Allenspach H, Hocker H, Beck-Sickinger AG (2000) Monitoring of the internalization of neuropeptide Y on neuroblastoma cell line SK-N-MC. *Eur J Biochem* 267: 5631–5637

150 Parker SL, Kane JK, Parker MS, Berglund MM, Lundell IA, Li MD (2001) Cloned neuropeptide Y (NPY) Y1 and pancreatic polypeptide Y4 receptors expressed in Chinese hamster ovary cells show considerable agonist-driven internalization, in contrast to the NPY Y2 receptor. *Eur J Biochem* 268: 877–886

151 Gicquiaux H, Lecat S, Gaire M, Dieterlen A, Mely Y, Takeda K, Bucher B, Galzi JL (2002) Rapid internalization and recycling of the human neuropeptide Y Y1 receptor. *J Biol Chem* 277: 6645–6655

152 Parker SL, Parker MS, Lundell I, Balasubramaniam A, Buschauer A, Kane JK, Yalcin A, Berglund MM (2002) Agonist internalization by cloned Y1 neuropeptide Y (NPY)

receptor in Chinese hamster ovary cells shows strong preference for NPY, endosome-linked entry and fast receptor recycling. *Regul Pept* 107: 49–62
153 Berglund MM, Schober DA, Statnick MA, McDonald PH, Gehlert DR (2003) The use of bioluminescence resonance energy transfer 2 to study neuropeptide Y receptor agonist-induced beta-arrestin 2 interaction. *J Pharmacol Exp Ther* 306: 147–156
154 Berglund MM, Schober DA, Esterman MA, Gehlert DR (2003) Neuropeptide Y Y4 receptor homodimers dissociate upon agonist stimulation. *J Pharmacol Exp Ther* 307: 1120–1126
155 Marklund U, Bystrom M, Gedda K, Larefalk A, Juneblad K, Nystrom S, Ekstrand J (2002) Intron-mediated expression of the human neuropeptide Y Y1 receptor. *Mol Cell Endocrinol* 188: 85–97
156 Holliday N, Lam CW, Tough IR, Cox HM (2005) Role of the C terminus in neuropeptide Y Y1 receptor desensitization and internalization. *Mol Pharmacol* 67: 655–664
157 Devi LA (2001) Heterodimerization of G-protein-coupled receptors: pharmacology, signaling and trafficking. *Trends Pharmacol Sci* 22: 532–537
158 Dinger MC, Bader JE, Kobor AD, Kretzschmar AK, Beck-Sickinger AG (2003) Homodimerization of neuropeptide y receptors investigated by fluorescence resonance energy transfer in living cells. *J Biol Chem* 278: 10562–10571
159 Schober DA, Berglund MM, Gehlert DR (2003) Neuropeptide Y (NPY) Y1 and Y5 receptors form constitutive dimers that elicit an enhanced response to Y5 agonists: *The Society for Neuroscience 33rd Annual Meeting*, New Orleans, Abstract 615.614
160 Gehlert DR, Gackenheimer SL, Schober DA (1992) [Leu^{31}-Pro^{34}] Neuropeptide Y identifies a subtype of ^{125}I-labelled peptide YY binding sites in the rat brain. *Neurochem Int* 21: 45–67
161 Dumont Y, Fournier A, St-Pierre S, Quirion R (1993) Comparative characterization and autoradiographic distribution of neuropeptide Y receptor subtypes in the rat brain. *J Neurosci* 13: 73–86
162 Dumont Y, Fournier A, St-Pierre S, Quirion R (1996) Autoradiographic distribution of [^{125}I]-Leu^{31}-Pro^{34}]PYY and [^{125}I]-$PYY_{3\text{-}36}$ binding sites in the rat brain evaluated with two newly developed Y1 and Y2 receptor radioligands. *Synapse* 22: 139–158
163 Gehlert DR, Gackenheimer SL (1997) Differential distribution of neuropeptide Y Y1 and Y2 receptors in rat and guinea-pig brains. *Neuroscience* 76: 215–224
164 Parker RM, Herzog H (1999) Regional distribution of Y-receptor subtype mRNAs in rat brain. *Eur J Neurosci* 11: 1431–1448
165 Landry M, Holmberg K, Zhang X, Hokfelt T (2000) Effect of axotomy on expression of NPY, galanin, and NPY Y1 and Y2 receptors in dorsal root ganglia and the superior cervical ganglion studied with double-labeling *in situ* hybridization and immunohistochemistry. *Exp Neurol* 162: 361–384
166 Wharton J, Gordon L, Byrne J, Herzog H, Selbie L, Moore K, Sullivan MHF, Elder MG, Moscoso G, Taylor KM et al (1993) Expression of the human neuropeptide tyrosine Y1 receptor. *Proc Natl Acad Sci* 90: 687–691
167 Goumain M, Voisin T, Lorinet AM, Laburthe M (1998) Identification and distribution

of mRNA encoding the Y1, Y2, Y4, and Y5 receptors for peptides of the PP-fold family in the rat intestine and colon. *Biochem Biophys Res Commun* 247: 52–56

168 Schober DA, Gackenheimer SL, Heiman ML, Gehlert DR (2000) Pharmacological characterization of (125)I-1229U91 binding to Y1 and Y4 neuropeptide Y/Peptide YY receptors. *J Pharmacol Exp Ther* 293: 275–280

169 Durkin MM, Walker MW, Smith KE, Gustafson EL, Gerald C, Branchek TA (2000) Expression of a novel neuropeptide Y receptor subtype involved in food intake: an *in situ* hybridization study of Y5 mRNA distribution in rat brain. *Exp Neurol* 165: 90–100

170 Dumont Y, Fournier A, Quirion R (1998) Expression and characterization of the neuropeptide Y Y5 receptor subtype in the rat brain. *J Neurosci* 18: 5565–5574

171 Dumont Y, Cadieux A, Doods H, Fournier A, Quirion R (2000) Potent and selective tools to investigate neuropeptide Y receptors in the central and peripheral nervous systems: BIB03304 (Y1) and CGP71683A (Y5). *Can J Physiol Pharmacol* 78: 116–125

172 Grove KL, Campbell RE, Ffrench-Mullen JM, Cowley MA, Smith MS (2000) Neuropeptide Y Y5 receptor protein in the cortical/limbic system and brainstem of the rat: expression on gamma-aminobutyric acid and corticotropin-releasing hormone neurons. *Neuroscience* 100: 731–740

173 Soll RM, Dinger MC, Lundell I, Larhammer D, Beck-Sickinger AG (2001) Novel analogues of neuropeptide Y with a preference for the Y1-receptor. *Eur J Biochem* 268: 2828–2837

174 Balasubramaniam A, Dhawan VC, Mullins DE, Chance WT, Sheriff S, Guzzi M, Prabhakaran M, Parker EM (2001) Highly selective and potent neuropeptide Y (NPY) Y1 receptor antagonists based on [Pro(30), Tyr(32), Leu(34)]NPY(28-36)-NH2 (BW1911U90). *J Med Chem* 44: 1479–1482

175 Parker EM, Balasubramaniam A, Guzzi M, Mullins DE, Salisbury BG, Sheriff S, Witten MB, Hwa JJ (2000) [D-Trp(34)] neuropeptide Y is a potent and selective neuropeptide Y Y(5) receptor agonist with dramatic effects on food intake. *Peptides* 21: 393–399

176 Poindexter GS, Bruce MA, LeBoulluec KL, Monkovic I, Martin SW, Parker EM, Iben LG, McGovern RT, Ortiz AA, Stanley JA et al (2002) Dihydropyridine neuropeptide y y(1) receptor antagonists. *Bioorg Med Chem Lett* 12: 379–382

177 Poindexter GS, Bruce MA, Breitenbucher JG, Higgins MA, Sit SY, Romine JL, Martin SW, Ward SA, McGovern RT, Clarke W et al (2004) Dihydropyridine neuropeptide Y Y1 receptor antagonists 2: bioisosteric urea replacements. *Bioorganic & Medicinal Chemistry* 12: 507–521

178 Hipskind PA, Lobb KL, Nixon JA, Britton TC, Bruns RF, Catlow J, Dieckman-McGinty DK, Gackenheimer SL, Gitter BD, Iyengar S et al (1997) Potent and selective 1,2,3-trisubstituted indole NPY Y-1 antagonists. *J Med Chem* 40: 3712–3714

179 Zimmerman DM, Cantrell BE, Smith EC, Nixon JA, Bruns RF, Gitter B, Hipskind PA, Ornstein PL, Zarrinmayeh H, Britton TC et al (1998) Structure-activity relationships of a series of 1-substituted-4- methylbenzimidazole neuropeptide Y-1 receptor antagonists. *Bioorg Med Chem Lett* 8: 473–476

180 Zarrinmayeh H, Zimmerman DM, Cantrell BE, Schober DA, Bruns RF, Gackenheimer

SL, Ornstein PL, Hipskind PA, Britton TC, Gehlert DR (1999) Structure-activity relationship of a series of diaminoalkyl substituted benzimidazole as neuropeptide Y Y1 receptor antagonists. *Bioorg Med Chem Lett* 9: 647–652

181 Zarrinmayeh H, Nunes AM, Ornstein PL, Zimmerman DM, Arnold MB, Schober DA, Gackenheimer SL, Bruns RF, Hipskind PA, Britton TC et al (1998) Synthesis and evaluation of a series of novel 2-[(4-chlorophenoxy)methyl]benzimidazoles as selective neuropeptide Y Y1 receptor antagonists. *J Med Chem* 41: 2709–2719

182 Martin SW, Romine JL, Chen L, Mattson G, Antal-Zimanyi IA, Poindexter GS (2004) Application of solution-phase parallel synthesis to preparation of trisubstituted 1,2,4-triazoles. *J Combinat Chem* 6: 35–37

183 Darrow JW, De Lombaert S, Blum C, Tran J, Giangiordano M, Griffith DA, Carpino PA (2001) Preparation of amino substituted pyrazolo[1,5-a]-1,5-pyrimidines and pyrazolo[1,5-a]-1,3,5-triazines as NPY receptor antagonists: *PCT Int Appl* Neurogen Corporation, USA; Pfizer Inc. Wo, p. 102

184 Darrow JW, De Lombaert S, Blum C, Tran J, Giangiordano M, Griffith DA, Carpino PA (2001) Preparation of certain alkylene diamine-substituted pyrazolo[1,5-a]-1,5-pyrimidines and pyrazolo[1,5-a]-1,3,5-triazines as selective modulators of NPY1 receptors: *PCT Int Appl* Neurogen Corporation, USA; Pfizer Inc.; De Lombaert, Stephane. Wo, p. 133

185 Horvath RF, Tran J, De LS, Hodgetts KJ, Carpino PA, Griffith DA (2001) Preparation of certain alkylene diamine-substituted heterocycles as NPY1 receptor inhibitors: *PCT Int Appl* Neurogen Corporation, USA; Pfizer, Inc.; De Lombaert, Stephane. Wo, p. 211

186 Doods H, Gaida W, Wieland HA, Dollinger H, Schnorrenberg G, Esser F, Engel W, Eberlein W, Rudolf K (1999) BIIE0246: a selective and high affinity neuropeptide Y Y(2) receptor antagonist. *Eur J Pharmacol* 384: R3–R5

187 Dumont Y, Cadieux A, Doods H, Pheng LH, Abounader R, Hamel E, Jacques D, Regoli D, Quirion R (2000) BIIE0246, a potent and highly selective non-peptide neuropeptide Y Y(2) receptor antagonist. *Br J Pharmacol* 129: 1075–1088

188 Andres CJ, Antal Zimanyi I, Deshpande MS, Iben LG, Grant-Young K, Mattson GK, Zhai W (2003) Differentially functionalized diamines as novel ligands for the NPY2 receptor. *Bioorg Med Chem Lett* 13: 2883–2885

189 Bonaventure P, Nepomuceno D, Mazur C, Lord B, Rudolph DA, Jablonowski JA, Carruthers NI, Lovenberg TW (2004) Characterization of N-(1-Acetyl-2,3-dihydro-1H-indol-6-yl)-3-(3-cyano-phenyl)-N-[1-(2-cyclopen tyl-ethyl)-piperidin-4yl]acrylamide (JNJ-5207787), a small molecule antagonist of the neuropeptide Y Y2 receptor. *J Pharmacol Exp Ther* 308: 1130–1137

190 Rueeger H, Rigollier P, Yamaguchi Y, Schmidlin T, Schilling W, Criscione L, Whitebread S, Chiesi M, Walker MW, Dhanoa D et al (2000) Design, synthesis and SAR of a series of 2-substituted 4-amino- quinazoline neuropeptide Y Y5 receptor antagonists. *Bioorg Med Chem Lett* 10: 1175–1179

191 Della-Zuana O, Revereault L, Beck-Sickinger A, Monge A, Caignard DH, Fauchere JL, Henlin JM, Audinot V, Boutin JA, Chamorro S et al (2004) A potent and selective NPY

Y(5) antagonist reduces food intake but not through blockade of the NPY Y(5) receptor. *Int J Obes Relat Metab Disord*

192 Block MH, Boyer S, Brailsford W, Brittain DR, Carroll D, Chapman S, Clarke DS, Donald CS, Foote KM, Godfrey L et al (2002) Discovery and optimization of a series of carbazole ureas as NPY5 antagonists for the treatment of obesity. *J Med Chem* 45: 3509–3523

193 Hammond M, Elliott RL, Gillaspy ML, Hager DC, Hank RF, LaFlamme JA, Oliver RM, Da Silva-Jardine PA, Stevenson RW, Mack CM et al (2003) Structure-activity relationships in a series of NPY Y5 antagonists: 3-amido-9-ethylcarbazoles, core-modified analogues and amide isosteres. *Bioorg Med Chem Lett* 13: 1989–1992

194 Blum CA, Zheng X, De Lombaert S (2004) Design, synthesis, and biological evaluation of substituted 2-cyclohexyl-4-phenyl-1H-imidazoles: potent and selective neuropeptide Y Y5-receptor antagonists. *J Med Chem* 47: 2318–2325

195 Elliott RL, Oliver RM, Hammond M, Patterson TA, Li S, Hargrove DM, Martin KA, Maurer TS, Kalvass JC, Morgan BP et al (2003) *In vitro* and *in vivo* characterization of 3-{2-[6-(2-tert-Butoxyethoxy)pyridin-3-yl]-1H-imidazol-4-yl}benzonitrile hydrochloride salt, a potent and selective NPY5 receptor antagonist. *J Med Chem* 46: 670–673

196 Rueeger H, Gerspacher M, Buehlmayer P, Rigollier P, Yamaguchi Y, Schmidlin T, Whitebread S, Nuesslein-Hildesheim B, Nick H, Cricione L (2004) Discovery and SAR of potent, orally available and brain-penetrable 5,6-dihydro-4H-3-thia-1-aza-benzo[e]azulen- and 4,5-dihydro-6-oxa-3-thia-1-aza-benzo[e]azulen derivatives as neuropeptide Y Y5 receptor antagonists. *Bioorg Med Chem Lett* 14: 2451–2457

197 Sato N, Takahashi T, Shibata T, Haga Y, Sakuraba A, Hirose M, Sato M, Nonoshita K, Koike Y, Kitazawa H et al (2003) Design and synthesis of the potent, orally available, brain-penetrable arylpyrazole class of neuropeptide Y5 receptor antagonists. *J Med Chem* 46: 666–669

198 Satoh Y, Hatori C, Ito H (2002) Novel potent antagonists of human neuropeptide Y-Y5 receptor. Part 4. Tetrahydrodiazabenzazulene derivatives. *Bioorg Med Chem Lett* 12: 1009–1011

199 Juanenea L, Galiano S, Erviti O, Moreno A, Perez S, Aldana I, Monge A (2004) Synthesis and evaluation of new hydrazide derivatives as neuropeptide Y Y5 receptor antagonists for the treatment of obesity. *Bioorg Med Chem* 12: 4717–4723

200 Aquino CJ, Ramanjulu JM, Heyer D, Daniels AJ, Palazzo F, Dezube M (2004) Synthesis and structure activity relationship of guanidines as NPY Y5 antagonists. *Bioorg Med Chem* 12: 2691–2708

201 Walker P, Munoz M, Martinez R, Peitsch MC (1994) Acidic residues in extracellular loops of the human Y1 neuropeptide Y receptor are essential for ligand binding. *J Biol Chem* 269: 2863–2869

202 Sautel M, Martinez R, Munoz M, Peitsch MC, Beck-Sickinger AG, Walker P (1995) Role of a hydrophobic pocket of the human Y1 neuropeptide Y receptor in ligand binding. *Mol Cell Endocrinol* 112: 215–222

203 Munch G, Walker P, Shine J, Herzog H (1995) Ligand binding analysis of human neuropeptide Y1 receptor mutants expressed in *E. coli. Receptors Channels* 3: 291–297

204 Sautel M, Rudolf K, Wittneben H, Herzog H, Martinez R, Munoz M, Eberlein W, Engel W, Walker P, Beck-Sickinger AG (1996) Neuropeptide Y and the nonpeptide antagonist BIBP 3226 share an overlapping binding site at the human Y1 receptor. *Mol Pharmacol* 50: 285–292

205 Du P, Salon JA, Tamm JA, Hou C, Cui W, Walker MW, Adham N, Dhanoa DS, Islam I, Vaysse PJ et al (1997) Modeling the G-protein-coupled neuropeptide Y Y1 receptor agonist and antagonist binding sites. *Protein Eng* 10: 109–117

206 Robin-Jagerschmidt C, Sylte I, Bihoreau C, Hendricksen L, Calvet A, Dahl SG, Benicourt C (1998) The ligand binding site of NPY at the rat Y1 receptor investigated by site-directed mutagenesis and molecular modeling [In Process Citation]. *Mol Cell Endocrinol* 139: 187–198

207 Kanno T, Kanatani A, Keen SL, Arai-Otsuki S, Haga Y, Iwama T, Ishihara A, Sakuraba A, Iwaasa H, Hirose M et al (2001) Different binding sites for the neuropeptide Y Y1 antagonists 1229U91 and J-104870 on human Y1 receptors. *Peptides* 22: 405–413

208 Munoz M, Sautel M, Martinez R, Sheikh SP, Walker P (1995) Characterization of the human Y1 neuropeptide Y receptor expressed in insect cells. *Mol Cell Endocrinol* 107: 77–86

209 Berglund MM, Fredriksson R, Salaneck E, Larhammar D (2002) Reciprocal mutations of neuropeptide Y receptor Y2 in human and chicken identify amino acids important for antagonist binding. *FEBS Lett* 518: 5–9

210 Sjödin P, Lundell I, Roschmann E, Philippi A, Desitter I, Roy S, Rosseau F, Hager J, Åkerberg H, Uhlen S et al (2004) Pharmacological and functional characterization of mutants of the human neuropeptide Y Y5 receptor and naturally occurring variants of the Y4 receptor: *The Society for Neuroscience 34th Annual Meeting*, San Diego, Abstract 317.319

211 Sainsbury A, Schwarzer C, Couzens M, Fetissov S, Furtinger S, Jenkins A, Cox HM, Sperk G, Hokfelt T, Herzog H (2002) Important role of hypothalamic Y2 receptors in body weight regulation revealed in conditional knockout mice. *Proc Natl Acad Sci* USA 99: 8938–8943

212 Sainsbury A, Baldock PA, Schwarzer C, Ueno N, Enriquez RF, Couzens M, Inui A, Herzog H, Gardiner EM (2003) Synergistic effects of Y2 and Y4 receptors on adiposity and bone mass revealed in double knockout mice. *Mol Cell Biol* 23: 5225–5233

213 Cabrele C, Wieland HA, Langer M, Stidsen CE, Beck-Sickinger AG (2001) Y-receptor affinity modulation by the design of pancreatic polypeptide/neuropeptide Y chimera led to Y(5)-receptor ligands with picomolar affinity. *Peptides* 22: 365–378

NPY processing in neuronal and non-neuronal tissues by proconvertases

Eric Grouzmann[1] *and Noureddine Brakch*[2]

[1]Lausanne University Hospital School, Division of Clinical Pharmacology and Toxicology, 1001 Lausanne, Switzerland; [2]Lausanne University Hospital School, Division of Angiology, Lausanne, Switzerland

Introduction

Since its discovery in 1982, several reports had established that NPY is an important regulator of whole body homeostasis (see the other chapters of this book [1, 2]). NPY is widely expressed in the central and peripheral nervous system and has been recognized to be one of the most abundant peptides in the brain [3]. It is not only found in the peripheral nervous system in nerve endings [4] but also in the adrenal medulla, the endocrine pancreas [5], platelets [6], the kidney [7] and the pituitary gland [8] and even in endothelial cells [9]. However, NPY immunoreactivity is relatively complex, including a precursor, several intermediate products resulting from the proteolysis of proNPY and even post-translational modification of the biologically active 36 amino acid by selective aminopeptidase which may switch the selectivity of NPY toward specific receptors. This chapter will review the nature of the enzymes involved in the processing of proNPY. In addition, the tissue-specific localization of these enzymes and how they affect the generation of NPY products will be addressed and the consequences on NPY functions discussed.

ProNPY proteolysis

The biochemical steps for the processing of proNPY and NPY are described in Figure 1. NPY is a 36 amino acid peptide derived from a 97 amino acid precursor, the pre-pro-neuropeptide Y [10]. After removal of the 28 amino acid signal sequence, the 69-amino acid pro-NPY undergoes cleavage at a single paired basic site (Lys^{38}-Arg^{39}) due to a specific family of prohormone converting enzymes called proconvertases. The resulting NPY1-39 is further processed by a ubiquitary carboxypeptidase-like enzyme (CPE) to yield NPY1-37, a glycine extended form of the peptide, which in turn is used as a substrate by an amidating enzyme (peptidyl-glycine-alpha-amidating monooxygenase; PAM) localized in neuroendocrine cells to yield the ami-

The NPY Family of Peptides in Immune Disorders, Inflammation, Angiogenesis and Cancer,
edited by Zofia Zukowska and Giora Z. Feuerstein

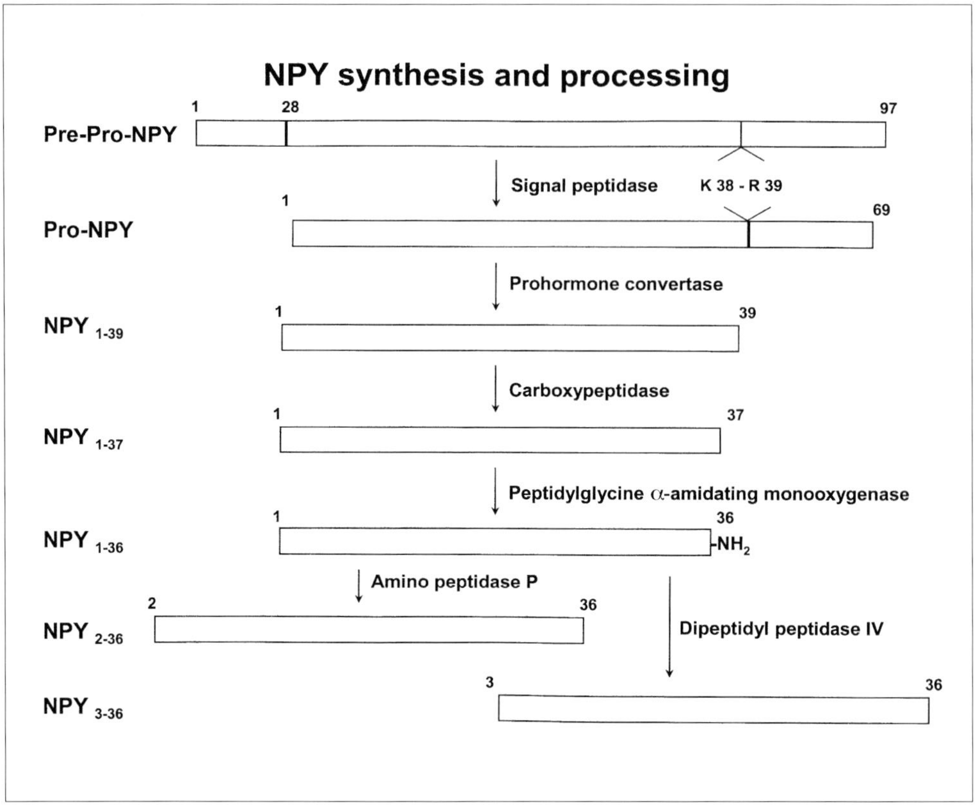

Figure 1
Synthesis and processing of neuropeptide Y

dated NPY1-36. A 30-amino acid carboxyl terminal peptide, the C-flanking peptide of NPY (CPON) is generated during the first step of processing [11]. The CPON is co-localized with NPY in the same tissues, but has no recognized biological functions identified so far. All these processing steps generate several NPY fragments (proNPY, NPY1-39, NPY1-37, and NPY1-36) called NPY immunoreactivity, of which only NPY1-36 is biologically active [12]. Unfortunately, the radioimmunoassay used to determine concentrations of the peptide in tissues or fluids often do not distinguish between the immunoreactive products of NPY. Carboxypeptidase and peptidyl-glycine-alpha-amidating monooxygenase are found in the whole body. The limiting step for proNPY processing in a given tissue is therefore dependent on the presence of the corresponding proconvertase at the proper location (subcellular compartment) and its capability to cleave this precursor.

Structure of convertases

The limited proteolysis at basic residues of peptides and proteins represents an important processing step to render them biologically active. A large number of researchers sought for at least three decades to identify prohormone and proprotein convertases. This is due to the fact that different endopeptidases having similar specificity are found in large amounts in lysosomes. These endopeptidases are difficult to separate from a given processing enzyme of interest during the preparation of secretory granules.

The first characterized prohormone-specific endoprotease was the product of the KEX2 gene isolated from yeast and responsible for cleavage at dibasic sites within the alpha-mating factor and killer toxin precursors [13]. This enzyme is a glycoprotein of 100 kDa belonging to the calcium-dependent serine-protease family. It has a high amino acid sequence homology with the bacterial subtilisin enzymes [14]. KEX2 is an integral membrane protein made up of a signal peptide followed by a pro-region, an extended subtilisin-like catalytic domain, a Ser/Thr rich sequence, a transmembrane hydrophobic domain and a short cytosolic tail [15]. The first mammalian homologue of KEX2, called furin, was found by computer homology and isolated as a gene upstream of the fes/fps proto-oncogenes. It has 50% amino acid identity with KEX2 in a 300-amino acid catalytic domain [16]. Furin exhibits a structure comparable to that of KEX2, including a signal sequence, a glycosylated and hydrophobic amino terminal domain, a transmembrane domain for anchoring in the membrane, and a highly charged carboxy-terminal region. Mutational studies have shown that the cleavage site of furin is Arg-X-Lys/Arg-Arg with Arg in the first and the fourth position being crucial for cleavage [17]. However, furin being localized in the Golgi apparatus is ubiquitously expressed in tissues and may therefore not be specific for prohormone processing. Using the conservated sequence around the subtilisin catalytic domain, several convertases were identified. These include PC1/3, PC2 PACE4, PC4, PC5/6 and PC7/8 [18–23]. In general, it was found that the catalytic segment of each convertase contains the sequence most conserved between all members of the family, with PC2 showing the least sequence similarity.

What is known about the localization of these enzymes?

Of all these convertases, only furin is ubiquitously expressed, while PACE4, PC5/6 and PC7/8 exhibit a widespread distribution in both regulated and constitutive cells. In contrast, PC1/3 and PC2 seem to be predominantly expressed in endocrine and neuronal cells and PC4 is exclusively synthesized within germ cells of testis [24].

Proteolytic cleavages can occur in a tissue-specific way; for instance, the proopiomelanocortin precursor is cleaved at the basic residue flanking ACTH to yield

ACTH and β-LPH in the anterior lobe of pituitary, whereas these two peptides are further processed in the intermediate lobe of pituitary in α-MSH and β-endorphin, respectively [25]. The specific localization of these proteolytic enzymes in neuroendocrine and non-endocrine tissue is not yet fully established.

This has been mainly addressed by *in situ* hybridization, or immunocytochemistry, looking for proconvertases and hormone precursor's co-localization. Day et al. have observed by *in situ* hybridization that PC1 and PC5 mRNAs were localized in paraventricular nucleus (PVN) and supraoptic nucleus (SON) magnocellular neurons, while PC2 mRNA was observed in both magnocellular and parvocellular PVN neurons as well as magnocellular SON neurons. On the other hand, PC5 is present in the PVN only in ocytocin producing cells [26]. These results support the hypothesis that processing enzymes are localized in tissues to perform the selective local production of selective bioactive peptides.

To identify which convertase is involved in proNPY cleavage several studies have been done based on heterologous expression of proNPY, co-expression of proNPY with prohormone convertases and *in vitro* proteolytic processing of proNPY.

Heterologous expression of proNPY

One of the most widely used methods for studying protein post-translational modification is heterologous expression in different kind of cells. In the present application mostly endocrine and neuronal cells are used. Cell transfection combined with systematic site-directed mutagenesis has been used to characterize the sequence dependency of the dibasic nature and its recognition by the endoproteases. Johansen et al. have shown that CHO cells are able to amidate NPY from a transfected C-peptide deleted form of the precursor, having as the terminal sequence Gly-Lys-Arg (i.e., NPY1-39), a sequence that does not require a dibasic endoprotease to be cleaved [27]. Dickerson et al. have shown that the proNPY processing is dependent on the sequence of the basic amino acids at the cleavage site, as well as on the cell type used [28, 29]. Indeed, GH_3 somatomammotrope cells that express PC2 when transfected with vectors encoding proNPY containing four different pairs of basic residues at the single endoproteolytic cleavage [Lys-Arg (wild type), Arg-Arg, Arg-Lys, and Lys-Lys] were able to process equally the four propeptides. In contrast, the corticotrope AtT-20 cells that express PC1/3, exhibit a strong hierarchy of cleavage pattern since native proNPY with a Lys-Arg pair and its Arg-Arg mutant are well cleaved, whereas a mutant with Arg-Lys is poorly cleaved and a Lys-Lys mutant is cleaved to a very small extent. Cell-type specificity was also observed in the removal patterns of the amino-terminal basic residue from the proNPY carboxylic-terminal peptide. Thus, GH_3 cells remove the NH_2-terminal basic residue less efficiently than AtT-20 cells, when processing the Arg-Arg mutants. In contrast, GH_3 cells remove about half the NH_2-terminal Lys from the COOH-terminal peptide produced from proNPY-Arg-Lys cleavage, while AtT-20 cells remove none of the NH_2-terminal Lys. However, these results are specific to proNPY since the Arg-Lys sites that are cleaved by At T-20 cells in proNPY remained unchanged with a proACTH/endorphin construction. The authors hypothesized that the lack of cleavage of proACTH/endorphin may be due to conformation differences, rather than the lack of appropriate enzyme. It is interesting to note that on proNPY cleavage site Arg residues are located at positions -1, -5, -7; this is completely different from the rules determined by Nakayama et al. for cleavage by furin. This author proposed rules for sequences that may govern the processing of NPY and particularly that the $Arg^{33}XArg^{35}XXArg^{38}$ sequence serves as a signal for the cleavage of the Arg^{38}-Lys^{39} mutant of proNPY by a KEX2-like endoprotease in AtT-20 cells [30]. Wulff et al. have studied the processing of proNPY and proPP in four neuroendocrine transfected cell lines expressing different precursor convertases [31]. They showed that proNPY is almost completely converted to NPY and that, in contrast, only 50% of proPP is converted to PP. Mutagenesis studies showed that on the N-terminal side of the dibasic processing site, the important difference in structure between PP and NPY, i.e., a proline in position 34 instead of a glutamine, is not responsible for the difference in processing efficiency. This suggests that the structural element essential for the efficient processing of proNPY is not located on the N-terminal side of the dibasic processing site. Paquet al. used time manipulation, temperature and intragranular pH to distinguish the action of PC1/3 and PC2. They found that PC1/3 mediated rapid processing of proNPY, this cleavage was characterized to be insensitive at 20°C, a temperature that blocked secretory pathway transport at the trans-golgi network. In contrast, PC2 mediated slower proNPY processing and was blocked at 20°C. These data clearly indicate that PC1/3 and PC2 do not perform their catalytic activity in the same local environment [32].

Coexpression of proNPY with prohormone convertases

In order to investigate whether other prohormone convertases were involved in proteolytic processing of proNPY our group had used a vaccinia virus derived expression system to co-express recombinant proNPY with each of the prohormone convertases PC1/3, PC2, furin and PACE4 in Neuro2A and NIH 3T3 cell lines as regulated neuroendocrine and constitutive prototype cell lines, respectively. The analysis of processed products (after separation by HPLC of NPY1-39, CPON and proNPY followed by a radioimmunoassay with NPY02, an anti-NPY monoclonal antibody that cross-reacts with proNPY and NPY1-39) shows that only PC1/3 generates NPY in NIH 3T3 cells, while in Neuro2A cells, both PC1/3 and PC2 are able to generate NPY. The convertases furin and PACE4 are unable to process proNPY in either cell line [33]. However, our conclusions were tempered by transfection experiments on neurons with antisense sequences coding for

PC1/3 or PC2 concluding that PC2 was the endoprotease responsible for proNPY in superior cervical ganglion [34].

In vitro proteolytic processing of proNPY

We have done *in vitro* processing assays with PC1/3 and PC2 using natural precursor [33]. proNPY was expressed in *E. coli*. The expressed fusion protein was purified by gel permeation and submitted to BrCN cleavage. The cleaved proNPY was purified by HPLC and identity of purified peptide was confirmed by electrospray mass spectrometry and Edman degradation. The enzyme activities of the proteases used were controlled with the fluorogenic substrate specific for these proteases, pGlu-Arg-Thr-Lys-Arg-MCA. The reaction mixture after digestion was separated by C18 HPLC and a sample of each fraction was assayed for NPY by RIA with NPY02. Under these conditions we were able to show that proNPY was completely cleaved by PC1/3, partially by PC2 but not by furin. As control reaction with chelating calcium with EDTA abolished the activities of both PC1/3 and PC2. The same amount of enzymatic activities was used, suggesting that the difference in activity is most probably due to substrate recognition. The significance of the difference in proNPY recognition between PC1/3 and PC2 was evaluated by determining kinetic parameters. PC1/3 and PC2 recognize proNPY with same range of Km (95 *versus* 69 μM), but with a great difference in V_{max} (47 *versus* 3 pmol/min).

To determine if structural conformation located in the NH2- and COOH-terminal portion of the precursor are involved in recognition short peptide proNPY namely proNPY20-49 and proNPY 28-43 have been used. The shortened sequences of proNPY are cleaved more efficiently than proNPY itself in the order proNPY 28-43 > (proNPY 20-49 > proNPY for PC1/3 and proNPY 20-49 > proNPY for PC2. In addition, the shortest peptide proNPY 28-43 is not cleaved by PC2 at all, while it is still cleaved by PC1/3. This confirms that PC1/3 and PC2 require different structural recognition elements for the cleavage of proNPY, as assessed by the V_{max}/Km ratio.

Based upon these observations it is proposed, first, that the constitutive secreted NPY does not result from cleavage carried out by ubiquitously expressed enzymes furin and PACE4. Second, that PC1/3 and PC2 are not equipotent in the cleavage of proNPY. Third, substrate peptide length might influence PC1/3 and PC2 processing activity. This could be a starting point to build molecules that could serve as specific proNPY processing inhibitors [33].

However, another enzyme, prohormone thiol protease (PTP) has been proposed to be more potent than PC1/3 and PC2 to cleave proNPY [35]. These authors have compared the cleavage of proNPY by PC1/3, PC2 and PTP. They used recombinant form of the three prohormones converting enzymes and a slightly modified proNPY sequence. The experiments were carried out in order to process 50% of the different precursors with the different enzymes. Interestingly, they found that compared to PC1/3 and PC2, PTP was the most potent enzyme to cleave proNPY. Therefore, at the present time we cannot exclude the existence of highly specific proNPY processing enzymes [35].

NPY and ProNPY presence in tumors, endocrine and non endocrine tissues

O'Hare and Schwartz reported the expression of NPY and its precursor in 16 human pheochromocytoma and three neuroblastoma tumors. An almost complete proteolytic processing of proNPY to mature NPY was observed in the tumors (median, 93%; range, 72–100%). A positive correlation between the processing efficiency and the NPY content was also observed. They also demonstrated, by both the elution position of the peptide in isoelectric focusing and its cross-reaction with an antiserum that recognizes the amidated C terminal region of NPY, that the peptide produced by the tumors is correctly processed and amidated [36]. The same authors studied the expression and precursor processing of NPY in 13 human and murine neuroblastoma and pheochromocytoma cell lines [37]. Among the 13 cell lines tested, eight produced a detectable immunoreactivity for NPY. In contrast with the results observed with the tumors, the cell content had a substantial amount of unprocessed proNPY (median, 57%; range, 33–72%). They concluded that many of these neuroendocrine cells express the NPY gene. The cells displayed, however, a partly impaired dibasic processing capacity while they generally amidated the products efficiently. Another study by Bjellerup et al. found that the degree of processing of proNPY to NPY in tumor tissue was lower in advanced neuroblastomas with regional or metastatic spread (stage 3 and 4) compared to the less aggressive stages 1, 2 and 4S tumors [38]. ProNPY processing of less than 50% was correlated with poor clinical outcome. N-Myc oncogene amplification was also correlated with a low degree of proNPY processing. Finally, it appears that a low degree of proNPY processing correlates with advanced clinical stage and poor outcome in neuroblastomas. Interestingly, Wulff et al. has shown that proNPY can be partially processed by non-endocrine cells such as Chinese hamster ovary (CHO) cells transfected with human NPY cDNA [39]. Although the majority of the immunoreactive NPY was found in the form of proNPY, as controlled by gel-filtration followed by specific radioimmunoassays, some processing occurred in the cells till the amidation (between 5 and 45% of the propeptide). However, the same authors showed that when proNPY, purified from a human neuroblastoma cell line, was incubated for 24 h with unconditioned tissue culture medium, it was cleaved to a molecular form of immunoreactive NPY eluting on a G50 column right before NPY indicating that the cleavage of proNPY can occur extracellularly [39]. Only a minor part of this material was amidated, indicating different kinds of proteolytic degradation. No attempt was made to further characterize these immunoreactive non-amidated forms of NPY. The authors postulated that the media could contain trypsin-like enzymes capable of cleaving the precursor at the monobasic level, either at Lys^{38} or the Arg^{39}, the natural processing site of proNPY, or at the Arg^{33} or the Arg^{35} in the NPY molecule itself. Dickerson et al. rule out cleavage of proNPY after secretion by incubating Tyr-labeled proNPY with protease inhibitors and in the absence of serum in the medium [40].

We have determined proNPY and NPY concentrations in the arcuate nucleus and the paraventricular nucleus of rats after 12 h of food deprivation using two-site specific enzyme immunoassays with monoclonal antibodies specific for NPY and proNPY. Although NPY mRNA has been shown to be elevated in these areas during starvation, amidated NPY content was unchanged, but proNPY concentrations significantly increased, reflecting the activation of the transcription of NPY mRNA. The discrepancies with already published data on an increase in NPY content in the arcuate nucleus and the paraventricular nucleus during starvation could be due to the fact that common RIA for NPY cross-react with proNPY, related fragments or apparented peptides of NPY. Therefore, we speculate that food starvation rather than increasing NPY synthesis induces its release (or those of proNPY) [41]. We have further investigated the content of proNPY and amidated NPY by ELISA and total NPY immunoreactivity (NPYir) by RIA in the hypothalamus and adrenal of mice. The hypothalamus contains 0.9 fmoles/mg of NPY, 0.1 fmoles/mg of proNPY and 14.8 fmoles/mg of NPYir; in contrast, the adrenal contains 1.1 fmoles/mg of NPY, 23.4 fmoles/mg of proNPY and 153.5 fmoles/mg of NPYir. Given that amidated NPY/amidated NPY + proNPY ratio may vary from 4 to 90%, depending on the tissue, we postulate that tissue-specific processing of proNPY in NPY could occur and that the biologically active amidated NPY in any case represents a minor portion of the total immunoreactive NPY content in these tissues [42]. More recently, NPY was also reported to increase endothelial cell stickiness for leukocytes [43]. Therefore, we examined the possibility that human umbilical venous endothelial cells (HUVECs) could also express NPY and that NPY could act as a modulator of inflammatory response. We found that a cytokine cocktail containing LPS, TNF-α, IFN-γ cause a biphasic increase in NPY and proNPY contents. We observed a first peak at 6–9 h and a second at 16 h with a proNPY and NPY cell contents elevated by six- and two-fold, respectively. The cytokine-induced increase in proNPY content strongly suggests a new synthesis of NPY and also the fact that non-neuronal cells are also able to not only produce a propeptide such as proNPY but also to process it into a biologically active form. ProNPY concentrations may follow mRNA expression of the peptide [44].

Processing of ProNPY in PC2 deficient mice

The involvement of PC1/3 and PC2 in neuropeptide proteolytic processing has been first established by using proconvertases antisense expression [45]. The availability of PC1/3 and PC2 null mice will allow us to confirm the association of a proconvertase to the proteolysis of a given prohormone. Ruthellen et al. have evaluated the cellular levels of several neuropeptides. Using selective NPY antibodies, which do not recognize proNPY, the authors have reported that NPY in the ileum was partially decreased (50%) in the PC2 deficient mice compared to control mice. This underlines the role of PC2 in ileum. In contrast, in the adrenal NPY was found to be increased by 2.5-fold. It was suggested that the increase in NPY may reflect compensatory mechanisms for general NPY production in the absence of PC2 activity. In the hypothalamus no change was observed which constitutes an interesting observation since the hypothalamus is rich in NPY. Also, NPY was not altered in the cortex or brain of PC2 deficient mice. These data show that the absence of PC2 only results in some tissue-selective changes in NPY levels [46].

Conclusion

The bioactivity of NPY depends on the correct processing by specific convertases of proNPY. This proteolysis is tissue-specific and can be regulated by various factors. Many conclusions are drawn on NPY regulation based on measurements by radioimmunoassay with antibodies that are not specific for the bioactive form. Since the NPY immunoreactivity is complex, it is essential to use well characterized assays to distinguish proNPY and its glycine-extended forms from the amidated form of NPY. The primary sequence of this precursor is simple but its processing toward the generation of biologically active ligand is rather complex. Several enzymes can be involved for its processing PC1/3, PC2, PTP and likely other not yet discovered enzymes. Its trimming also involves several enzymes, namely carboxypeptidase, PAM, DPPIV and Amp. In addition, there are many difficulties encountered in its measurement. These together permit only experienced researchers to address questions regarding the physiological, physiopathological and biosynthesis of this precursor. However, it remains to be established if genetic diseases could be associated to a defect in a proconvertase that would result in NPY abnormal concentrations in man.

References

1 Pedrazzini T, Pralong F, Grouzmann E (2003) Neuropeptide Y: the universal soldier. *Cell Mol Life Sci* 60(2): 350–377

2 Silva AP, Cavadas C, Grouzmann E (2002) Neuropeptide Y and its receptors as potential therapeutic drug targets. *Clin Chim Acta* 326(1–2): 3–25

3 Tatemoto K, Carlquist M, Mutt V (1982) Neuropeptide Y a novel brain peptide with structural similarities to peptide YY and pancreatic polypeptide. *Nature* 296: 659–660

4 Ekblad E, Edvinsson L, Wahlestedt C, Udman R, Hakanson R, Sundler F (1984) Neu-

ropeptide Y co-exists and co-operates with noradrenaline in perivascular nerve fibers. *Regul Peptides* 8: 225–235
5 Jamal H, Jones PM, Byrne J, Suda K, Ghatei MA, Kanse SM, Bloom SR (1991) Peptide contents of neuropeptide Y, vasoactive intestinal peptide, and calcitonin gene-related peptide and their messenger ribonucleic acids after dexamethasone treatment in the isolated rat islets of Langerhans. *Endocrinology* 129: 3372–3380
6 Ericsson A, Schalling M, Mc Intyre KR, Lundberg JM, Larhammar D, Seroogy K, Hökfelt T, Persson H (1987) Detection of neuropeptide Y and its mRNA in megakaryocytes: enhanced levels in certain autoimmune mice. *Proc Natl Acad Sci USA* 84: 5587–5591
7 Grouzmann E, Alvarez-Bolado G, Meyer C, Osterheld MC, Burnier M, Brunner HR, Waeber B (1994) Localization of Neuropeptide Y and its C-terminal flanking peptide in human renal tissue. *Peptides* 15 (8): 1377–1382
8 Gehlert DR, Gackenheimer SL, Millington MR, Manning AB, Chronwall BM (1994) Localization of neuropeptide Y immunoreactivity and [^{125}I] peptide YY binding sites in the human pituitary. *Peptides* 15 (4): 651–656
9 Zukowska-Grojec Z, Karwatowska-Prokopczuk E, Rose W, Rone J, Movafagh S, Ji H, Yeh Y, Chen WT, Kleinman HK, Grouzmann E et al (1998) Neuropeptide Y: a novel angiogenic factor from the sympathetic nerves and endothelium. *Circ Res* 83(2): 187–195
10 Minth CD, Bloom SR, Polak JM, Dixon JE (1984) Cloning, characterization, DNA sequence of human cDNA encoding neuropeptide tyrosine *Proc Natl Acad Sci USA* 81: 4577–4581
11 Allen JM, Polak JM, Bloom SR (1985) Presence of the predicted C flanking peptide of neuropeptide Y (CPON) in tissue extract. *Neuropeptides* 6: 95–100
12 Wahlestedt C, Yanaihara N, and Hakanson R (1986) Evidence for different pre- and post-junctional receptors for neuropeptide Y and related peptides. *Regul Pept* 13: 307–318
13 Fuller RS, Sterne RE, Thorner J (1988) Enzymes required for yeast prohormone processing. *Annu Rev Physiol* 50: 345–362
14 Mizuno K, Nakamura T, Oshima T, Tanaka S, Matsuo H (1988) Yeast KEX2 gene encodes an endopeptidase homologous to subtilisin-like serine proteases. *Biochem Biophys Res Commun* 156: 246–254
15 Fuller RS, Brake A, Thorner J (1989) Yeast prohormone processing enzyme (KEX2 gene product) is a Ca^{2+} dependent serine protease. *Proc Natl Acad Sci USA* 86: 1434–1438
16 Roebroek AJM, Schalken JA, Leunissen JAM, Onnekink C, Bloemers HPJ, Van De Ven WJM (1986) Evolutionary conserved closed linkage of the c-fes/fps proto-oncogene and genetic sequences encoding a receptor-like protein. *EMBO J* 5: 2197–2202
17 Hatsuzawa K, Nagahama M, Takahashi S, Takada K, Murakami K, Nakayama K (1992) Purification and characterization of furin, a Kex2-like processing endoprotease, produced in chinese hamster ovary cells. *J Biol Chem* 267: 16094–16099
18 Smeekens SP, Steiner DF (1990) Identification of human insulinoma cDNA encoding a novel mammalian protein structurally related to the yeast dibasic processing protease kex2. *J Biol Chem* 265: 2997–3000

19 Seidah NG, Marcinkiewicz M, Benjannet S, Gaspar L, Beaubien G, Mattei MG, Lazure C, Mbikai M, Chretien M (1991) Cloning and primary sequence of a mouse candidate prohormone convertase PC1 homologous to PC2, furin, and kex2: distinct chromosomal localization and messenger RNA distribution in brain and pituitary compared to PC2. *Mol Endocrinol* 5: 111–122

20 Nakayama K, Kim WS, Torii S, Hosaka M, Nakagawa T, Ikemizu J, Baba T, Murakami K (1992) Identification of the fourth member of the mammalian endoprotease family homologous to the yeast kex2 protease. *J Biol Chem* 267: 5897–5900

21 Lusson L, Vieau D, Hamelin J, Day R, Chretien M, Seidah NG (1993) cDNA structure of the mouse and rat subtilisin/kexin like PC5: a candidate proprotein convertase expressed in endocrine and nonendocrine cells. *Proc Natl Acad Sci USA* 90: 6691–6695

22 Nakagawa T, Hosaka M, Torii S, Watanabe T, Murakami K, Nakayama K (1992) Identification and functional expression of the mammalian kex2-like processing endoprotease family: its striking structural similarity to PACE4. *J Biochem* 113: 132–135

23 Tsuji A, Hine C, Mori K, Tamai Y, Higashine K, Nagamune H, Matsuda Y (1994) A novel member, PC7, of the mammalian kexin-like protease family: homology to PACE4A, its brain-specific expression and identification of isoforms. *Biochem Biophys Res Commun* 202 (3): 1452–1459

24 Seidah NG, Chretien M, Day R (1994) The family of subtilisin/kexin like pro-protein and pro-hormone convertases: divergent or shared functions. *Biochimie* 76: 197–209

25 Rhodes CJ, Thorne BA, Lincoln B, Nielsen E, Hutton JC, Thomas G (1993) Processing of proopiomelanocortin by insulin secretory granule proinsulin processing endopeptidases. *J Biol Chem* 268: 4267–4275

26 Dong W, Seidel B, Marcinkiewicz M, Chretien M, Seidah NG, Day R (1997) Cellular localization of the prohormone convertases in the hypothalamic paraventricular and supraoptic nuclei: selective regulation of PC1 in corticotrophin-releasing hormone parvocellular neurons mediated by glucocorticoids. *J Neurosci* 17(2): 563–575

27 Johansen TE, O'Hare MMT, Wulff BS, Schwartz TW (1991) CHO cells synthesize amidated neuropeptide Y from a C-peptide deleted form of the precursor. *Endocrinology* 129 (1): 553–555

28 Dickerson IM, Mains RE (1990) Cell-type specific posttranslational processing of peptides by different pituitary cell lines. *Endocrinology* 127 (1): 133–140

29 Dickerson IM, Dixon JE, Mains RE (1990) Biosynthesis and posttranslational processing of site-directed endoproteolytic cleavage mutants of pro-neuropeptide Y in mouse pituitary cells. *J Biol Chem* 265 (5): 2642–2649

30 Nakayama K, Watanabe T, Nakagawa T, Kim WS, Nagahama M, Hosaka M, Hatsuzawa K, Kondo-Hashiba K, Murakami K (1992) Consensus sequence for precursor processing at mono-arginyl sites. *J Biol Chem* 267 (23): 16335–16340

31 Wulff BS, Johansen TE, Dalboge H, O'Hare MMT, Schwartz TW (1993) Processing of two homologous precursors, pro-neuropeptide Y and pro-pancreatic polypeptide, in transfected cell lines expressing different precursor convertases. *J Biol Chem* 268 (18): 13327–13335

32 Paquet L, Zhou A, Chang EY, Mains RE (1996) Peptide biosynthetic processing: distinguishing prohormone convertases PC1 and PC2. *Mol Cell Endocrinol* 120(2): 161–168

33 Brakch N, Rist B, Beck-Sickinger AG, Goenaga J, Wittek R, Bürger E, Brunner HR, Grouzmann E (1998) Role of prohormone convertases in pro-neuropeptide y processing: coexpression and *in vitro* kinetic investigations. *Biochemistry* 36(51): 16309–16320

34 Paquet L, Massie B, Mains RE (1996) Proneuropeptide Y processing in large dense-core vesicles: manipulation of prohormone convertase expression in sympathetic neurons using adenoviruses. *J Neurosci* 16(3): 964–973

35 Hook VY, Schiller MR, Azaryan AV (1996) The processing proteases prohormone thiol protease, PC1/3 and PC2, and 70-kDa aspartic proteinase show preferences among proenkephalin, proneuropeptide Y, and proopiomelanocortin substrates. *Arch Biochem Biophys* 328(1): 107–114

36 O'Hare MMT, Schwartz TW (1989) Expression and precursor processing of neuropeptide Y in human pheochromocytoma and neuroblastoma tumors *Cancer Res* 49: 7010–7014

37 O'Hare MMT, Schwartz TW (1989) Expression and precursor processing of neuropeptide Y in human and murine neuroblastoma and pheochromocytoma cell lines. *Cancer Res* 49: 7015–7019

38 Bjellerup P, Theodorsson E, Jörnvall H, Kogner P (2000) Limited neuropeptide Y precursor processing in unfavourable metastatic neuroblastoma tumours. *British J Cancer* 83: 171–176

39 Wulff BS, O'Hare MMT, Boel E, Theill LE, Schwartz TW (1990) Partial processing of the neuropeptide Y precursor in transfected CHO cells. *FEBS Lett* 261 (1): 101–105

40 Dickerson IM, Mains RE (1990) Cell-type specific posttranslational processing of peptides by different pituitary cell lines. *Endocrinology* 127 (1): 133–140

41 Bjenning C, Grouzmann E, Oksenshteyn E, Leibowitz SF (1997) Effect of maternal nutrition and food deprivation on hypothalamic NPY and proNPY levels in the neonate and weanling rat. *Abst Neuroscience meeting*, New Orleans

42 Gaertner H, Léchenne B, Imboden H, Michalet S, Stöcklin R, Monod M, Grouzmann E (2004) Production Of native proneuropeptide Y in *E. coli* and its quantification in tissues. *7th International NPY Meeting* Coimbra

43 Sung CP, Arleth AJ, Feuerstein GZ (1991) Neuropeptide Y upregulates the adhesiveness of human endothelial cells for leukocytes. *Circ Res* 68(1): 314–318

44 Silva AP, Cavadas C, Baisse-Agushi B, Spertini O, Brunner HR, Grouzmann E (2003) NPY, NPY receptors, and DPP IV activity are modulated by LPS, TNF-alpha and IFN-gamma in HUVEC. *Regul Pept* 116(1–3): 71–79

45 Bloomquist BT, Eipper BA, Mains RE (1991) Prohormone-converting enzymes: regulation and evaluation of function using antisense RNA. *Mol Endocrinol* 5(12): 2014–2024

46 Miller R, Toneff T, Vishnuvardhan D, Beinfeld M, Hook VY (2003) Selective roles for the PC2 processing enzyme in the regulation of peptide neurotransmitter levels in brain and peripheral neuroendocrine tissues of PC2 deficient mice. *Neuropeptides* 37(3): 140–148

Dipeptidyl-peptidase IV and aminopeptidase P: molecular switches of NPY/PYY receptor affinities

Rolf Mentlein

University of Kiel, Department of Anatomy, Olshausenstraße 40, 24098 Kiel, Germany

Introduction

Like other signal-transmitting substances, neuropeptides and hormones have to be inactivated after their release and interaction with their receptors. Otherwise, a constant stimulus results that permits no regulated information transfers. For neuropeptides, proteolytic degradation by cell-surface peptidases is the major inactivation route; other processes like diffusion and receptor-mediated endocytosis followed by lysosomal degradation appear to be minor processes [1–3]. For circulating small peptide hormones, degradation by endothelial cell-surface or soluble peptidases in the blood as well as renal elimination are both important inactivation routes. The relative contributions of the one or other mechanism depend on the anatomical site as well as at the time scale.

Cell-surface peptidases as peptide-inactivation system

Degradation of bioactive peptides can be accomplished by a variety of proteases. However, a special group of oligopeptidases acts preferentially on small neuropeptides or hormones [1, 3]. These oligopeptidases or peptidases are specialized proteases cleaving 3–100 residue oligopeptides (the former term 'peptidase' is now also used synonymously to protease or proteinase that cleave also larger peptides/proteins). These above-defined oligopeptidases are mainly cell-surface peptidases (Table 1 gives important examples mentioned further in the text), but often soluble forms exist in body fluids that are mostly generated by shedding of the membrane-anchored forms. Cell-surface peptidases are exo- or endo-peptidases with a certain specificity regarding the cleavage sites, but not with respect to a single regulatory peptide (see Tab. 1; an exception is pyroglutamyl-peptidase II that is specific for thyrotropin-releasing hormone). They have a wide tissue distribution and are mostly involved in the inactivation, but sometimes also in the activation of bioactive pep-

The NPY Family of Peptides in Immune Disorders, Inflammation, Angiogenesis and Cancer,
edited by Zofia Zukowska and Giora Z. Feuerstein

Table 1 - Overview of the main cell surface peptidases involved in the inactivation of regulatory peptides

Peptidase	EC	Catalytic type	Specificity	Localization (examples)
Aminopeptidase N APN (CD13)	3.4.11.2	Metallo	Xaa-↓-Xaa-...	Endothelial cells, epithelial cells, macrophages
Aminopeptidase A APA	3.4.11.7	Metallo	Glu/Asp-↓-Xaa...	B cells
Aminopeptidase P APP	3.4.11.9	Metallo	Xaa-↓-Pro-...	Endothelial cells
Pyroglutamyl-peptidase II	3.4.19.6	Metallo	Glp-↓-His-Pro-NH_2	Neurons, pituitary cells
Dipeptidyl-peptidase IV DPP IV (CD26)	3.4.14.5	Serine	Xaa-Pro/Ala-↓-Yaa...	Endothelial cells, epithelial cells, T cells, hepatocytes
Carboxypeptidase M CPM	3.4.17.12	Metallo	...Xaa-↓-Arg/Lys	Epithelial cells, neurons, smooth muscle cells
Angiotensin-converting enzyme ACE (CD143)	3.4.15.1	Metallo	...Xaa-↓-Yaa-Zaa	Endothelial cells, epithelial cells, neurons, fibroblasts, monocytes
Neprilysin Neutral endopeptidase NEP (CD10)	3.4.24.11	Metallo	...Xaa-↓-Yaa...	Epithelial cells, neurons, Schwann cells, lymphocytes, smooth muscle cells

For more details and further cell-surface peptidases see [3] or consult http://merops.sanger.ac.uk/ or http://brenda.bc.uni-koeln.de/ (or further internet resources). Methods to monitor the degradation of bioactive peptides by these cell-surface peptidases or other proteases are reviewed in [2].

tides that serve in cell communication. Some of them are in the focus of recent and actual pharmacological research, since specific inhibitors are widely used or under development as antidiabetic or antihypertensive drugs (see [3] for a recent review).

In contrast to neuropeptides/peptide hormone receptors, their affinity is not in the physiological nano- or pico-molar range. Instead, their K_m-values range from 1–50 μM. That means they do not cleave their substrates under saturated V_{max} conditions (respectively k_{cat} = V_{max}/mol enzyme). At concentrations lower than their K_m-values, there is a linear dependence of the reaction rate v from the substrate concentration [5], and the slope is given by the rate constant k_{cat}/K_m. High rate constants correspond to high cleavage rates under physiological concentrations and, e.g., doubling of peptide concentration results in doubled reaction rates [3, 4].

Peptidases involved in the degradation of neuropeptide Y (NPY) and peptide YY (PYY)

By their structure, neuropeptide Y (NPY), peptide YY (PYY) and pancreatic polypeptide (PP) are protected against degradation by many non-specialized proteases: carboxypeptidase attack is hindered by the amidation at the C-terminus, and non-specialized aminopeptidases do not degrade sequences with proline residues as found in positions 2, 5 and 8 of the N-terminal structure. Furthermore, the attack of endopeptidases is restricted by the PP-fold secondary structure. In fact, non-specialized amino- and carboxypeptidases including aminopeptidase N, aminopeptidase W, leucyl aminopeptidase (EC 3.4.11.1, soluble), carboxypeptidase M and angiotensin-converting enzyme do not cleave NPY and PYY [5–7]. On the other hand some specialized peptidases attack NPY and PYY and generate either new forms or degrade them to inactive fragments (Fig. 1).

Aminopeptidases as NPY/PYY convertases

Two specialized aminopeptidases are able to remove the first or the first two amino acids from the N-terminus of NPY and PYY: Dipeptidyl-peptidase IV (DPP IV; CD26; Xaa-Pro-dipeptidylaminopeptidase; adenosine deaminase binding protein-2; EC 3.4.14.5) and aminopeptidase P (APP; X-Pro aminopeptidase; EC 3.4.11.9). Both are selective for the second N-terminal amino acid, either cleaving after (DPP IV) or in front (APP) of a penultimate proline (or alanine) residue. Purified DPP IV liberates the ultimate Tyr-Pro-dipeptides to yield NPY(3-36) or PYY(3-36); purified APP liberates only the terminal Tyr-residue to generate NPY(2-36) or PYY(2-36) [5–8].

Human DPP IV is a dimeric 240 kDa glycoprotein composed of two 120 kDa subunits. Its mRNA codes for a type II integral transmembrane protein of 766 amino acids which is anchored by a 22 residue hydrophobic membrane-spanning domain preceded by a short, 6-residue intracellular sequence at the N-terminus. DPP IV belongs to the serine class of proteases, clan SC, family S9, with a Ser-630, Asp-708, and His-740 catalytic triad, inversely arranged as compared to the trypsin-like serine proteases ("nonclassical serine protease"). The enzyme has a wide tissue distribution; it is found on the cell surface of many epithelial cells, endothelial cells, activated T cells, thymocytes, hepatocytes and specialized fibroblasts [3, 4]. NPY and PYY belong to the best peptide substrates for DPP IV in terms of k_{cat}/K_m. However, several other neuropeptides, peptide hormones and chemokines with N-termi-

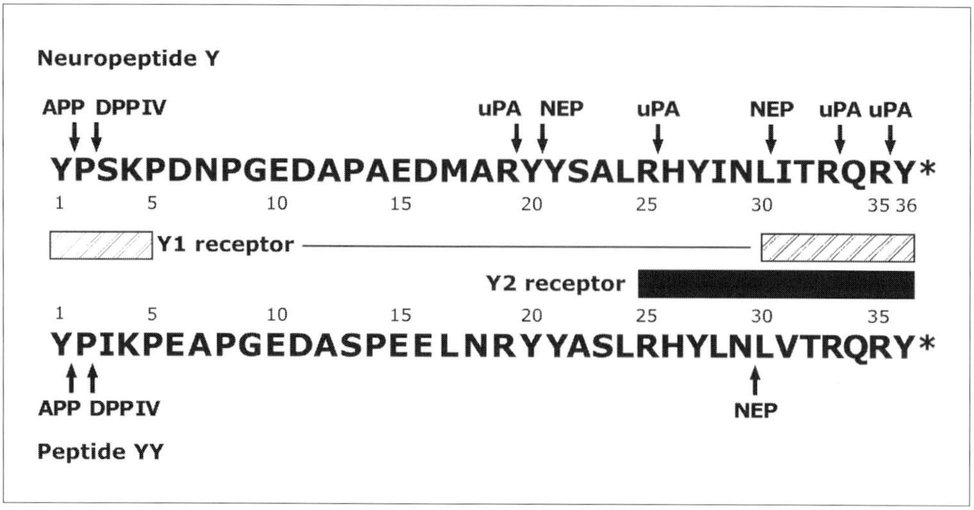

Figure 1
Cleavage of NPY and PYY by different proteases. The exopeptidases aminopeptidase P (APP) and dipeptidyl-peptidase IV (DPP IV) release N-terminal Tyr or Tyr-Pro affecting the Y1 receptor binding affinities, whereas the endopeptidases neutral endopeptidase (neprilysin, NEP) and urokinase-type plasminogen activator (urokinase, uPA) target intrapeptide sequences and completely abolish the binding affinities.

nal Xaa-Pro- or Xaa-Ala-sequence are also good substrates for DPP IV. So far, its most important physiological function appears to be the destruction of the insulinotropic hormone, glucagon-like peptide 1 (GLP-1) that is released after meals by endocrine L cells of the gut and stimulates insulin secretion from pancreatic β cells (incretin effect). DPP IV inhibitors have been shown to improve glucose tolerance in normal and diabetic animals and in human diabetic patients [9]. DPP IV-knockout mice show a normal phenotype, but are refractory to the development of obesity and hyperinsulinemia [10]. However, inactivation of other hormones beside GLP-1 might also contribute to these effects.

Human APP exists in two forms encoded by different genes: a membrane-bound form (AP-P2) encoded by a gene of the human X chromosome and a cytosolic form encoded by a gene on human chromosome 10 [11, 12], both with nearly identical substrate specificities (also with respect to NPY and PYY [8, 13–16]). The membrane-bound form is located as an ectoenzyme via a glycosyl-phosphatidylinositol anchor on the plasma membrane of endothelial and epithelial cells whereas the soluble form is a cytosolic enzyme found in leucocytes, platelets and astroglial cells of the brain [13–15]. The cytosolic form displays 43% amino acid identity and 63% similarity to the membrane-bound form [16]. Both forms are metalloproteases (manganese-dependent) and belong to the peptidase clan MG [12, 17]. Beside NPY and PYY, bradykinin (Arg-Pro-Pro-Gly-Phe-Ser-Pro-Phe-Arg) appears to be the most important substrate for APP and is partially inactivated to desArg-bradykinin [6, 15].

NPY and PYY are the best natural substrates for DPP IV and are moderate substrates for APP [3–8]. N-terminal truncation by DPP IV or APP does not abolish the biological activities of NPY and PYY to all Y-receptor subtypes: The truncated forms still bind to Y2 and the Y5 receptors (Y5: dipeptide-truncated forms with less activity), but they are inactive at Y1 receptors [18]. In particular, DPP IV-truncated NPY and PYY have been shown to bind with equal potency in binding assays to Y2, but not to Y1 receptors [19, 20]. In accordance, the DPP IV-truncated forms lack Y1 biological potencies, e.g., Y1-mediated inhibition of vagally stimulated acid secretion and exocrine pancreatic secretion by PYY [20–22].

The biological importance of peptidases involved in the proteolytic truncation of NPY and PYY may be estimated from their occurrence in the body.

Occurrence of proteolytic NPY and PYY fragments

In fact, not only full-length NPY and PYY are found in tissues and body fluids, but also in considerable amounts in their N-terminal-truncated forms. NPY(3-36) has been isolated from porcine brain (35% of total NPY immunoreactivity) and human pancreatic tumors [23, 24]. In human cerebrospinal fluid NPY(3-36) is the most important immunoreactive component beside the intact peptide [25]. PYY(3-36) accounts for 37% of total PYY immunoreactivity in human plasma samples collected in the fasted state, and even 54% of that in the postprandial state [19, 26]. Furthermore, intact PYY and PYY(3-36) can be isolated from rabbit intestine in nearly equal amounts [27].

These data suggest that DPP IV-cleavage of NPY and PYY are frequent modifications of both peptides *in vivo*. APP might act in more specialized locations.

Biological importance of aminopeptidase-mediated NPY/PYY conversion

Regarding the physiological role of this process, DPP IV and APP can be regarded as NPY/PYY convertases switching their receptor selectivity by abolishing Y1 effects and generating Y2 and partially Y5 agonists [28]. So far, the physiological importance has been investigated in a few examples for NPY-mediated effects.

NPY is a co-transmitter with noradrenaline in sympathetic perivascular nerves. Here, NPY acts as a proliferative and contraction factor for smooth muscle cells via Y1 receptors as well as an angiogenic factor on endothelial cells via non-Y1 receptors (Fig. 2). Truncation of NPY by APP on smooth muscle cells [6] or DPP IV on

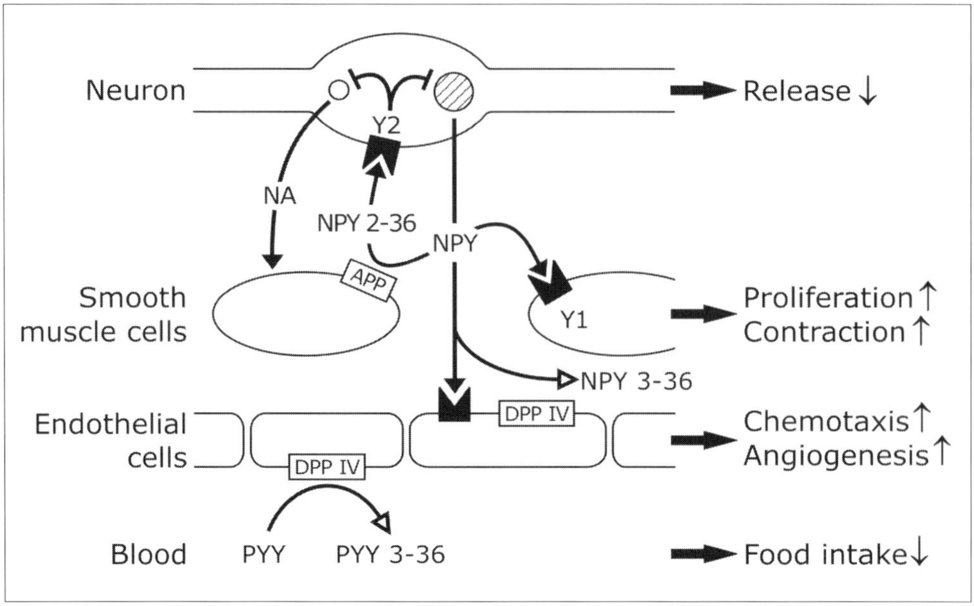

Figure 2
Schematic examples of the roles of aminopeptidase P (APP) and dipeptidyl-peptidase IV (DPP IV) in the switch of receptor affinities of neuropeptide Y (NPY) and peptide YY (PYY). NPY and noradrenaline (NA) are transmitters of sympathetic nerves surrounding blood vessels. NPY stimulates the proliferation and contraction of vascular smooth muscle cells via Y1 receptors, chemotaxis of endothelial cells and subsequent angiogenesis. The Y1 receptor-mediated effects are abolished upon proteolytic truncation by APP (on smooth muscle cells) or DPP IV on endothelial cells, whereas the Y2-mediated effects, e.g., presynaptic inhibition of NPY and NA release, remain unchanged. PYY is converted in the blood to the Y2-agonist PYY(3-36) that inhibits food intake.

endothelial cells [5] abolishes all Y1, but not Y2 effects. Whereas blood pressure effects or vasoconstriction by DPP IV-inhibitors have not yet been reported, a critical role of DPP IV has been demonstrated in angiogenesis, in particular in endothelial cell migration in response to wound healing [28, 29]. The role of DPP IV in angiogenesis is, however, not fully elucidated and involves probably other factors, e.g., DPP IV adhesion to extracellular matrix components.

Peripheral NPY can potentate local inflammatory responses via Y1 and Y5 receptors. As a consequence, stabilization of full-length NPY by application of the DPP IV inhibitor Ile-thiazolidide has been shown to enhance inflammatory paw oedema in a rat model [30]. A further interesting aspect of peripheral NPY action is its role in the regulation of malignant cells. NPY and NPY(3-36) can be isolated from human pancreatic endocrine tumors [24]. Furthermore, breast carcinoma cells express Y1 and Y2 receptors, highly metastatic ones in particular Y1 receptors [31]. In SK-N-MC cells NPY reduced tumor growth via Y1 receptors. Consequently, DPP IV/APP regulation of NPY-regulated tumor growth and immune reactions should be considered.

In the central nervous system, NPY is one of the most abundant neuropeptides. NPY is involved in a variety of neurological functions such as memory, anxiety, anti-convulsion, food intake and proliferation of neuronal precursor cells. In DPP IV activity-deficient (not protein-deficient) Fischer 344 rat substrains a reduced basal and NPY-evoked behavioural stress response was observed [32]. NPY administration induced higher anxiolytic-like effects (increased social interaction time in the social interaction test) and sedative-like effects (decreased motor activity in the elevated plus maze) in DPP IV activity-deficient rats [32].

DPP IV-deficient rats and mice show also a reduced body weight, especially after a high-fat diet. However, for satiety effects observed after DPP IV-inhibitor application or in DPP IV-deficient rats or mice not only the stabilization of central NPY, but also of peripheral hormones like PYY, GLP-1 and others might be responsible. PYY(3-36) has been claimed to be a physiological satiety factor which is low in obese patients [33].

Thus, regulation of the biological potency of NPY and PYY appear to be important switches of receptor and consequently biological activity. More experimental data with DPP IV inhibitors and knockout animals should reveal the importance under normal and pathological conditions.

Endopeptidases as inactivation enzymes

Apart from these exopeptidases, some endopeptidases are known to cleave NPY and PYY (Fig. 1). Neprilysin (= neutral endopeptidase, NEP, CD10) cleaves NPY at the Tyr^{20}-Tyr^{21} and Leu^{30}-Ile^{31} bonds and PYY preferentially at the Asn^{29}-Leu^{30} bond [7, 16]. Urokinase (urokinase-type plasminogen activator, uPA; a mainly soluble, cell-secreted protease) and related trypsin-like enzymes cleave NPY after arginine residues in positions 19, 25, 33 and 35 [34]. This fragmentation was also observed with cultivated microglial cells and neurons, but not with astroglial cells [34]. In binding studies with rabbit kidney membranes, a cleavage of NPY by endopeptidase-2 (meprin A; EC 3.4.24.18), a membrane-bound metalloproteinase, was observed [35].

These cleavages of internal peptide-bonds (endopeptidase cleavage) result in the complete inactivation of NPY and PYY at all receptor subtypes. For NPY, endopeptidase cleavage is probably the most important final inactivation route in the brain, for peripheral NPY and circulating PYY diffusion and renal elimination might be of similar relevance.

Conclusion and perspectives

Cell-surface and soluble peptidases regulate the biological potency of neuropeptides, peptide hormones and chemokines by activation, partial inactivation and complete inactivation. So far, the activation of angiotensin I to II is the best known example for activation (conversion) of a peptide hormone and convenient drug-target. Stabilization of natriuretic peptides against inactivation by neutral endopeptidase has been thoroughly investigated as a target for the treatment of hypertension. Stabilization of GLP-1 and other incretins (hormones increasing insulin secretion postprandially) against inactivation by DPP IV is currently being evaluated as a target for the treatment of type 2 diabetes (DPP IV inhibitors are now in clinical phase III studies). The differential regulation of the receptor selectivity of NPY and PYY by DPP IV and APP has not yet been fully investigated for side effects and new applications of inhibitors. Of both ectopeptidases, DPP IV appears to be of higher importance in humans. Presently, regulation of NPY-mediated angiogenesis and inflammation as well as central NPY- and PYY-regulated food intake and anxiety are the most important targets to be considered. Clearly, more experimental data with DPP IV knockout animals and DPP IV inhibitors, especially those entering the brain, are needed.

Acknowledgements
I thank Prof. Dr. Stephan von Hörsten, Medical University of Hannover, Germany, for critical reading and stimulating discussion and Clemens Franke for drawing the figures.

References

1 Kenny AJ, Hooper NM (1991) Peptidases involved in the metabolism of bioactive peptides. In: JH Henrickson (ed): *Degradation of bioactive substances: physiology and pathophysiology*. CRC Press, Boca Raton, 47–79
2 Mentlein R, Lucius R (1997) Methods for the investigation of neuropeptide catabolism and stability *in vitro*. *Brain Res Brain Res Protoc* 1: 237–246
3 Mentlein R (2004) Cell-surface peptidases. Int Rev Cytol 235: 165–213
4 Mentlein R (1999) Dipeptidyl-peptidase IV (CD26) – role in the inactivation of regulatory peptides. *Regul Peptides* 85: 9–24
5 Mentlein R, Dahms P, Grandt D, Krüger R (1993) Proteolytic processing of neuropeptide Y and peptide YY by dipeptidyl peptidase IV. *Regul Peptides* 49: 133–144
6 Mentlein R, Roos T (1996) Proteases involved in the metabolism of angiotensin II, bradykinin, calcitonin gene-related peptide (CGRP), and neuropeptide Y by vascular smooth muscle cells. *Peptides* 17: 709–720
7 Medeiros MD, Turner AJ (1996) Metabolism and functions of neuropeptide Y. *Neurochem Res* 21: 1125–1132

8 Orawski AT, Simmons WH (1995) Purification and properties of membrane-bound aminopeptidase P from rat lung. *Biochemistry* 34: 11227–11236

9 Mentlein R (2005) Therapeutic assessment of GLP-1 agonists compared to DP IV inhibitors as potential anti-diabetic drugs. *Expert Opinion Invest Drugs* 14: 57–64

10 Conarello SL, Li Z, Ronan J, Roy RS, Zhu L, Jiang G, Liu F, Woods J, Zycband E, Moller DE, Thornberry NA, Zhang BB (2003) Mice lacking dipeptidyl peptidase IV are protected against obesity and insulin resistance. *Proc Natl Acad Sci USA* 100: 6825–6830

11 Sprinkle TJ, Stone AA, Venema RC, Denslow ND, Caldwell C, Ryan JW (1998) Assignment of the membrane-bound human aminopeptidase P gene (XPNPEP2) to chromosome Xq25. *Genomics* 50: 114–116

12 Sprinkle TJ, Caldwell C, Ryan JW (2000) Cloning, chromosomal sublocalization of the human soluble aminopeptidase P gene (XPNPEP1) to 10q25.3 and conservation of the putative proton shuttle and metal ligand binding sites with XPNPEP2. *Arch Biochem Biophys* 378: 51–56

13 Harbeck HT, Mentlein R (1991) Aminopeptidase P from rat brain. Purification and action on bioactive peptides. *Eur J Biochem* 198: 451–458

14 Hooper NM, Hryszko J, Turner AJ (1990) Purification and characterization of pig kidney aminopeptidase P. A glycosyl-phosphatidylinositol-anchored ectoenzyme. *Biochem J* 267: 509–515

15 Simmons WH, Orawski AT (1992) Membrane-bound aminopeptidase P from bovine lung. Its purification, properties, and degradation of bradykinin. *J Biol Chem* 267: 4897–4903

16 Medeiros MD, Turner AJ (1994) Processing and metabolism of peptide-YY: pivotal roles of dipeptidylpeptidase-IV, aminopeptidase-P, and endopeptidase-24.11. *Endocrinology* 134: 2088–2094

17 Cottrell GS, Hooper NM, Turner AJ (2000) Cloning, expression, and characterization of human cytosolic aminopeptidase P: a single manganese(II)-dependent enzyme. *Biochemistry* 39: 15121–15128.

18 Berglund MM, Hipskind PA, Gehlert DR (2003) Recent developments in our understanding of the physiological role of PP-fold peptide receptor subtypes. *Exp Biol Med (Maywood)* 228: 217–244

19 Grandt D, Teyssen S, Schimiczek M, Reeve JR Jr, Feth F, Rascher W, Hirche H, Singer MV, Layer P, Goebell H, Ho FJ, Eysselein VE (1992) Novel generation of hormone receptor specificity by amino terminal processing of peptide YY. *Biochem Biophys Res Commun* 186: 1299–1306

20 Grandt D, Schimiczek M, Rascher W, Feth F, Shively J, Lee TD, Davis MT, Reeve JR, Michel MC (1996) Neuropeptide Y 3-36 is an endogenous ligand selective for Y2 receptors. *Regul Peptides* 67: 33–37

21 Grandt D, Siewert J, Sieburg B, al Tai O, Schimiczek M, Goebell H, Layer P, Eysselein VE, Reeve JR Jr, Muller MK (1995) Peptide YY inhibits exocrine pancreatic secretion in isolated perfused rat pancreas by Y1 receptors. *Pancreas* 10: 180–186

22 Lloyd KC, Grandt D, Aurang K, Eysselein VE, Schimiczek M, Reeve JR Jr (1996) Inhibitory effect of PYY on vagally stimulated acid secretion is mediated predominantly by Y1 receptors. *Am J Physiol* 270: G123–127

23 Grandt D, Schimiczek M, Rascher W, Feth F, Shively J, Lee TD, Davis MT, Reeve JR Jr, Michel MC (1996) Neuropeptide Y 3-36 is an endogenous ligand selective for Y2 receptors. *Regul Peptides* 67: 33–37

24 Shaw C, Cormican K, Thim L, Maule AG, Sloan JM, Buchanan KD (1993) Neuropeptide Y and neuropeptide Y 3-36: isolation from human pancreatic endocrine tumours. *Regul Peptides* 45: 387–394

25 Nilsson C, Westman A, Blennow K, Ekman R (1998) Processing of neuropeptide Y and somatostatin in human cerebrospinal fluid as monitored by radioimmunoassay and mass spectrometry. *Peptides* 19: 1137–1146

26 Grandt D, Schimiczek M, Beglinger C, Layer P, Goebell H, Eysselein VE, Reeve JR (1994) Two molecular forms of peptide YY (PYY) are abundant in human blood: characterization of a radioimmunoassay recognizing PYY 1-36 and PYY 3-36. *Regul Peptides* 51: 151–159

27 Grandt D, Schimiczek M, Struk K, Shively J, Eysselein VE, Goebell H, Reeve JR Jr (1994) Characterization of two forms of peptide YY, PYY(1-36) and PYY(3-36), in the rabbit. Peptides 15: 815–820

28 Zukowska-Grojec Z, Karwatowska-Prokopczuk E, Rose W, Rone J, Movafagh S, Ji H, Yeh Y, Chen WT, Kleinman HK, Grouzmann E, Grant DS (1998) Neuropeptide Y: a novel angiogenic factor from the sympathetic nerves and endothelium. *Circ Res* 83: 187–195

29 Ghersi G, Chen W, Lee EW, Zukowska Z (2001) Critical role of dipeptidyl peptidase IV in neuropeptide Y-mediated endothelial cell migration in response to wounding. *Peptides* 22: 453–458

30 Dimitrijevic M, Stanojevic S, Vujic V, Kovacevic-Jovanovic V, Beck-Sickinger A, Demuth H, von Hörsten S (2002) Effect of neuropeptide Y on inflammatory paw edema in the rat: involvement of peripheral NPY Y1 and Y5 receptors and interaction with dipeptidyl-peptidase IV (CD26). *J Neuroimmunol* 129: 35–42

31 Reubi JC, Gugger M, Waser B, Schaer JC (2001) Y(1)-mediated effect of neuropeptide Y in cancer: breast carcinomas as targets. *Cancer Res* 61: 4636–4641

32 Karl T, Hoffmann T, Pabst R, von Hörsten S (2003) Behavioral effects of neuropeptide Y in F344 rat substrains with a reduced dipeptidyl-peptidase IV activity. *Pharmacol Biochem Behav* 75: 869–879

33 Gale SM, Castracane VD, Mantzoros CS (2004) Energy homeostasis, obesity and eating disorders: recent advances in endocrinology. *J Nutr* 134: 295–298

34 Ludwig R, Feindt J, Lucius R, Petersen A, Mentlein R (1996). Metabolism of neuropeptide Y and calcitonin gene-related peptide by cultivated neurons and glial cells. *Brain Res Mol Brain Res* 37: 181–191

35 Price JS, Kenny AJ, Huskisson NS, Brown MJ (1991) Neuropeptide Y (NPY) metabolism by endopeptidase-2 hinders characterization of NPY receptors in rat kidney. *Br J Pharmacol* 104: 321–326

The NPY family of peptides in immune disorders and inflammation

NPY, NPY receptors and DPPIV in innate immunity and autoimmune disorders

Sammy Bedoui and Stephan von Hörsten

Hannover Medical School,Department of Functional and Applied Anatomy, OE 4120, Carl-Neuberg Str. 1, 30625 Hannover, Germany

Introduction

It is now well established that the immune system and the nervous system communicate functionally with each other [1]. A major pathway for the bidirectional interaction between the nervous and the immune system is provided by the sympathetic nervous system (SNS) [2]. Though earlier studies focused mostly on the catecholamines, more recent work has established that other sympathetic transmitters, such as neuropeptide Y (NPY), also play a significant role within the neuroimmune crosstalk (see Fig. 1) [3].

NPY containing nerves and NPY receptors in lymphoid organs

A physiological role for NPY in the neuroimmune crosstalk depends on specific anatomical and functional prerequisites. One such prerequisite is the presence of NPY containing nerve fibers in primary and secondary lymphoid organs from which NPY can be released within the local immunological microenvironment. Importantly, cells of the immune system should also possess NPY receptors enabling a direct communication between locally present NPY and immune cells.

The presence of sympathetic fibers innervating lymphoid organs is well documented [1]. Several of these not only contain catecholamines, the prototypical transmitter of the SNS, but also store NPY. Lundberg et al. [4] found several NPY+ fibers innervating the cat spleen and Weihe et al. [5] investigated the presence of NPY+ sympathetic nerves supplying lymph nodes in a variety of mammalian species (guinea pig, rat, cat, pig, mouse and human). Both visceral and somatic lymph nodes are richly innervated by NPY+ fibers and some fibers were also found to branch off into the lymphoid parenchyma. Electron microscopy studies revealed that NPY+ nerve terminals form close contacts with lymphocytes

The NPY Family of Peptides in Immune Disorders, Inflammation, Angiogenesis and Cancer, edited by Zofia Zukowska and Giora Z. Feuerstein

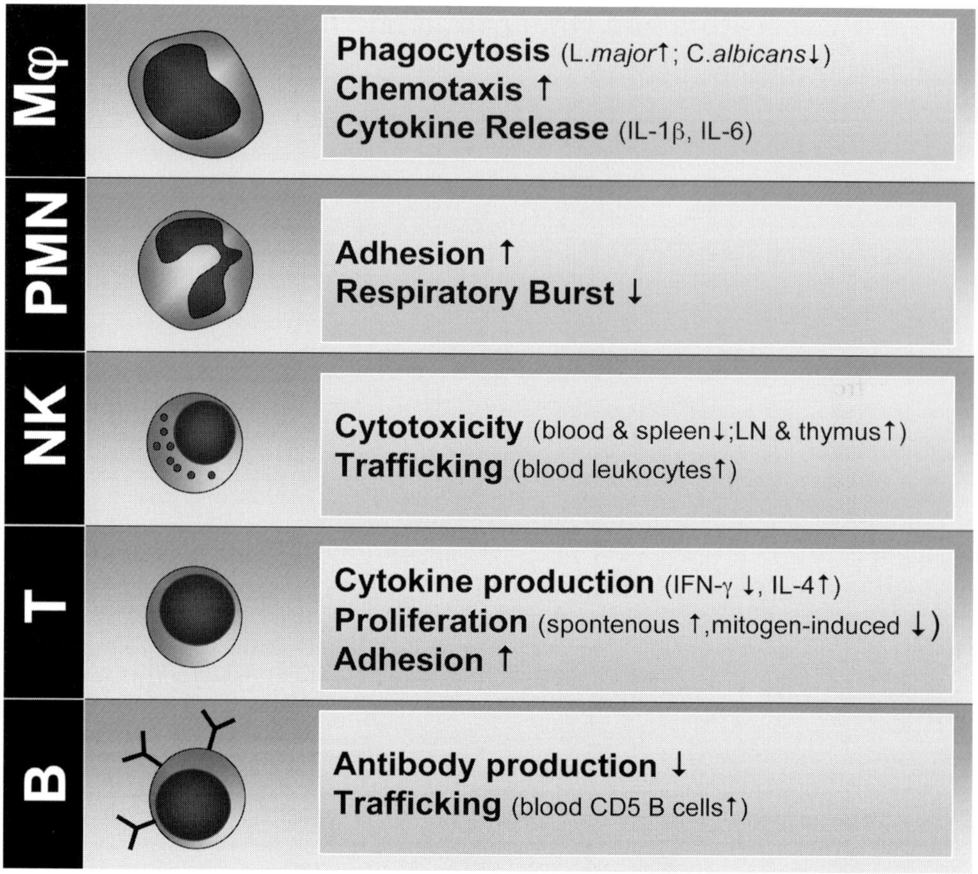

Figure 1
Overview on the effect of NPY on various immune cells

and macrophages in the marginal zone of the rat spleen [6]. NPY+ nerves also innervate other important lymphoid organs, such as the thymus and the bone marrow. Though the majority of NPY+ fibers were found to surround blood vessels, several fibers also ramified among immune cells within these organs [7, 8]. The functional significance of these findings is further substantiated by studies demonstrating that electrical stimulation of pig and murine spleens induces the release of NPY from sympathetic nerves into the local immunological microenvironment [9, 10]. Thus, the presence of NPY containing fibers in lymphoid organs and the release of NPY within the local microenvironment allows for a direct interaction between NPY and immune cells.

NPY receptors in the immune system

If NPY truly functions as a transmitter in the neuroimmune crosstalk, leukocytes should be equipped with functional NPY receptors. Initial investigations have revealed a high density of the Y1 receptor in the spleen of different species [11]. Even though much stronger signals were obtained in the red pulp and blood vessels, moderate signals are also present in lymphocyte rich areas of the white pulp [4]. This first indication for the expression of the Y1 receptor on immune cells was further substantiated by de la Fuente et al. [12], who demonstrated binding sites for radiolabeled NPY on peritoneal murine macrophages. A more detailed investigation on this matter was conducted by Petitto et al. [13]. They successfully cloned a Y1 receptor from human splenic lymphocytes, which displayed a 100% sequence homology upon comparison to the Y1 receptor cloned from the frontal cortex. Using RT-PCR and radioligand binding studies, they demonstrated the expression of the Y1 receptor protein on the surface of splenic lymphocytes. Functional Y1 receptors are also expressed on murine T lymphocytes [14] and rat peripheral blood mononuclear cells [15].

Dipeptidyl peptidase IV (DPPIV, EC 3.4.14.5, CD26): distribution in the immune system and relevance of the NPY-DPPIV-axis

DPPIV is an ectopeptidase with a triple functional role: CD26 is involved in catalyzing the release of Xaa-Pro dipeptides from circulating hormones and chemokines [16, 17], in T cell dependent immune responses [18, 19], and in cell adhesion [17, 20]. The role of NPY in immune regulation cannot be fully understood without a deeper understanding of its degradation and associated changes in receptor specificity. Both are accomplished by DPPIV, an ubiquitous cell-surface or soluble peptidase that is involved in the activation, partial or complete inactivation of bioactive peptides. Together with aminopeptidase N (APP), DPPIV is involved in the post-secretory regulation of the biological activity of peripheral and central NPY and PYY via N-terminally truncation thereby generating metabolites of altered receptor selectivity. N-terminally truncated NPY(3-36) as well as PYY(3-36) lose their affinity and potency to Y1 and partially Y5 receptors, but not to Y2 receptors and are both found in serum and many tissues. Most likely, therefore, NPY-conversion plays a role in many NPY-modulated processes including angiogenesis, inflammation, food intake and anxiolysis. Apart from its widespread distribution in organs and body fluids, activated T lymphocytes also express high levels of DPPIV/CD26. In these cells, the ectoenzyme DPPIV has been shown to play a crucial role in T cell activation and specific inhibitors of DPPIV have been used to suppress DNA synthesis as well as cytokine production (IL-2, IFN-γ, etc.) of stimulated human and mouse T cells [18, 19].

Innate immune function

Phagocytosis

Phagocytosis of extracellular particles and microorganisms represents a major function of monocytes and neutrophils. This complex phenomenon, involving engulfment of the pathogen and subsequent enzymatic degradation within the phagocyte, is mediated through surface molecules that are directed against highly conserved microbial components, such as lipopolysaccharide (LPS) or formyl-methionylleucylphenylalanine (fMLP). An alternative means initiating phagocytosis is opsonization by antibodies or complement fragments bound to the surface of a pathogen. There is convincing evidence that NPY profoundly interferes with phagocytosis of both monocytes and neutrophils. Notably, the effect of NPY on phagocytosis by monocytes depends on the type of microorganism studied, as NPY increases phagocytosis of *Candida albicans* [12], whereas the engulfment of *Leishmania major* is inhibited [21]. The different results can probably be attributed to the various pathophysiological mechanisms by which microbes interact with monocytes. With phagocytosis of *Leishmania* representing a critical step for their replication and proliferation intracellularly, one may argue that the inhibition of phagocytosis of *Leishmania* may still be protective for the organism. NPY also potently modulates the ability of neutrophils to phagocytoze gram-negative bacteria. Using fluorescence-labeled *Escherichia coli*, we recently studied the effect of NPY and various NPY specific agonists and antagonists on human neutrophil phagocytosis [22]. NPY itself modulated phagocytosis in a bimodal manner. When stimulated with low concentrations, NPY significantly inhibited phagocytosis of *E. coli*. However, with increasing concentrations of NPY this effect changed, eventually resulting in a significant enhancement of phagocytosis. In order to assess which NPY receptor subtype is responsible for the modulation of phagocytosis, we also tested a panel of different receptor specific agonists and antagonists. The results demonstrated that NPY-mediated modulation of phagocytosis is mediated via the Y5 receptor subtype. In the light of these findings, new significance might be led to the demonstration that DPPIV is upregulated on the surface of phagocytes following ingestion of latex beats [23]. Accelerated cleavage of NPY and resulting higher concentrations of Y2/5 preferring fragments, might represent an endogenous mechanism by which phagocytosis is augmented in a NPY-dependent fashion. This possibility, however, needs to be checked experimentally.

From the studies discussed here it becomes evident that NPY profoundly modulates the ability of both monocytes and neutrophils for phagocytosis. However, most remains unclear regarding the underlying mechanisms. Which receptor subtypes modulate monocyte-mediated phagocytosis? What intracellular events govern these effects? Does phagocytosis involve a specific mechanism of NPY and DPPIV interaction?

Production of highly reactive oxygen derivates

A remarkable feature of the host defense apparatus is its ability to deploy a group of highly reactive oxidizing reagents, including oxidizing radicals and singlet oxygen. The importance of this so called "respiratory burst (RB)" became obvious when the syndrome chronic granulomatous disease, characterized by predisposition to bacterial and fungal infections, was shown to be associated with decreased oxygen consumption [24]. However, the capacity for bacterial killing carries with it an implicit capacity for host tissue destruction and must therefore be tightly regulated. In addition to numerous immunological and non-immunological mechanisms controlling the RB, mediators of the SNS are also shown to be potent modulators of the RB. During phagocytosis, the respiratory burst is strongly activated. Whereas NPY has no influence on the respiratory burst of resting murine peritoneal monocytes, the stimulation of the respiratory burst resulting from activation of phagocytosis is directly increased by NPY [12]. More recently we have extended these investigations regarding the involvement of different NPY receptors and the role of DPPIV. The RB of rat peritoneal macrophages induced by phorbol-12-myristate-13-acetate (PMA) stimulation is potentiated not only by NPY, but also by NPY13-36, a selective Y2 receptor agonist. Supporting the role of Y2 receptors in PMA-induced RB, the action of NPY can be abrogated by co-administration of a selective Y2 receptor antagonist [25]. Interestingly, a functional involvement of DPPIV was observed as well, since a DPPIV inhibitor attenuated the NPY-induced potentiation of the RB in PMA-stimulated macrophages. This observation implies that under circumstances of regular DPPIV activity, most of the NPY is cleaved and therefore predominantly activates the Y2 receptor. Notably, NPY also modifies the RB when administered to the central nervous system. Blood neutrophils immediately isolated from i.c.v NPY treated rats exhibited a decrease in zymosan and PMA-induced oxygen radical formation [26–28]. When the same assay was performed 24 h after i.c.v administration of NPY an increase in oxygen radical formation was observed. Further studies revealed that these effects are mediated via the Y1 receptor, since the results were mimicked using an Y1 receptor agonist and were blocked by an Y1 receptor antagonist [29]. Thus, at this stage a very complex receptor interplay is involved in the modulation of innate immune functions via NPY-DPPIV-axis *in vivo* (see Fig. 2).

Natural killer cell activity

Host response to viruses and tumors are controlled to a large extent by natural killer (NK) cells. NK cell activity is also modulated, among several other mechanisms, by physical and psychological stress, such as exercise or mood disorders [30]. However the underlying mechanisms of this interaction are not fully understood. Most likely these alterations are mediated via the SNS. This suggestion is underlined by sev-

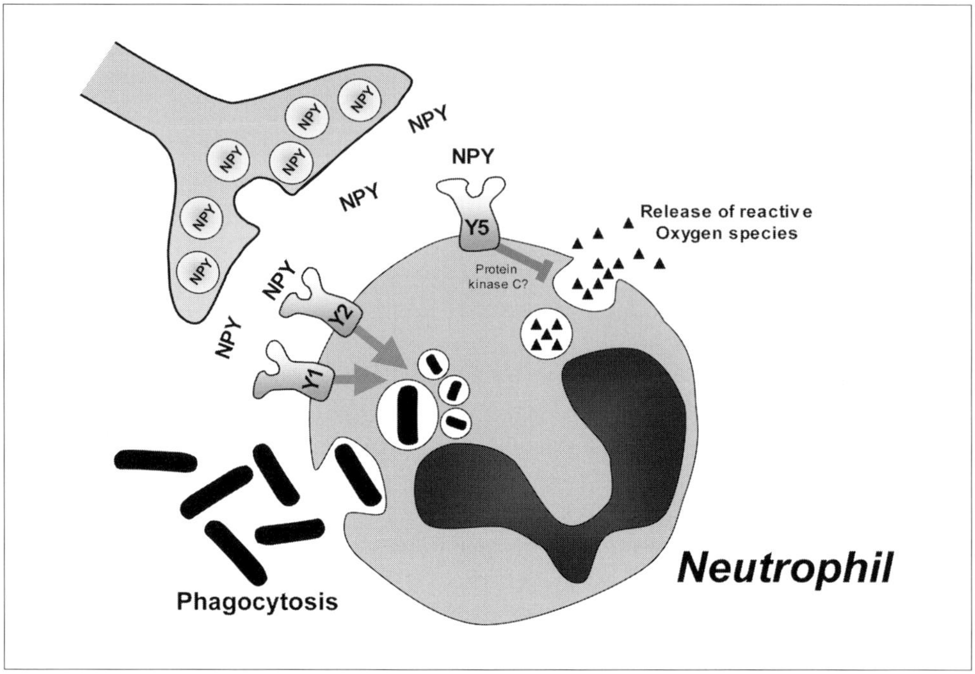

Figure 2
Interplay between nerve-derived NPY and neutrophil granulocytes. Neutrophils express the NPY receptors Y1, Y2 and Y5. Activation of these receptors results in differential functional consequences. Parallel activation of both the Y1 and the Y2 receptor modulates phagocytosis of gram-negative in a complex manner (see text). Conversely, the production and secretion of highly reactive oxygen species can be inhibited by activating the Y5 receptor.

eral findings. Chemical sympathectomy is shown to increase NK cell activity [31], whereas SNS activation inhibits NK cell activity [32]. Furthermore, bilateral lesions in the hypothalamic preoptic nucleus result in suppressed NK cell activity [33] and these effects can be abrogated by resection of sympathetic splenic nerves [34]. Interestingly, a growing number of studies indicate that sympathetic control on NK cell activity also involves NPY.

The first detailed study on the association of NPY levels and NK activity was performed in patients undergoing bereavement or other severely threatening life events. It was found that the decrease in NK activity resulting from such events [35, 36] is inversely correlated with NPY plasma levels. Increased NPY levels result in decreased NK cell activity, and *vice versa* [37]. These findings were further emphasized by *in vitro* experiments demonstrating a direct and dose-dependent inhibition of NK cell activity by NPY. This latter study clearly demonstrated the specificity of the findings, since anti-NPY sera abrogated the effects and the impaired NK activity was not due to effects of NPY on the target cells themselves [38]. Interestingly, these suppressive effects on NK activity can not only be induced by direct interaction of NPY with these cells, also intracerebroventricular (i.c.v) application is able to induce a similar depression in NK cell function. However, these indirect effects seem to have a rebound effect, since NK activity increased after 24 h [26, 27]. The peripheral distribution of NK cells is also modulated by NPY. This is of particular importance since immune reactions and inflammations involve the migration and trafficking of leukocytes to their site of action. Recently, we have shown that intravenous application of NPY dose-dependently alters NK cell distribution in the blood [39]. The effects were bimodal: a low dose NPY injection resulted in decreased numbers of NK cells, whereas high dose NPY was able to mobilize these cells. Interestingly, the distribution of blood NK cells can be also altered by i.c.v administration of NPY. Shortly after injection of NPY a significant increase in NK cell numbers was detected [26, 27]. The role of a differential expression of adhesion molecules induced or suppressed by NPY or effects on chemokines and their receptors are still unknown. However, considerable evidence is available for a further involvement of DPPIV, the NPY degrading enzyme, on NK cell activity, suggesting a further loop in the modulation of innate immunity via the DPPIV-NPY-axis.

Previous work on the role of DPPIV on NK cells has demonstrated that IL-2 stimulation increased DPPIV expression on a subpopulation of these cells [18, 40]. However, NK cytotoxicity of CD26 positive cells was not different in comparison to DPPIV negative NK cells [40]. Since, in addition, DPPIV inhibitors had no effect on NK cell function but instead suppressed DNA synthesis and cell cycle progression of NK cells, it was concluded that CD26 is involved in the regulation of NK cell proliferation, whereas natural cytotoxicity seems to be regulated independently [40]. In partial contrast, using classical chromium release NK cytotoxicity assays, we have repeatedly observed a blunted cytotoxicity splenic NK cell (*ex vivo*) against tumor cell targets co-cultured *in vitro* using NK cells from rats deficient in CD26 [20, 41].

Cytokine release

A major repository of cytokines released at inflammatory sites is provided by macrophages. This cytokine release is modulated by the SNS, as the interaction of catecholamines (CA) and NPY in the sympathetic nerve-macrophage interplay was recently examined [42] using electrical stimulation of ex vivo spleen slices [10]. Administration of exogenous NPY inhibits IL-6 release from splenic macrophages. This process is mediated via the Y1 receptor, because administration of the Y1 receptor antagonist BIBP 3226 abrogates this action. Interestingly, co-administration of NPY and norepinephrine (NE) resulted in an even stronger inhibition of the IL-6 release, indicating a potentiation of NE-mediated effect. Because NE mainly acts through α-ARs, but at higher concentrations also involves β-adrenergic signaling, the effect of the β-AR agonist isoproterenol was also investigated. Isoproterenol increases the release of IL-6 dose-dependently and thus operates in an opposite manner to a α_2-AR agonist. Surprisingly, the combined action of NPY and isoproterenol intensifies the increased IL-6 release. One possible explanation for the potentiating effects of NPY can be a modulation of intracellular cyclic adenosine monophosphate (cAMP) formation. Since NPY is shown to induce both an increase and an inhibition of cAMP formation [43], we speculate that the NPY induced effects depend on the catecholaminergic pathway that is activated. In case of β-adrenergic stimulation NPY augments the increase of cAMP, whereas upon α_2-AR activation NPY further inhibits cAMP formation. In conclusion, NPY potentiates the outcome of adrenergic stimuli in the nerve-macrophage interplay, depending on which AR subtype is activated [42]. This provides evidence for co-transmission between NPY and CA during neuroimmune interactions.

Gut-derived human macrophages and lymphocytes proliferate in response to NPY treatment [44]. IL-1β released from macrophages mediates this effect, since it is abolished by depletion of adherent cells and restored by the addition of exogenous IL-1β. This conclusion is supported by the *in vitro* capacity of NPY to increase IL-1β production from whole blood [45] and murine monocytes_ [46]. The effects of NPY on the release of cytokines, such as IL-1β and IL-6, and the interaction of NPY with the release of oxidative reagents may be of major importance for the host defense. Furthermore, after being secreted from monocytes these cytokines diffuse throughout the body. Since systemically released cytokines can induce a wide spectrum of actions, it is of major importance to investigate to what extend locally NPY-evoked cytokine release contributes to systemic levels of cytokines.

Histamine release by mast cells

Another innate branch of the immune system is formed by mast cells and their mediators are crucial components in the first line defense of immune responses against microbes. Murine connective tissue mast cells respond to NPY by releasing histamine [47], whereas NPY has no effect on human mucosa mast cells [48]. The effect on connective tissue mast cells appears to be different from other NPY evoked effects on the immune system. Investigations focusing on the involvement of NPY receptors reveal that degranulation in rat peritoneal mast cells can be evoked by C-terminal fragments of NPY, which, however, do not have proper affinities for NPY receptors [49]. This receptor-independent effect is most likely due to positive charges at the C-terminal end of centrally truncated NPY analogues [50], and may represent a receptor-independent effect of NPY.

The findings on the interaction between NPY and mast cells suggest unspecific effects on the release of histamine that are associated with structural components of NPY receptor-specific analogues. Future development of drugs based on NPY analogues should carefully consider this receptor-independent mechanism.

Autoimmunity

An important feature of T cells is their ability to develop into distinct subtypes that are characterized by specific cytokine patterns. The Th1 phenotype predominantly secretes cytokines, such as IL-2 and IFN-γ, whereas the production of IL-4 is an important feature of T cells biased towards a Th2 subtype [51]. This Th1/Th2 balance is of critical significance for autoimmune disorders, such as multiple sclerosis (MS), rheumatoid arthritis (RA), and systemic lupus erythematosus (SLE): in MS and RA Th1 cells predominate and in SLE a Th2 bias is present. Though more recent findings argue against a clear cut distinction towards either the Th1 or Th2 phenotype, this concept is still valuable in approaching general features of these disorders.

Autoimmune models of multiple sclerosis (MS): Experimental allergic encephalomyelitis (EAE)

In MS, a disease characterized by the destruction of myelin sheaths, impaired sympathetic functions are indicated both experimentally and clinically. The course of EAE, an animal model for MS, is much more severe when sympathetic nerves are chemically disrupted by the application of 6-hydroxydopamine (6-OHDA) [52]. Conversely, clinical signs and the pathology of EAE can be significantly reduced by treating the animals with β-adrenergic agonists [53]. Clinical observations support the concept of sympathetic dysfunction in MS as a variety of sympathetic dysfunctions have been reported to occur in MS patients. For example pupillary disturbances [54], impaired sympathetic skin response [55] and cardiovascular abnormalities [56] have been described. An important indication that the sympathetic defect is also relevant for the communication between the SNS and the immune system during MS is the finding that the density of adrenergic receptors on lymphocytes from MS patients is altered. The upregulation of β-adrenoreceptors on T cells from MS patients [57] is believed to reflect the loss of sympathetic influence, since chemical sympathectomy results in a similar receptor upregulation [58]. Hence, a sympathetic dysfunction is accompanied by altered responsiveness of lymphocytes. This makes a solitary action of catecholamines as mediators for the sympathetic influence unlikely. In support of the concept that the sympathetic influence on the immune system is alleviated, decreased concentrations of NPY in the cerebrospinal fluid and plasma of MS patients have been documented [59]. These findings were confirmed by others and it was additionally found that a more severe form of MS is associated with even lower cerebrospinal fluid NPY levels than observed in milder forms [60]. In an attempt to approach this concept more functionally, we recently examined whether administration of exogenous NPY would have any impact on the clinical course of EAE. The results demonstrate a dose-dependent suppression of actively and also passively-induced EAE [14]. Pharmacological assessment of the involved receptor revealed that the suppression is mediated via the Y1 receptor. Comparing T cell responses and anti-MOG35-55 antibody titers from treated *versus* control animals we found that the suppression is due to a Y1 receptor-mediated induction of a Th2 shift of autoreactive lymphocytes *in vivo*. This study proofs that NPY, serving as a sympathetic transmitter, exerts a protective role in EAE. The potential relevance of NPY for MS was very recently emphasized in a genetic study revealing a vulnerability locus on human chromosome 7q15, a locus that actually encodes NPY [61].

Notably, the interaction between NPY and DPPIV might also be of significant importance. The result of N-terminal truncation of NPY by DPPIV is a peptide fragment that has significantly altered receptor affinities, as the degradation product NPY3-36 loses its ability to activate the Y1 receptor and becomes a selective agonist to the Y2 and Y5 receptor [17]. Increased degradation of NPY by DPPIV under autoimmune conditions may thus account for insufficient activation of Y1 receptors, which in turn favors the development of Th1 responses. Interestingly, this possibility is consistent with the recent report that the pharmacological inhibition of CD26 suppresses the clinical course of EAE via the induction of a Th2 shift [62]. It remains to be investigated whether the suppressive action of DPPIV inhibition results from increased Y1 receptor activation and the corresponding Th2 shift.

Rheumatoid arthritis

An interesting feature of rheumatoid arthritis, an autoimmune disorder characterized by joint inflammation and destruction, is the loss of sympathetic nerve fibers. Despite the presence of sympathetic fibers in all layers of healthy joints, sympathetic nerves are absent from the synovium of RA joints [63]. To study the significance of such a loss of sympathetic nerves, chemical depletion of sympathetic transmitters by 6-OHDA was performed in typical animal models of arthritis. Depleting sympathetic nerves in lymph nodes draining the hind limb of rats was recently shown to aggravate joint inflammation and increase paw diameter in a model of adjuvans-induced arthritis [64]. Conversely, continued treatment using salbutamol, a β-adrenergic agonist, inhibits collagen-induced arthritis in mice [65]. These findings strongly suggest a protective role for the SNS in RA that is probably mediated by catecholamines. A more detailed look into the involvement of catecholamines in RA and experimental models reveals some interesting observations. Studies comparing the density of adrenoreceptors on synovial lymphocytes from RA patients and healthy controls revealed a significantly reduced expression of adrenergic receptors on the surface of lymphocytes from RA patients [66]. Furthermore, the suppressive effect of catecholamines on the proliferation of these cells is reduced in synovial lymphocytes of RA patients. Finally, there is a direct correlation between disease severity and the number of adrenergic binding sites: patients with high systemic disease activity have lower numbers of adrenoreceptors than patients with low disease activity [67]. These findings indicate that in addition to the loss of sympathetic innervation, the responsiveness towards catecholamines is also decreased in RA joint tissues.

Several lines of evidence suggest that NPY is also involved in the regulation of RA. A specific loss of NPY+ containing fibers is found in the synovium from RA patients [68, 69]. Investigations to determine the concentration of NPY in inflamed arthritic joints have revealed some unexpected observation. Despite the lack of NPY+ containing nerves, different studies reported elevated NPY levels in RA joints, such as the knee, the ankle or the temporomandibular joint [70, 71]. One study also found a correlation between NPY levels in the temporomandibular joint and plasma NPY levels [72]. The question should be addressed how NPY levels can be elevated when a dramatic reduction of sympathetic innervation is present. Ahmed et al. [73] investigated possible mechanisms underlying elevated NPY levels in the inflamed ankle joint of arthritic Lewis rats. If increased activity of NPY+ fibers would account for this phenomenon, altered NPY levels in the dorsal root ganglia should be expected. However, NPY levels in the dorsal root ganglia revealed no changes. To investigate extraneuronal sources of NPY, they determined the number of megakaryocytes in the bone marrow of the tibial bone and found a dramatic increase of NPY+ megakaryocytes. These findings suggest the interesting possibility that the increased concentration of NPY could be derived from megakaryocytes, a cell type that contains high amounts of NPY. This suggestion is further supported by a study seeking to determine the source of NE in RA synovial tissues. Miller et al. [63] reported that despite the loss of sympathetic fibers the concentration of norepinephrine is increased in synovial tissues from RA patients. The study revealed that the additional NE amount is produced by CD163+ macrophages residing in the synovium. Hence, in analogy to the possibility that megakaryocytes release NPY in inflamed RA tissues, extraneuronal NE is released by local macrophages in the synovium of RA patients.

The endogenous regulation of DPPIV activity and expression levels on certain immune cell types appears to be altered in RA as well. Muscat and colleagues analyzed the expression of DPPIV on T lymphocytes isolated from the synovial fluid of patients with either active or inactive joint inflammation and found that T lymphocytes from patients with active inflammation express higher amounts of DPPIV on their surface [74]. Supporting the significance of DPPIV expression in RA, a recent study has demonstrated that administration of a fungi-derived DPPIV inhibitor improves clinical features, such as paw swelling, in adjuvans arthritis [75]. This observation is associated with decreased T lymphocyte proliferation and decreased IL-12 secretion. As inhibition of DPPIV results in increased levels of uncleaved, native NPY, DPPIV inhibition can be regarded an indirect means to increase Y1 receptor stimulation. Accordingly, it may well be that the suppression of joint inflammation and T lymphocyte suppression is at least partially a results of increased Y1 receptor activation by endogenous NPY. So far, there is no experimental evidence available regarding the interaction between DPPIV and NPY in models of autoimmune arthritis. Clearly, experiments are needed to investigate such a possibility.

Systemic lupus erythomatodes (SLE)

SLE is driven by autoreactive T cells that are biased towards the Th2 subtype. Th2 biased T helper cells activate B cells to produce autoantibodies, which in turn mediate typical tissue destruction. Contrasting sympathetic dysfunctions in Th1 driven autoimmune disorders, such as MS and RA, observations in SLE patients indicate increased sympathetic activity. Prolonged pupillary reaction and increased maximal pupillary areas, as well as increases in blood pressure and heart rate are found in SLE patients [76, 77]. Comparing stress-induced alterations in the composition of leukocytes in the peripheral blood of SLE patients and healthy controls, Jacobs et al. [78] recently observed an interesting difference between SLE patients and healthy subjects. Whereas in the control group the number of both IFN-γ (Th1) and IL-4 (Th2) producing cells increased upon stress, in SLE patients only IL-4 producing cells increased, suggesting a selective inhibition of stress-induced Th1 responses in SLE patients. Strikingly, alterations in NPY contents have been reported in models of SLE. Ericsson et al. [79] compared the expression of NPY mRNA in the spleen, the bone marrow and the peripheral blood in different mouse strains and found largely elevated NPY levels in mice that develop SLE. Furthermore, it was demonstrated that the onset of glomerulonephritis in lupus-prone mice is accompanied by increased quantities of NPY in the kidney [80]. In line with the association of Th1-driven autoimmune disorders with increased DPPIV activity, several investigators have documented that the serum activity of DPPIV is decreased in SLE [81], implying that Th2-driven autoimmunity is associated with decreased DPPIV activity. Suppressed cleavage of native NPY shifts the balance between preferential binding of DPPIV-cleaved fragments to Y2/Y5 receptors on the one hand, and activation of Y1 receptors by fully intact, uncleaved NPY one the other, towards latter. Such NPY-mediated Th1 suppression may explain as to why Th2 predominate in SLE. This possibility needs to be addressed in future studies.

Table 1 - Involvement of NPY in various autoimmune disorders. The table is modified after Bedoui et al. 2003

Disease	Organ/Compartment	NPY alteration	Literature
Fibromyalgia	Plasma	NPY ↓	[82]
Inflammatory bowel disease	Colon	NPY+ nerves ↑	[83]
	Myenteric plexus	NPY+ nerves ↑	[84]
	Platelets	No differences	[85]
	Serum	Loss of correlation with cortisol	[86]
Lichen planus	Oral mucosa	NPY+ nerves ↓ in inflamed areas	[87]
Multiple sclerosis	Cerebrospinal fluid	NPY ↓	[59]
	Thymus	NPY ↑ after IFN therapy	[7]
Rheumatoid arthritis	Synovium	NPY+ nerves ↓ in superficial layer	[68]
	Synovial fluid (Knee, ankle, TMJ)	NPY ↑	[70, 71]
	Synovial fluid (TMJ)	NPY ↓ GC treatment	[88]
	Synovial fluid (Ankle)	NPY ↑	[73]
Sjögren syndrome	Saliva	NPY ↑	[89]
	Labial glands	NPY+ nerves ↓	[90]
	Labial glands	NPY+ nerves ↓ in inflamed areas	[91]
Systemic lupus erythematosus	Spleen, BM, blood	NPY mRNA ↑	[79]
	Inflamed kidneys	NPY ↑	[80]
Type 1 diabetes	Plasma	NPY ↓ after exercise	[92]

Concluding remarks

NPY is released in the central and the peripheral nervous system and acts via multiple pathways on immune cells, which express receptors for NPY. As also described in this book, in the future, at least three potential avenues of research will utilize this knowledge on the role of NPY in neuroimmunomodulation: (a) modulation of autoimmune disease (see Tab. 1), (b) vascular pathology/repair, and (c) usage of NPY receptors on tumors for labeling/therapy. However, this perspective becomes challenging as the degradation of NPY by dipeptidyl-peptidase IV and associated changes in receptor specificity will complicate the understanding on the role of NPY at the local microenvironment during inflammation and/or host defense.

References

1 Elenkov IJ, Wilder RL, Chrousos GP, Vizi ES (2000) The sympathetic nerve – an integrative interface between two supersystems: the brain and the immune system. *Pharmacol Rev* 52: 595–638

2 Kohm AP, Sanders VM (2001) Norepinephrine and beta 2-adrenergic receptor stimulation regulate $CD4^{+}$ T and B lymphocyte function *in vitro* and *in vivo*. *Pharmacol Rev* 53: 487–525

3 Bedoui S, Kawamura N, Straub RH, Pabst R, Yamamura T, von Hörsten S (2003) Relevance of neuropeptide Y for the neuroimmune crosstalk. *J Neuroimmunol* 134: 1–11

4 Lundberg JM, Martinsson A, Hemsen A, Theodorsson-Norheim E, Svedenhag J, Ekblom B, Hjemdahl P (1985) Co-release of neuropeptide Y and catecholamines during physical exercise in man. *Biochem Biophys Res Commun* 133: 30–36

5 Weihe E, Muller S, Fink T, Zentel HJ (1989) Tachykinins, calcitonin gene-related peptide and neuropeptide Y in nerves of the mammalian thymus: interactions with mast cells in autonomic and sensory neuroimmunomodulation? *Neurosci Lett* 100: 77–82

6 Romano TA, Felten SY, Felten DL, Olschowka JA (1991) Neuropeptide-Y innervation of the rat spleen: another potential immunomodulatory neuropeptide. *Brain Behav Immun* 5: 116–131

7 Cavallotti D, Artico M, Iannetti G, Cavallotti C (2002) Occurrence of adrenergic nerve fibers in human thymus during immune response. *Neurochem Int* 40: 211–221

8 Tabarowski Z, Gibson-Berry K, Felten SY (1996) Noradrenergic and peptidergic innervation of the mouse femur bone marrow. *Acta Histochem* 98: 453–457

9 Lundberg JM (1996) Pharmacology of cotransmission in the autonomic nervous system: integrative aspects on amines, neuropeptides, adenosine triphosphate, amino acids and nitric oxide. *Pharmacol Rev* 48: 113–178

10 Straub RH (2004) Complexity of the bi-directional neuroimmune junction in the spleen. *Trends Pharmacol Sci* 25: 640–646

11 Gehlert DR, Gackenheimer SL (1996) Unexpected high density of neuropeptide Y Y1 receptors in the guinea pig spleen. *Peptides* 17: 1345–1348

12 De la Fuente M, Bernaez I, Del Rio M, Hernanz A (1993) Stimulation of murine peritoneal macrophage functions by neuropeptide Y and peptide YY. Involvement of protein kinase C. *Immunology* 80: 259–265

13 Petitto JM, Huang Z, McCarthy DB (1994) Molecular cloning of NPY-Y1 receptor cDNA from rat splenic lymphocytes: evidence of low levels of mRNA expression and [125I]NPY binding sites. *J Neuroimmunol* 54: 81–86

14 Bedoui S, Miyake S, Lin Y, Miyamoto K, Oki S, Kawamura N, Beck-Sickinger A, von Hörsten S, Yamamura T (2003) Neuropeptide Y (NPY) suppresses experimental autoimmune encephalomyelitis: NPY1 receptor-specific inhibition of autoreactive Th1 responses *in vivo*. *J Immunol* 171: 3451–3458

15 Bedoui S, Lechner S, Gebhardt T, Nave H, Beck-Sickinger AG, Straub RH, Pabst R, von Hörsten S (2002) NPY modulates epinephrine-induced leukocytosis via Y-1 and Y-5

receptor activation *in vivo*: sympathetic co-transmission during leukocyte mobilization. *J Neuroimmunol* 132: 25–33

16 De Meester I, Korom S, Van Damme J, Scharpe S (1999) CD26, let it cut or cut it down. *Immunol Today* 20: 367–375

17 Mentlein R (1999) Dipeptidyl-peptidase IV (CD26) – role in the inactivation of regulatory peptides. *Regul Pept* 85: 9–24

18 Kähne T, Lendeckel U, Wrenger S, Neubert K, Ansorge S, Reinhold D (1999) Dipeptidyl peptidase IV: a cell surface peptidase involved in regulating T cell growth (review). *Int J Mol Med* 4: 3–15

19 Korom S, De Meester I, Stadlbauer TH, Chandraker A, Schaub M, Sayegh MH, Belyaev A, Haemers A, Scharpe S, Kupiec-Weglinski JW (1997) Inhibition of CD26/dipeptidyl peptidase IV activity *in vivo* prolongs cardiac allograft survival in rat recipients. *Transplantation* 63: 1495–1500

20 Shingu K, Helfritz A, Zielinska-Skowronek M, Meyer-Olson D, Jacobs R, Schmidt RE, Mentlein R, Pabst R, von Horsten S (2003) CD26 expression determines lung metastasis in mutant F344 rats: involvement of NK cell function and soluble CD26. *Cancer Immunol Immunother* 52: 546–554

21 Ahmed AA, Wahbi A, Nordlind K, Kharazmi A, Sundqvist KG, Mutt V, Liden S (1998) *in vitro Leishmania major* promastigote-induced macrophage migration is modulated by sensory and autonomic neuropeptides. *Scand J Immunol* 48: 79–85

22 Bedoui SK, Zwinkmann D, Gebhardt T, Jacobs R, Sowa E, Heine J, von Hörsten S (2005) Neuropeptide Y receptor-specifically modulates human neutrophil function. *Submitted*

23 De Bruijn ML, Peterson PA, Jackson MR (1996) Induction of heat-stable antigen expression by phagocytosis is involved in *in vitro* activation of unprimed CTL by macrophages. *J Immunol* 156: 2686–2692

24 Holmes B, Quie PG, Windhorst DB, Good RA (1966) Fatal granulomatous disease of childhood. An inborn abnormality of phagocytic function. *Lancet* 1: 1225–1228

25 Dimitrijevic M, Stanojevic S, Vujic V, Kovacevic-Jovanovic V, Beck-Sickinger A, Demuth H, von Horsten S (2002) Effect of neuropeptide Y on inflammatory paw edema in the rat: involvement of peripheral NPY Y1 and Y5 receptors and interaction with dipeptidyl-peptidase IV (CD26). *J Neuroimmunol* 129: 35–42

26 von Hörsten S, Ballof J, Helfritz F, Nave H, Meyer D, Schmidt RE, Stalp M, Klemm A, Tschernig T, Pabst R (1998) Modulation of innate immune functions by intracerebroventricularly applied neuropeptide Y: dose and time dependent effects. *Life Sci* 63: 909–922

27 von Hörsten S, Nave H, Ballof J, Helfritz F, Meyer D, Schmidt RE, Stalp M, Exton NG, Exton MS, Straub RH et al (1998) Centrally applied NPY mimics immunoactivation induced by non-analgesic doses of met-enkephalin. *Neuroreport* 9: 3881–3885

28 Song C, Earley B, Leonard BE (1996) The effects of central administration of neuropeptide Y on behavior, neurotransmitter, and immune functions in the olfactory bulbectomized rat model of depression. *Brain Behav Immun* 10: 1–16

29 von Hörsten S, Exton NG, Exton MS, Helfritz F, Nave H, Ballof J, Stalp M, Pabst R (1998) Brain NPY Y1 receptors rapidly mediate the behavioral response to novelty and a compartment-specific modulation of granulocyte function in blood and spleen. Brain Res 806: 282–286
30 Irwin M, Smith TL, Gillin JC (1992) Electroencephalographic sleep and natural killer activity in depressed patients and control subjects. Psychosom Med 54: 10–21
31 Reder A, Checinski M, Chelmicka-Schorr E (1989) The effect of chemical sympathectomy on natural killer cells in mice. *Brain Behav Immun* 3: 110–118
32 Cunnick JE, Lysle DT, Armfield A, Rabin BS (1988) Shock-induced modulation of lymphocyte responsiveness and natural killer activity: differential mechanisms of induction. *Brain Behav Immun* 2: 102–113
33 Katafuchi T, Ichijo T, Take S, Hori T (1993) Hypothalamic modulation of splenic natural killer cell activity in rats. J Physiol 471: 209–221
34 Hori T, Katafuchi T, Take S, Shimizu N, Niijima A (1995) The autonomic nervous system as a communication channel between the brain and the immune system. Neuroimmunomodulation 2: 203–215
35 Kiecolt-Glaser JK, Ricker D, George J, Messick G, Speicher CE, Garner W, Glaser R (1984) Urinary cortisol levels, cellular immunocompetency, and loneliness in psychiatric inpatients. Psychosom Med 46: 15–23
36 Irwin M, Smith TL, Gillin JC (1987) Low natural killer cytotoxicity in major depression. Life Sci 41: 2127–2133
37 Irwin M, Brown M, Patterson T, Hauger R, Mascovich A, Grant I (1991) Neuropeptide Y and natural killer cell activity: findings in depression and Alzheimer caregiver stress. Faseb J 5: 3100–3107
38 Nair MP, Schwartz SA, Wu K, Kronfol Z (1993) Effect of neuropeptide Y on natural killer activity of normal human lymphocytes. *Brain Behav Immun* 7: 70–78
39 Bedoui S, Kuhlmann S, Nave H, Drube J, Pabst R, von Hörsten S (2001) Differential effects of neuropeptide Y (NPY) on leukocyte subsets in the blood: mobilization of B-1-like B-lymphocytes and activated monocytes. *J Neuroimmunol* 117: 125–132
40 Bühling F, Kunz D, Reinhold D, Ulmer AJ, Ernst M, Flad HD, Ansorge S (1994) Expression and functional role of dipeptidyl peptidase IV (CD26) on human natural killer cells. Nat Immun 13: 270–279
41 Karl T, Chwalisz WT, Wedekind D, Hedrich HJ, Hoffmann T, Jacobs R, Pabst R, von Hörsten S (2003) Localization, transmission, spontaneous mutations, and variation of function of the Dpp4 (Dipeptidyl-peptidase IV; CD26) gene in rats. *Regul Pept* 115: 81–90
42 Straub RH, Schaller T, Miller LE, von Horsten S, Jessop DS, Falk W, Schölmerich J (2000) Neuropeptide Y cotransmission with norepinephrine in the sympathetic nerve-macrophage interplay. J Neurochem 75: 2464–2471
43 Michel MC, Beck-Sickinger A, Cox H, Doods HN, Herzog H, Larhammar D, Quirion R, Schwartz T, Westfall T (1998) XVI. International Union of Pharmacology recom-

mendations for the nomenclature of neuropeptide Y, peptide YY, and pancreatic polypeptide receptors. *Pharmacol Rev* 50: 143–150
44 Elitsur Y, Luk GD, Colberg M, Gesell MS, Dosescu J, Moshier JA (1994) Neuropeptide Y (NPY) enhances proliferation of human colonic lamina propria lymphocytes. *Neuropeptides* 26: 289–295
45 Hernanz A, Tato E, De la Fuente M, de Miguel E, Arnalich F (1996) Differential effects of gastrin-releasing peptide, neuropeptide Y, somatostatin and vasoactive intestinal peptide on interleukin-1 beta, interleukin-6 and tumor necrosis factor-alpha production by whole blood cells from healthy young and old subjects. *J Neuroimmunol* 71: 25–30
46 De la Fuente M, Del Rio M, Medina S (2001) Changes with aging in the modulation by neuropeptide Y of murine peritoneal macrophage functions. *J Neuroimmunol* 116: 156–167
47 Arzubiaga C, Morrow J, Roberts LJ 2nd, Biaggioni I (1991) Neuropeptide Y, a putative cotransmitter in noradrenergic neurons, induces mast cell degranulation but not prostaglandin D2 release. *J Allergy Clin Immunol* 87: 88–93
48 Bischoff SC, Schwengberg S, Lorentz A, Manns MP, Bektas H, Sann H, Levi-Schaffer F, Shanahan F, Schemann M (2004) Substance P and other neuropeptides do not induce mediator release in isolated human intestinal mast cells. *Neurogastroenterol Motil* 16: 185–193
49 Shen GH, Grundemar L, Zukowska-Grojec Z, Hakanson R, Wahlestedt C (1991) C-terminal neuropeptide Y fragments are mast cell-dependent vasodepressor agents. *Eur J Pharmacol* 204: 249–256
50 Cross LJ, Beck-Sickinger AG, Bienert M, Gaida W, Jung G, Krause E, Ennis M (1996) Structure activity studies of mast cell activation and hypotension induced by neuropeptide Y (NPY), centrally truncated and C-terminal NPY analogues. *Br J Pharmacol* 117: 325–332
51 Abbas AK, Murphy KM, Sher A (1996) Functional diversity of helper T lymphocytes. *Nature* 383: 787–793
52 Chelmicka-Schorr E, Checinski M, Arnason BG (1988) Chemical sympathectomy augments the severity of experimental allergic encephalomyelitis. *J Neuroimmunol* 17: 347–350
53 Chelmicka-Schorr E, Kwasniewski MN, Thomas BE, Arnason BG (1989) The beta-adrenergic agonist isoproterenol suppresses experimental allergic encephalomyelitis in Lewis rats. *J Neuroimmunol* 25: 203–207
54 de Seze J, Arndt C, Stojkovic T, Ayachi M, Gauvrit JY, Bughin M, Saint Michel T, Pruvo JP, Hache JC, Vermersch P (2001) Pupillary disturbances in multiple sclerosis: correlation with MRI findings. *J Neurol Sci* 188: 37–41
55 Yokota T, Matsunaga T, Okiyama R, Hirose K, Tanabe H, Furukawa T, Tsukagoshi H (1991) Sympathetic skin response in patients with multiple sclerosis compared with patients with spinal cord transection and normal controls. *Brain* 114 (Pt 3): 1381–1394
56 Nordenbo AM, Boesen F, Andersen EB (1989) Cardiovascular autonomic function in multiple sclerosis. *J Auton Nerv Syst* 26: 77–84

57 Karaszewski JW, Reder AT, Anlar B, Arnason GW (1993) Increased high affinity beta-adrenergic receptor densities and cyclic AMP responses of CD8 cells in multiple sclerosis. *J Neuroimmunol* 43: 1–7

58 Miles K, Quintans J, Chelmicka-Schorr E, Arnason BG (1981) The sympathetic nervous system modulates antibody response to thymus-independent antigens. *J Neuroimmunol* 1: 101–105

59 Maeda K, Yasuda M, Kaneda H, Maeda S, Yamadori A (1994) Cerebrospinal fluid (CSF) neuropeptide Y- and somatostatin-like immunoreactivities in man. *Neuropeptides* 27: 323–332

60 Gallai V, Sarchielli P, Firenze C, Trequattrini A, Paciaroni M, Usai F, Franceschini M, Palumbo R (1994) Neuropeptide Y plasma levels and serum dopamine-beta-hydroxylase activity in MS patients with and without abnormal cardiovascular reflexes. *Acta Neurol Belg* 94: 44–52

61 Coppin H, Ribouchon MT, Fontaine B, Edan G, Clanet M, Roth MP (2004) A vulnerability locus to multiple sclerosis maps to 7p15 in a region syntenic to an EAE locus in the rat. *Genes Immun* 5: 72–75

62 Steinbrecher A, Reinhold D, Quigley L, Gado A, Tresser N, Izikson L, Born I, Faust J, Neubert K, Martin R et al (2001) Targeting dipeptidyl peptidase IV (CD26) suppresses autoimmune encephalomyelitis and up-regulates TGF-beta 1 secretion *in vivo*. *J Immunol* 166: 2041–2048

63 Miller LE, Justen HP, Schölmerich J, Straub RH (2000) The loss of sympathetic nerve fibers in the synovial tissue of patients with rheumatoid arthritis is accompanied by increased norepinephrine release from synovial macrophages. *Faseb J* 14: 2097–2107

64 Lorton D, Bellinger D, Duclos M, Felten SY, Felten DL (1996) Application of 6-hydroxydopamine into the fatpads surrounding the draining lymph nodes exacerbates adjuvant-induced arthritis. *J Neuroimmunol* 64: 103–113

65 Malfait AM, Malik AS, Marinova-Mutafchieva L, Butler DM, Maini RN, Feldmann M (1999) The beta2-adrenergic agonist salbutamol is a potent suppressor of established collagen-induced arthritis: mechanisms of action. *J Immunol* 162: 6278–6283

66 Baerwald CG, Laufenberg M, Specht T, von Wichert P, Burmester GR, Krause A (1997) Impaired sympathetic influence on the immune response in patients with rheumatoid arthritis due to lymphocyte subset-specific modulation of beta 2-adrenergic receptors. *Br J Rheumatol* 36: 1262–1269

67 Baerwald C, Graefe C, von Wichert P, Krause A (1992) Decreased density of beta-adrenergic receptors on peripheral blood mononuclear cells in patients with rheumatoid arthritis. *J Rheumatol* 19: 204–210

68 Mapp PI, Kidd BL, Gibson SJ, Terry JM, Revell PA, Ibrahim NB, Blake DR, Polak JM (1990) Substance P-, calcitonin gene-related peptide- and C-flanking peptide of neuropeptide Y-immunoreactive fibres are present in normal synovium but depleted in patients with rheumatoid arthritis. *Neuroscience* 37: 143–153

69 Pereira da Silva JA, Carmo-Fonseca M (1990) Peptide containing nerves in human syn-

ovium: immunohistochemical evidence for decreased innervation in rheumatoid arthritis. *J Rheumatol* 17: 1592–1599

70 Larsson J, Ekblom A, Henriksson K, Lundeberg T, Theodorsson E (1991) Concentration of substance P, neurokinin A, calcitonin gene-related peptide, neuropeptide Y and vasoactive intestinal polypeptide in synovial fluid from knee joints in patients suffering from rheumatoid arthritis. *Scand J Rheumatol* 20: 326–335

71 Appelgren A, Appelgren B, Kopp S, Lundeberg T, Theodorsson E (1993) Relation between the intra-articular temperature of the temporomandibular joint and the presence of neuropeptide Y-like immunoreactivity in the joint fluid. A clinical study. *Acta Odontol Scand* 51: 1–8

72 Holmlund A, Ekblom A, Hansson P, Lind J, Lundeberg T, Theodorsson E (1991) Concentrations of neuropeptides substance P, neurokinin A, calcitonin gene-related peptide, neuropeptide Y and vasoactive intestinal polypeptide in synovial fluid of the human temporomandibular joint. A correlation with symptoms, signs and arthroscopic findings. *Int J Oral Maxillofac Surg* 20: 228–231

73 Ahmed M, Bjurholm A, Theodorsson E, Schultzberg M, Kreicbergs A (1995) Neuropeptide Y- and vasoactive intestinal polypeptide-like immunoreactivity in adjuvant arthritis: effects of capsaicin treatment. *Neuropeptides* 29: 33–43

74 Muscat C, Bertotto A, Agea E, Bistoni O, Ercolani R, Tognellini R, Spinozzi F, Cesarotti M, Gerli R (1994) Expression and functional role of 1F7 (CD26) antigen on peripheral blood and synovial fluid T cells in rheumatoid arthritis patients. *Clin Exp Immunol* 98: 252–256

75 Tanaka S, Murakami T, Nonaka N, Ohnuki T, Yamada M, Sugita T (1998) Anti-arthritic effects of the novel dipeptidyl peptidase IV inhibitors TMC-2A and TSL-225. *Immunopharmacology* 40: 21–26

76 Gluck T, Oertel M, Reber T, Zietz B, Scholmerich J, Straub RH (2000) Altered function of the hypothalamic stress axes in patients with moderately active systemic lupus erythematosus. I. The hypothalamus-autonomic nervous system axis. *J Rheumatol* 27: 903–910

77 Nakajima A, Sendo W, Tsutsumino M, Koseki Y, Ichikawa N, Akama H, Taniguchi A, Terai C, Hara M, Kashiwazaki S (1998) Acute sympathetic hyperfunction in overlapping syndromes of systemic lupus erythematosus and polymyositis. *J Rheumatol* 25: 1638–1641

78 Jacobs R, Pawlak CR, Mikeska E, Meyer-Olson D, Martin M, Heijnen CJ, Schedlowski M, Schmidt RE (2001) Systemic lupus erythematosus and rheumatoid arthritis patients differ from healthy controls in their cytokine pattern after stress exposure. *Rheumatology (Oxford)* 40: 868–875

79 Ericsson A, Schalling M, McIntyre KR, Lundberg JM, Larhammar D, Seroogy K, Hokfelt T, Persson H (1987) Detection of neuropeptide Y and its mRNA in megakaryocytes: enhanced levels in certain autoimmune mice. *Proc Natl Acad Sci USA* 84: 5585–5589

80 Bracci-Laudiero L, Aloe L, Stenfors C, Theodorsson E, Lundeberg T (1998) Development of systemic lupus erythematosus in mice is associated with alteration of neu-

ropeptide concentrations in inflamed kidneys and immunoregulatory organs. *Neurosci Lett* 248: 97–100

81 Kobayashi H, Hosono O, Mimori T, Kawasaki H, Dang NH, Tanaka H, Morimoto C (2002) Reduction of serum soluble CD26/dipeptidyl peptidase IV enzyme activity and its correlation with disease activity in systemic lupus erythematosus. *J Rheumatol* 29: 1858–1866

82 Crofford LJ, Pillemer SR, Kalogeras KT, Cash JM, Michelson D, Kling MA, Sternberg EM, Gold PW, Chrousos GP, Wilder RL (1994) Hypothalamic-pituitary-adrenal axis perturbations in patients with fibromyalgia. *Arthritis Rheum* 37: 1583–1592

83 Bjorck S, Jennische E, Dahlstrom A, Ahlman H (1997) Influence of topical rectal application of drugs on dextran sulfate-induced colitis in rats. *Dig Dis Sci* 42: 824–832

84 Belai A, Boulos PB, Robson T, Burnstock G (1997) Neurochemical coding in the small intestine of patients with Crohn's disease. *Gut* 40: 767–774

85 Fagerstam JP, Whiss PA, Strom M, Andersson RG (2000) Expression of platelet P-selectin and detection of soluble P-selectin, NPY and RANTES in patients with inflammatory bowel disease. *Inflamm Res* 49: 466–472

86 Straub RH, Herfarth H, Falk W, Andus T, Scholmerich J (2002) Uncoupling of the sympathetic nervous system and the hypothalamic-pituitary-adrenal axis in inflammatory bowel disease? *J Neuroimmunol* 126: 116–125

87 Nissalo S, Hietanen J, Malmstrom M, Hukkanen M, Polak J, Konttinen YT (2000) Disorder-specific changes in innervation in oral lichen planus and lichenoid reactions. *J Oral Pathol Med* 29: 361–369

88 Alstergren P, Appelgren A, Appelgren B, Kopp S, Lundeberg T, Theodorsson E (1995) Co-variation of neuropeptide Y, calcitonin gene-related peptide, substance P and neurokinin A in joint fluid from patients with temporomandibular joint arthritis. *Arch Oral Biol* 40: 127–135

89 Santavirta N, Konttinen YT, Tornwall J, Segerberg M, Santavirta S, Matucci-Cerinic M, Bjorvell H (1997) Neuropeptides of the autonomic nervous system in Sjogren's syndrome. *Ann Rheum Dis* 56: 737–740

90 Feher E, Zelles T, Nagy G (1999) Immunocytochemical localisation of neuropeptide-containing nerve fibres in human labial glands. *Arch Oral Biol* 44 (Suppl 1): S33–37

91 Konttinen YT, Sorsa T, Hukkanen M, Segerberg M, Kuhlefelt-Sundstrom M, Malmstrom M, Polak JM (1992) Topology of innervation of labial salivary glands by protein gene product 9.5 and synaptophysin immunoreactive nerves in patients with Sjogren's syndrome. *J Rheumatol* 19: 30–37

92 Ahlborg G, Lundberg JM (1996) Exercise-induced changes in neuropeptide Y, noradrenaline and endothelin-1 levels in young people with type I diabetes. *Clin Physiol* 16: 645–655

NPY and phagocytic cell functions

Mónica De la Fuente and Sonia Medina

Department of Physiology (Animal Physiology II), Faculty of Biological Sciences, Complutense University, 28040 Madrid, Spain

General effects of NPY on macrophages

The phagocytic cell is one of the outstanding cells involved in the non-specific immune response of the host faced with infectious and inflammatory processes. For this function, phagocytic cells carry out the phagocytic process as their most characteristic property. This process involves several functions such as adherence to tissues, mobility to the infectious focus (chemotaxis), ingestion of the foreign material (phagocytosis) and digestion and destruction of this ingested material. These mechanisms are one of the most effective means for the production of free radicals, the first of which is the superoxide anion. The phagocytic function of these cells initiates many other biological activities that comprise the whole spectrum of the immune response. Moreover, phagocytes, especially macrophages, play an important regulatory and functional role in the immune system. Thus, these cells process antigens that are presented to T lymphocytes, kill tumor cells, and serve as accessory cells for lymphocytes [1].

As regards to the functional connection between the two homeostatic systems, namely the nervous and immune systems, that is carried out by molecular mediators and receptors common to both systems, the presence of NPY receptors on immune cells has been described. Presently, six types of NPY receptors are known, with only five of them cloned in different physiological systems. Although little is known about the presence of NPY receptors on phagocytic cells, it has been described that the central NPY Y1 receptor subtype is involved in the modulation of granulocyte function [2] and it has been identified on rat and guinea pig spleen cells [3]. In relation to this, our group has suggested the presence of high and low affinity NPY receptors in murine peritoneal macrophages [4]. In a study using agonists or antagonists of Y1, Y2 and Y3 NPY receptor subtypes, the possible role of these receptor subtypes mediating NPY effects on the different steps of the phagocytic process have also been proposed [5]. Recently, it has been demonstrated that NPY Y1, Y2, and Y5 receptors have different responses in rat peritoneal macrophage function during

The NPY Family of Peptides in Immune Disorders, Inflammation, Angiogenesis and Cancer,
edited by Zofia Zukowska and Giora Z. Feuerstein

NPY stimulation [6]. The role of the Y2 receptor also has further effects on the adhesion of rat monocytes [7]. Therefore, the immunomodulatory properties of NPY are highly dependent on the subtype of NPY receptor that is expressed on the immune cells and on the functional interaction between different subtypes of NPY receptors.

Although NPY exerts critical effects on various immunological parameters, including important actions on both innate and acquired immunity, the aim of this chapter is to outline the involvement of NPY on phagocytic cells, especially on macrophage functions.

It is important to point out that macrophages are distributed in different locations throughout the body and that the innervation by sympathetic nervous fibers, which release NPY along with other neurotransmitters, is different in each of those places. For this reason, the modulation exerted *in vivo* by NPY on macrophage functions will be different depending on the source of them and it is possible, as commented below, that the same occurs in the *in vitro* studies with macrophages. Among the different sources of macrophages, we have chosen in our laboratory those from the peritoneum because they can be obtained easily and without sacrificing the animals.

The knowledge at this moment about the influence of NPY on the different steps of the phagocytic process of peritoneal macrophages, as well as its way of action at the intracellular level is analyzed and summarized in Figure 1.

Adherence

The first step of the phagocytic process involves the adherence of phagocytic cells to tissue substrate before migration to the site of inflammation. Thus, adherence is a crucial event in the entire immune response shared by lymphocytes and phagocytes [8]. NPY affects the abilities of leukocytes to adhere to the endothelium, enhancing the adhesiveness to extracellular molecules *in vitro*, which is due to an increased overall affinity of the adhesion molecule receptors on leukocytes [9, 10]. Recently, Nave et al. [7] have found that NPY enhances the adherence of rat peritoneal macrophages activated with lipopolysaccharides (LPS) *in vitro* and *in vivo*, with NPY Y2 receptor mediating this effect. Our group has described an increase or no effect of this neuropeptide on the adhesion of murine peritoneal macrophages, depending on the time of incubation of macrophages with NPY [4, 5]. It has been reported that NPY increased the adhesion of lymphocytes to fibronectin through the activation of integrin B1 [10], but to our knowledge no studies on macrophages have been performed in relation to the possible activation of adhesion molecules by NPY. For a better understanding of the underlying mechanisms of this phenomenon, future investigations should carefully examine the involvement of adhesion molecules in NPY-induced leukocyte adhesion.

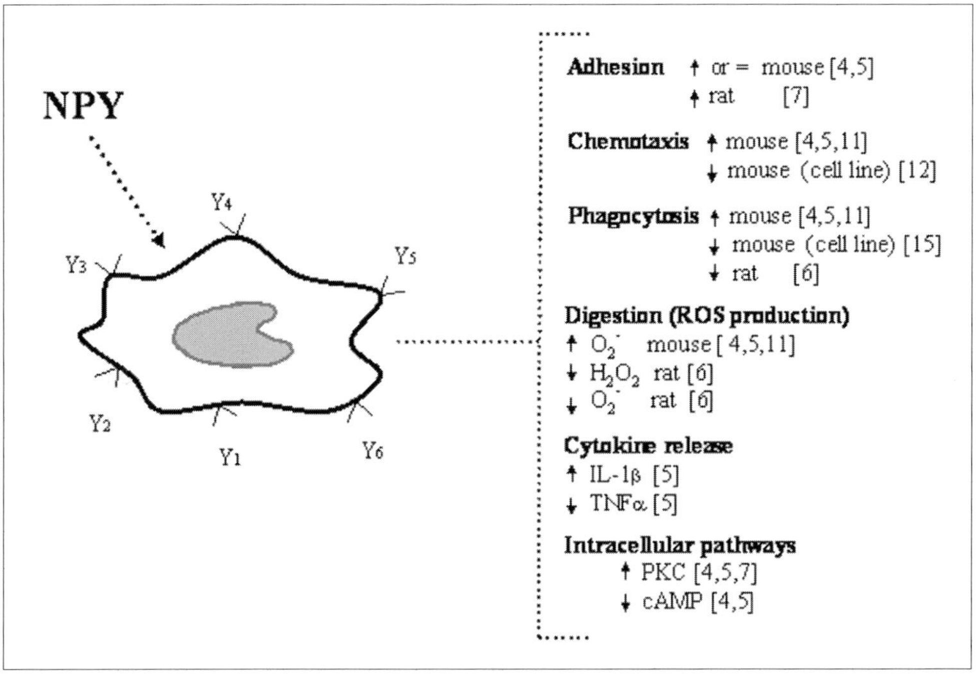

Figure 1
Effects of NPY on several functions of macrophages. ↑ Increase, ↓ decrease, = no effect

Chemotaxis

Phagocytes show great mobility, both spontaneous and directed towards a chemical gradient (chemotaxis) produced from an infectious focus. This property, extensively studied in leukocytes, in addition to adherence is also shared by lymphocytes and phagocytic cells. We have observed an increase of spontaneous mobility and chemotaxis of peritoneal murine macrophages in response to NPY [4, 5, 11]. However, an inhibitory effect of NPY on macrophage chemotaxis induced by *Leishmania major* has also been reported [12]. These opposite results could be explained because of the different cells employed, that is, murine peritoneal macrophages in our studies and a murine monocyte/macrophage cell line Raw 264.7 in the studies performed by Ahmed et al. [12].

NPY profoundly alters chemotaxis, a critical step of phagocytic cells, directly, and additionally modulates the outcome of other chemotactic stimuli. Thus, in an *in vitro* assay with human monocytes, NPY increased the chemotaxis of these phagocytic cells, but the fMLP, a strong chemoattractant agent obtained from bacterial walls, exerted a two- to three-fold stronger effect in comparison to NPY [13];

however, the combined action of fMLP and NPY did not result in synergistic effects, since NPY inhibited the fMLP-induced chemotaxis of human monocytes [14].

Phagocytosis

The phagocytic capacity of macrophages is the most representative and relevant function of these cells. Not much is known about the effect of NPY on this function, but we have found that the ingestion capacity of foreign cells (*Candida albicans*) of inert particles (latex beads) was increased by NPY [4, 5, 11]. Moreover, we have confirmed that the action of NPY on macrophage phagocytosis depends on the presence of other cells such as lymphocytes [5], a fact that will be extensively explained below. However, contradictory effects have also been reported. Thus, an inhibitory effect of NPY on phagocytosis of the *Leishmania major* parasites by a monocyte/macrophage cell line has been described [15]. In rat peritoneal macrophages stimulated with zymosan, NPY seems to inhibit phagocytosis [6]. As commented before and will be discussed in more detail, any change in the experimental design leads to different results of NPY's effects on immune cell functions.

Digestion

The digestion capacity is needed for the destruction of the ingested material and, in macrophages, for the antigenic presentation to lymphocytes. This activity is carried out in different ways, but the most relevant is known as the respiratory burst. In this process the presence of a phagocytic stimulus activates the enzyme NADPH oxidase, which catalyzes a reaction that produces superoxide anion (O_2^-), an oxygen free radical which is a precursor of the active microbiocidal oxidants [16]. With regard to this activity, several *in vitro* studies on NPY and peritoneal macrophages have shown that this neuropeptide increases the superoxide anion levels of these cells in response to latex beads as a phagocytic stimulus [4, 5, 11]. It has also been reported that NPY increased hydrogen peroxide production in PMA-stimulated rat peritoneal macrophages with Y1 and Y2 receptor subtypes mediating the stimulation, while the Y5 receptor subtype showed suppression of that peroxide production [6]. In the same study, it has been reported that NPY inhibits superoxide anion release when macrophages are stimulated with zymosan [6].

Cytokine release and pro- and anti-inflammatory functions

The immunomodulatory activity of neuropeptides and NPY is mediated, at least in part, through their effects on the production of cytokines. Thus, the modulatory effect of NPY on macrophage functions could be interpreted in the context of the effect of this neuropeptide on the levels of inflammatory/anti-inflammatory cytokines that it produces. Moreover, the macrophages are a major source of several cytokines released at inflammatory sites. It has been reported that NPY inhibits IL-6 release from spleen macrophages via the Y1 receptor [17]. NPY has also been shown to decrease TNF-α production, a clearly proinflammatory cytokine [5] which can be interpreted as a positive modulation since this cytokine is involved in many diseases and inflammatory processes. With respect to IL-1β, a proinflammatory cytokine necessary for immune functions such as the proliferative response of lymphocytes, NPY seems to play a homeostatic role to keep the amount of this cytokine at optimal levels depending on the age of the animal studied [5].

Moreover, NPY shows other anti-inflammatory actions such as the above mentioned, i.e., it decreases leukocyte infiltration to tissues during experimental endotoxemia, which could reduce the damage of those tissues [7], and decreases oxygen free radical levels in rat peritoneal macrophages, preventing a possible oxidative/inflammatory state [6]. However, we have observed that NPY did not exert any effect on the levels of nitric oxide or prostaglandin E2 (PGE2) in lipopolysaccharide-stimulated peritoneal leukocytes. However, when NPY was co-administered with norepinephrine, an inhibition of both PGE2 and nitric oxide (NO) levels was observed, giving these neurotransmitters an anti-inflammatory role (unpublished data).

Since systemically released cytokines are involved in many actions, some anti- and others proinflammatory in nature, it is of major importance that the extent to which local NPY-evoked cytokine release contributes to the systemic effects of cytokines will be further investigated.

Action mechanisms

The final effect of a peptide, such as NPY on a cell, depends on the intracellular pathway that it activates. With respect to the two most relevant intracellular signals, several studies suggest that agents that increase intracellular levels of cAMP and decrease diacylglycerol and inositol triphosphate levels, with subsequent lower protein kinase C (PKC) activation, inhibit the functions of a variety of immune cells [18, 19]. Nevertheless, low levels of cAMP and high protein kinase C (PKC) levels accompany an increase of these functions [18]. NPY has been shown to induce both an increase and an inhibition of cAMP formation [20]. We have described that NPY decreases cAMP levels, stimulates PKC and increases phagocytosis [4, 5, 18]. The study of PKC activity in murine macrophages showed that NPY produced a significant increase in the membrane fraction activity of PKC and a loss of activity in the cytosolic fraction, which seems to indicate an activation of the enzyme by translocation from the cytosol to the membranes [4]. In this way, it has been proposed recently that NPY increases the oxidative burst in phorbol myristate acetate-stimulated rat peritoneal macrophages involving the activation of PKC [6].

The current knowledge of the effect of NPY on phagocytic cell function is very limited and more research is needed for extending and clarifying its role on these immune cells. The data are also, in many cases, contradictory and with respect to this fact it is useful in each experimental design to consider the involvement of a great number of factors which could exert a relevant influence on the effects that NPY shows on the immune cell functions, especially on macrophage functions such as those that have been dealt with in this section. Next, we will discuss several of these factors.

Factors that could be involved in the effect of NPY on macrophage functions

The results shown until now on the modulation capacity of NPY on phagocytic cells have been obtained *in vitro*. However, *in vivo*, NPY is released together with other neuropeptides and hormones. Furthermore, other factors that must be considered are the interactions among the different immune cellular types that are found in each immune location. Thus, a determined effect of NPY could be different if the study were performed with an isolated cellular population or when other cells were present (total populations). Likewise, since there are changes with aging in the neuropeptide modulation of immune cells, the age of the subject from which the immune cells are obtained is an important factor to take into account when these kinds of studies are carried out. In addition to the above, the cellular activation state, the magnitude and duration of the NPY stimulus, as well as the biological rhythms, are also important factors, hardly considered in these types of studies, which affect the results of NPY effects on immune cell functions. Next, some of these factors will be analyzed, but others could be equally important.

Presence of other neurotransmitters

In vivo, the sympathetic nervous system (SNS) activation results not only in the release of catecholamines such as norepinephrine and epinephrine, but also in the release of NPY and other neuropeptides [21]. This pattern of releasing catecholamines and NPY also occurs in sympathetic nerve fibers terminating in lymphoid tissues [22]. Since NPY is found in sympathetic nerves in the periphery, where it is stored either alone in small vesicles or in large vesicles in combination with catecholamines [23]. Thus, *in vivo*, when the SNS is activated, the immune cells will receive NPY together with other neurotransmitters. Moreover, in the blood a great number of different hormonal factors reach the immune cells, further affecting the immunomodulation of NPY [24]. Our group has performed several investigations

Table 1 - Effects of co-administration on NPY and norepinephrine on several macrophage functions

	PGE_2	NO	IL-6	Phagocytosis	Chemotaxis	Adhesion
NPY	=	=	↓	↑	↑↑	↓
NE	=	=	↓	↑	↑	=
NPY+NE	↓	↓	↓↓	↑↑	↑	↓

NE, norepinephrine; NPY, neuropeptide Y; PGE_2, prostaglandin E_2; NO, nitric oxide; IL-6, interleukin 6; = no effect; ↑ increase; ↓ decrease

in order to study if the co-administration of NPY with norepinephrine (NE) mimicked the naturally occurring co-release of both SNS neurotransmitters. The effects were different when compared to those obtained with either NE or NPY alone on several leukocyte functions [25]. As regards the modulation of NPY and NE, separately and jointly, on different functions of phagocytic cells, as well as their effects on several cytokine levels (unpublished data); the results are summarized in Table 1. In reference to the NO and PGE_2 release, an inhibitory effect was observed only with the combination of both neurotransmitters. However, with respect to the assay of phagocytosis, the coadministration promoted the effect observed with each of them alone. This effect is not observed on chemotaxis, a macrophage function in which NPY alone exerted the highest effect. With respect to the adhesion to fibronectin, inhibitory effects were seen when NPY was administered alone and co-administered with NE. With regard to the co-administration of NPY and NE on IL-6 release by spleen macrophages, a stronger inhibition has been shown when compared to the effect of NPY alone [17].

According to all these results, the idea that we want to emphasize is that it is very important to carry out the *in vitro* experiments reproducing as much as possible the *in vivo* conditions, since different effects can be found on immune cells, as seen in macrophages, depending on the presence of one or more chemical mediators. Thus, the study of specific combinations of neurotransmitters mimicking the *in vivo* events could contribute to a better understanding of that situation.

Presence of other cells

The communication among the different cellular types is an important factor to take into consideration when analyzing the influence of a neurotransmitter on any cellular function such as macrophage functions. Addressing this aspect, our group has demonstrated that when leukocyte populations were depleted of adherent cells, the effect of NPY on leukocyte chemotaxis disappeared. Thus, adherent cells seem to be directly involved with NPY's effects [26]. With respect to macrophage phagocytosis, different NPY effects on this function were observed depending on whether NPY was added to macrophages in the presence or absence of B lymphocytes [5]. According to these data, in recent experiments (unpublished data), our group has studied the influence of different cell types of the mouse peritoneal suspension on the phagocytosis of macrophages. Figure 2 (A) shows the phagocytosis index (number of latex beads ingested by 100 macrophages) of peritoneal macrophages from mice, from: a total peritoneal cell suspension, isolated macrophages, macrophages plus T lymphocyte, and macrophages plus B lymphocytes. The highest phagocytic index was found when both macrophages and B lymphocytes were present. On the contrary, the lowest index was observed when macrophages and T lymphocytes were together. The presence of the total peritoneal population (TP), similar to "*in vivo*" situations, showed a higher phagocytosis index than that found in isolated macrophages and in macrophages plus T lymphocytes. When supernatants obtained from B lymphocytes, T lymphocytes and either B lymphocytes plus T lymphocytes cultures were added to macrophages, no differences on the phagocytosis indexes were found, indicating that cellular interactions are necessary. Therefore, these data seem to indicate that the presence of T and B lymphocytes is necessary for macrophage function, as seen here in phagocytic function. With this basis we evaluated the NPY effects on the four situations named above. The results obtained are represented in Figure 2 (B). No NPY effects on phagocytosis were observed with isolated macrophages, neither with macrophages and T lymphocytes. Nevertheless, NPY increased the phagocytosis index when macrophages plus B lymphocytes were present, while this neuropeptide decreased the phagocytic index when the assay was performed with a total peritoneal population. These results (unpublished data) suggest that B lymphocytes have a stimulatory role on phagocytosis (Fig. 2A), as well as on the effects exerted by NPY on this function (Fig. 2B). However, T lymphocytes need the presence of B lymphocytes to exert their action. Thus, it is possible that in the total populations, more similar to "*in vivo*" situations, the NPY effects observed on that phagocytosis could be due to interactions between T and B lymphocytes. Thus, different results on the effect of NPY on macrophage function, even in apparently identical experiments, could be explained by the presence or absence of other immune cells.

Aging

Aging may be defined as a progressive, endogenous and irreversible accumulation of adverse changes that increases vulnerability to disease and finally to death. It is well known that aging is accompanied by a decline of physiological systems and

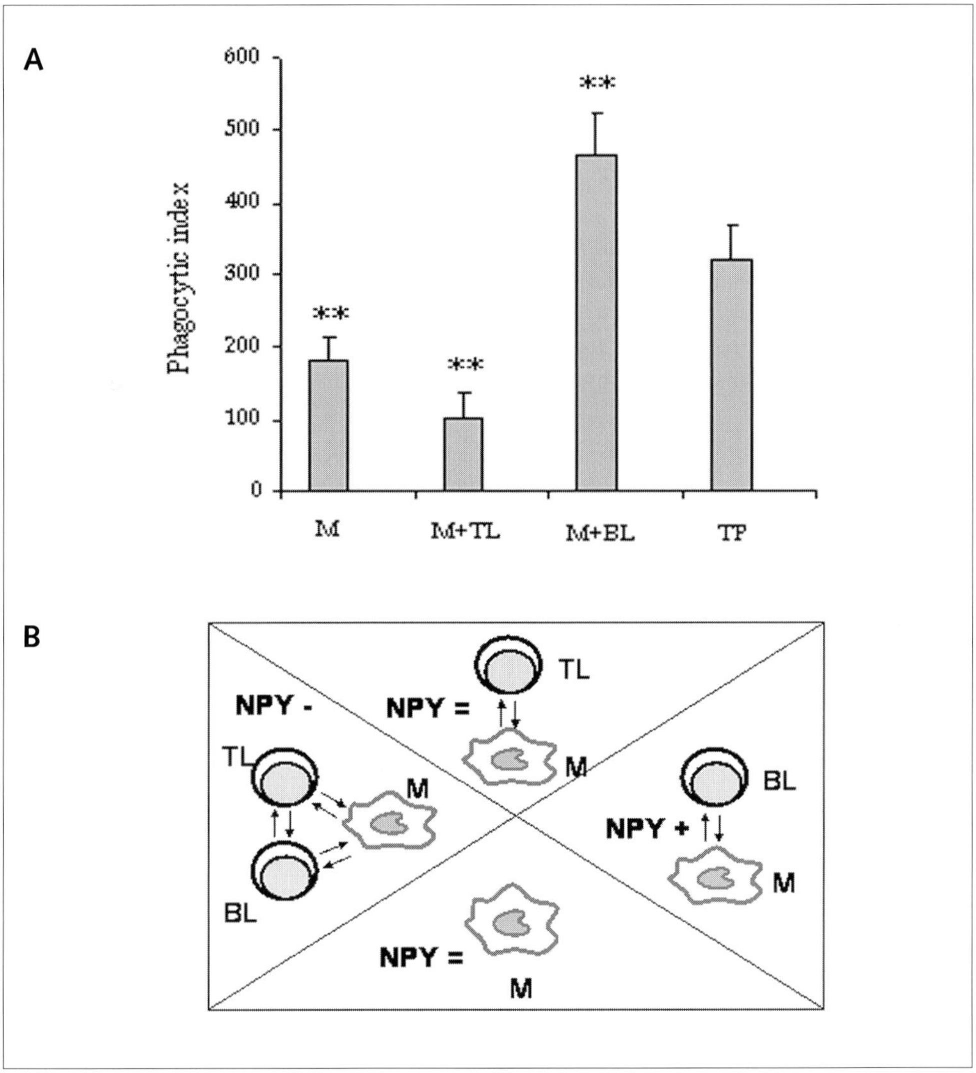

Figure 2
*(A) Phagocytic index (number of latex beads phagocytozed by 100 peritoneal macrophages) with or without the presence of T and B lymphocytes. M: macrophages; M+TL: macrophages plus T lymphocytes. M+BL: macrophages plus B lymphocytes. TP: total population. Each value is the mean±standard error of eight experiments, each value being the mean of duplicate assays. **p > 0.001 with respect to the corresponding value in total population.*
(B) Effect of NPY on peritoneal macrophage phagocytosis in assays with or without other types of cells from murine peritoneal suspension. TL, T lymphocyte; BL, B lymphocyte; M, macrophage. +, increase, –, decrease, = no NPY effects on phagocytic index.

impairment of nervous and immune functions [27, 28]. Moreover, the changes with aging in the communication between the immune and nervous system have been reported as a possible cause of physiological senescence [29]. To our knowledge, the age-depending changes in the effect of neuropeptides on the immune function, a subject scarcely studied, is not considered in this kind of work. Therefore, it is of great importance to consider the age of the subjects in the experimental studies since, even without any stimulus, there are differences in the immune functions depending on the age, and the modulation exerted by nervous mediators on immune cell functions will depend on the age of the animals which were the source of these cells. According to this, we have found, firstly, that macrophage functions, and the above commented different step of the phagocytic process, change with aging and, secondly, that the NPY regulation of those macrophage functions depend on the animal's age. Moreover, we have suggested that NPY could exert a regulatory role by maintaining the function of phagocytes at physiologically optimal levels [5]. Table 2 summarizes the changes with age of the regulatory action of NPY on several functions of the phagocytic process of peritoneal macrophages from mice. It can be observed that in cells from adult mice, NPY stimulated all the steps of the phagocytic process. On the contrary, NPY decreased chemotaxis and phagocytosis, increased adherence and did not show any effects on the superoxide anion production (digestion capacity) when the cells were from old animals. In relation to TNF-α NPY decreased the levels of this cytokine at both ages. With respect to IL-1β level, NPY increased the levels of this cytokine in adult mice, but decreased them in the old. As regards to the action pathways related to phagocytosis capacity in macrophages from adult mice, NPY increased PKC activity and, at the same time, decreased the cAMP levels, and consequently stimulated the ingestion capacity of macrophages. Nevertheless, the action pathway of NPY was the opposite in these same cells from old mice, in which NPY decreased PKC activity and increased cAMP levels, inhibiting the phagocytosis capacity [5].

In view of the above results, we could propose that, in the future, the age of the animals should be taken into consideration when analyzing the effects of any nervous mediators on immune cell functions.

State of cell activation and length of time of the stimulus

The state of activation of the cells and the duration of the applied stimulus are relevant when the effects of NPY on phagocytic cell functions are evaluated. Moreover, we must bear in mind that depending on the duration of sympathetic nerve activation and the magnitude of this activation, thus is the composition of the nervous fiber transmitter output, being dominated by either catecholamines or NPY. It is known that NPY is preferentially released under conditions of elevated SNS activity, whereas the release of catecholamines dominates under moderate stimulation

Table 2 - Effect of age on NPY actions on macrophage functions

	Adhesion	Chemotaxis	Phagocytosis	Digestion	IL-1β	TNF-α	cAMP	PKC
Adult	↑	↑	↑	↑	↑	↓	↓	↑
Old	↑↑↑	↓	↓	=	↓	↓	↑	↓

IL-1β, interleukin-1β; TNF-α, tumor necrosis factor-α; cAMP, cyclic adenosine monophosphate; PKC, protein kinase C. = no effect; ↑ increase; ↓ decrease

[30]. Thus, it has been shown that blood polymorphonuclear cells recently isolated from intracerebroventricular (i.c.v) NPY-treated rats exhibit a decrease in zymosan- and PMA-induced oxygen radical formation. However, when the same assay was performed 24 h after i.c.v administration of NPY, an increase in oxygen radical formation was observed [31, 32]. Therefore, the duration of the stimulus produces important changes in the effect analyzed. Our group has also observed that the time of incubation of NPY with macrophages is relevant when analyzing the NPY effect on adhesion capacity. Thus, no effects were observed in the macrophage adhesion of adult animals with 5 and 10 min of incubation [4, 5], but NPY increased this parameter [4] with longer times (20 or 30 min) of incubation. Besides, regarding macrophage phagocytosis assays with peritoneal adherent population of mice, we were able to observe (Fig. 3) that incubation of NPY together with the phagocytic stimulus (latex beads) for 30 min, resulted in an increase of the phagocytosis index. However, pre-incubation of macrophages with NPY for 30 min, before adding the phagocytic stimulus, resulted in a decrease of phagocytosis. Moreover, when carrying out this last experiment by performing a wash between the addition of NPY and the phagocytosis stimulus, the results were similar to the first case (i.e., NPY and phagocytosis stimulus at the same time). Thus, the incubation conditions are of a great importance when analyzing NPY effects on any immune activity and more concretely on macrophage functions.

Although the concentration of NPY used in the different experiments is an important factor to regard, in general, a broad range of concentrations are used. Thus, we have studied a range between 10^{-14} M to 10^{-6} M of NPY, with the majority of effects shown in the range of 10^{-11} M to 10^{-9} M [4, 5, 18].

Biological rhythms

Other important factors influencing the effect of nervous mediators, such as NPY, on the immune parameters are biological rhythms, namely circadian and circannual rhythms, given that all cells and biological systems show rhythms of different fre-

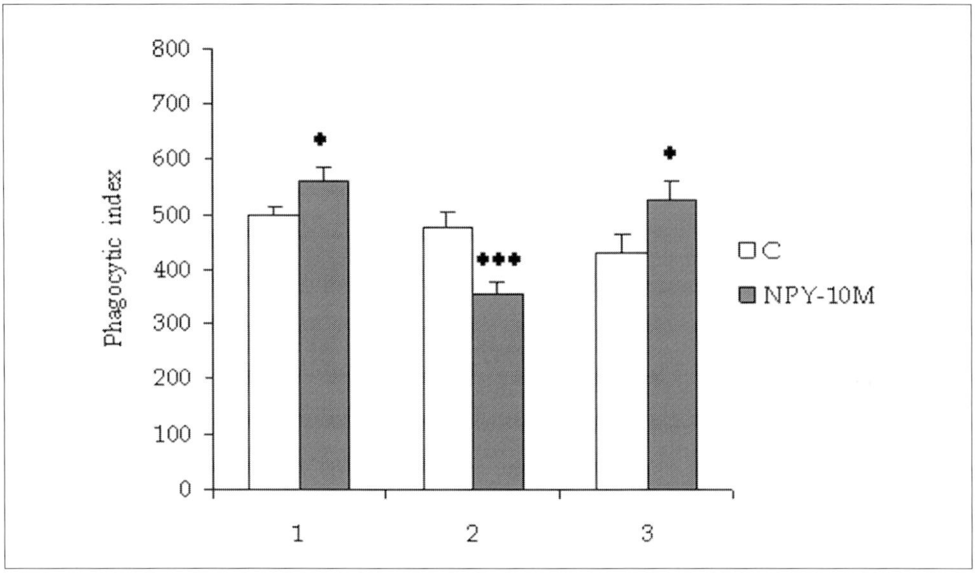

Figure 3
*NPY effects on phagocytic index of peritoneal macrophages. Assay performed in adherent population with different conditions. 1. NPY + latex (30 min); 2. NPY (30 min) + latex (30 min); 3. NPY (30 min) + washing + latex (30 min). Each column represents the media ± standard error of eight experiments performed in duplicate. *$p < 0.05$; ***$p < 0.001$ with respect to the corresponding control value.*

quencies which can be observed in single cells, in networks of tissues and organs, as well as in whole organisms and even in populations. Basically, it is necessary to distinguish between the exogenous rhythms, which only reflect the biological response to the environmental input, and the endogenous one, which is inborn in the organism, persists in a milieu deprived of all external influences, and is self-sustained. It is known that there are important physiological variations, as regard to the number and proportion of leukocytes with the circadian rhythm. Thus, granulocytes and macrophages, as well as natural killer cells, extra-thymic T cells, γδ T cells, and CD8+ subset show an increase in the light period of day, whereas other immune cells show an increase at night. Humans are active and show sympathetic nerve dominance in the daytime, a period in which their leukocytes may be under the regulation of the autonomic nervous system. Interestingly, granulocyte and lymphocyte subsets, with the daytime rhythm, were found to carry a high density of adrenergic receptors [33]. It has also been described that in BALB/c mice, the maximum inert-particle ingestion capacity of the peritoneal macrophages occurred during the night.

However, when animals are subjected to stress, they present a loss of the rhythm and the macrophage phagocytic capacity was greater at all hours of the day [34]. Another leukocyte property that changes with circadian rhythm is the expression of adhesion molecules [35]. Moreover, the time of sleep or the disruption of sleep, influence the activity of immune cells [36], as well as the expression of adhesion molecules and the immune cell migration, with the nocturnal changes in the levels of circulating norepinephrine being one of the factors that mediates these effects [36, 37]. NPY has also been involved in circadian rhythms [38], showing changes in its blood levels, depending on the time of day that it is analyzed. Moreover, it has been suggested that the NPY rhythm is dependent on the environmental light:dark cycle [39]. Thus, time of day, or more correctly, stage of rhythm, needs to be taken into account for understanding the effect of NPY on immune cells, in particular, phagocytic cells. The findings to date suggest that "when" something is done should be considered, just as importantly as "what" and "how much", for interpretation and treatment of many conditions and illnesses, thereby adding a fourth dimension (time) to our understanding of the physiological effects of a nervous mediator on immune functions [40].

Acknowledgement
This work has been supported in part by a grant from MCYT of Spain. The authors would like to thank Dr Hernanz for his suggestions about the text and especially Dr Puerto for carrying out several of the experiments shown in the chapter.

References

1 Unanue, ER (1989) Macrophages, antigen-presenting cells and the phenomena of antigen handling and presentation. In: WE Paul (ed): *Fundamental Immunology*. Raven Press, New York, 95
2 Hörsten SV, Exton NG, Exton MS, Helfritz F, Nave H, Ballof J, Stalp M, Pabst R (1998) Brain NPY Y_1 receptors rapidly mediate the behavioral response to novelty and a compartment-specific modulation of granulocyte function in blood and spleen. *Brain Res* 806: 282–286
3 Gehlert DR, Gackenheimer SL (1996) Unexpected high density of neuropeptide Y Y_1 receptors in guinea pig spleen. *Peptides* 17: 1345–1348
4 De la Fuente M, Bernaez I, Del Río M, Hernanz A (1993) Stimulation of murine peritoneal macrophage functions by neuropeptide Y and peptide YY. Involvement of protein kinase C. *Immunology* 80: 259–265
5 De la Fuente M, Del Rio M, Medina S (2001) Changes with aging in the modulation by neuropeptide Y of murine peritoneal macrophage functions. *J Neuroimmunol* 116: 156–167

6 Dimitrijevic M, Stanojevic S, Vujic V, Beck-Sickinger A, von Hörsten S (2005) Neuropeptide Y and its receptor subtypes specifically modulate rat peritoneal macrophage functions *in vitro*: counter regulation through Y1 and Y2/5 receptors. *Regul Pept* 124: 163–172

7 Nave H, Bedoui S, Moenter F, Steffens J, Felies M, Gebhardt T, Straub RH, Pabst R, Dimitrijevic M, Stanojevic S et al (2004) Reduced tissue inmigration of monocytes by neuropeptide Y during endotoxemia is associated with Y2 receptor activation. *J Neuroimmunol* 155: 1–12

8 Mackay CR, Imhof BA (1993) Cell adhesion in the immune system. *Immunol Today* 14: 99–102

9 Sung CP, Arleth AJ, Feuerstein GZ (1991) Neuropeptide Y upregulates the adhesiveness of human endothelial cells for leukocytes. *Circ Res* 68: 314–318

10 Levite M, Cahalon L, Hershkoviz R, Steinman L, Lider O (1998) Neuropeptides, via specific receptors, regulate T cell adhesion to fibronectin. *J Immunol* 160: 993–1000

11 De la Fuente M, Medina S, Del Rio M, Ferrandez MD, Hernanz A (2000) Effect of aging on the modulation of macrophage functions by neuropeptides. *Life Sci* 67: 2125–2135

12 Ahmed AA, Wahbi A, Nordlind K, Kharazmi A, Sundqvist KG, Mutt V, Liden S (1998) *In vitro Leishmania major* promastigote-induced macrophage migration is modulated by sensory and autonomic neuropeptides. *Scan J Immunol* 48: 79–85

13 Straub RH, Mayer M, Kreutz M, Leeb S, Shölmerich J, Falk W (2000) Neurotransmitters of the sympathetic nerve terminal are powerful chemoattractants for monocytes. *J Leukoc Biol* 67: 553–558

14 Dureus P, Louis D, Grant AV, Bilfinger TV, Stefano GB (1993) Neuropeptide Y inhibits human and invertebrate immunocyte chemotaxis, chemokinesis, and spontaneous activation. *Cell Moll Neurobiol* 13: 541–546

15 Ahmed AA, Wahbi AH, Nordlin K (2001) Neuropeptides modulate a murine monocyte/macrophage cell line capacity for phagocytosis and killing of *Leishmania major* parasites. *Immunopharmacol Immunotoxicol* 23: 397–409

16 Segal AW, Abo A (1993) The biochemical basis of NADPH oxidase of phagocytes. *Trends Biochem Sci* 18: 43–47

17 Straub RH, Schaller T, Miller LE, von Horsten S, Jessop DS, Falk W, Sholmerich J (2000) Neuropeptide Y cotransmission with norepinephrine in the sympathetic nerve-macrophage interplay. *J Neurochem* 75: 2464–2471

18 De la Fuente M (1999) Modulation of immune function by neuropeptides. *Curr Trends* Immunol 2: 111–122

19 Vang T, Torgersen KM, Sundvold V, Saxena M, Levy FO, Skalhegg BS, Hansson V, Mustelin T, Tasken K (2001) Activation of the COOH-terminal Src kinase (Csk) by cAMP-dependent protein kinase inhibits signalling through the T cell receptor. *J Exp Med* 193: 497–507

20 Michel MC, Beck-Sickinger A, Cox H, Doods HN, Herzog H, Larhammar D, Quirion R, Schwartz T, Westfall T (1998) XVI International Union of Pharmacology recom-

mendations for the nomenclature of neuropeptide Y, peptide YY, and pancreatic polypeptide receptors. *Pharmacol Rev* 50: 143–150

21 Lundberg JM, Rudehill A, Sollevi A, Fried G, Wallin G (1989) Corelease of neuropeptide Y and noradrenaline from pig spleen *in vivo*: importance of subcellular storage, nerve impulse frequency and pattern, feedback regulation and resupply by axonal transport. *Neuroscience* 28: 475–486

22 Felten DL, Felten SY, Carlson SL, Olschowka JA, Livnat S (1985) Noradrenergic and peptidergic innervation of lymphoid tissues. *J Immunol* 135: 755–765

23 Ekblad E, Edvinsson L, Wahlestedt C, Uddman R, Hakanson R, Sundler F (1984) Neuropeptide Y co-exists and co-operates with noradrenaline in perivascular nerve fibers. *Regul Pept* 8: 225–235

24 Elenkov IJ, Wilder RL, Chrousos GP, Vizi ES (2000) The sympathetic nerve – an integrative interface between two supersystems: the brain and the immune system. *Pharmacol Rev* 52: 595–638

25 Puerto M, Guayerbas N, Alvarez P, De la Fuente M (2005) Modulation of neuropeptide Y and norepinephrine on several leucocyte functions in adults, old and very old mice. *J Neuroimmunol* 165 (1–2): 33–40

26 Medina S, Del Río M, Hernanz A, De la Fuente M (2000) The NPY effects on murine leukocyte adherence and chemotaxis change with age. *Regul Pept* 95: 35–45

27 Pawelec G, Barnett Y, Forsey R, Frasca D, Globerson A, McLeod J, Caruso C, Franceschi C, Fulop T, Gupta S et al (2002) T cells and aging. *Front Biosci* 7: 1056–1183

28 De la Fuente M (2002) Effects of antioxidants on immune system ageing. *Eur J Clin Nutr* 56: 5–8

29 Fabris N (1991) Neuroendocrine-immune interactions: a theoretical approach to aging. *Arch Gerontol Geriatr* 12: 219–230

30 Hökfelt T (1991) Neuropeptides in perspective: the last ten years. *Neuron* 7: 867–879

31 von Horsten S, Nave H, Ballof J, Helfritz F, Meyer D, Schmidt RE, Stalp M, Exton NG, Exton MS, Straub RH et al (1998) Centrally applied NPY mimics immunoactivation induced by non-analgesic doses of met-enkephalin. *Neuroreport* 9: 3881–3885

32 von Horsten S, Ballof J, Helfritz F, Nave H, Meyer D, Schmidt RE, Stalp M, Klemm A, Tschernig T, Pabst R (1998) Modulation of innate immune functions by intracerebroventricularly applied neuropeptide Y: dose and time dependent effects. *Life Sci* 63: 909–922

33 Suzuki S, Toyabe S, Moroda T, Tada T, Tsukahara A, Iiai T, Minagawa M, Maruyama S, Hatakeyama K, Endoh K et al (1997) Circadian rhythm of leucocytes and lymphocytes subsets and its possible correlation with the function of the autonomic nervous system. *Clin Exp Immunol* 110: 500–508

34 Barriga C, Martin MI, Tabla R, Ortega E, Rodriguez AB (2001) Circadian rhythm of melatonin, corticosterone and phagocytosis: effect of stress. *J Pineal Res* 30: 180–187

35 Niehaus G, Ervin E, Patel A, Khanna K, Vanek VW, Fagan DL (2002) Circadian varia-

tion in cell-adhesion molecule expression by normal human leukocytes. *Can J Physiol Pharmacol* 80: 935–940

36 Redwine L, Dang J, Irwin M (2004) Cellular adhesion molecule expression, nocturnal sleep, and partial night sleep deprivation. *Brain Behav Immun* 18: 333–340

37 Irwin M, Clark C, Kennedy B, Gillin JC, Ziegler M (2003) Nocturnal catecholamines and immune function in insomniacs, depressed patients, and control subjects. *Brain Behav Immun* 17: 365–372

38 Baraban SC (1998) Neuropeptide Y and limbic seizures. *Rev Neurosci* 9: 117–128

39 Djeridane Y, Eurin J, Touitou Y (2004) Daily variation in the concentration of neuropeptide Y in the rat atrium: effects of age and photoperiodic conditions. *Peptides* 25: 1153–1157

40 Sothern RB, Roitman-Johnson B (2001) Biological rhythms and immune function. In: R Ader, DL Felten, N Cohen (eds): *Psychoneuroimmunology*. Academic Press, London, 445–479

NPY in allergic and respiratory disorders

Basile N. Landis, Isabelle Plouin-Gaudon and Jean-Silvain Lacroix

Unité de Rhinologie-Olfactologie, Service d'Oto-Rhino-Laryngologie, Hôpitaux Universitaires de Genève, Rue Micheli-du-Crest 24, 1211Geneva, Switzerland

Introduction

The respiratory tract can be divided into upper airway including the nasal cavities, the pharynx and larynx, and the lower airway comprising the tracheobronchial tract and the respiratory portion of the lungs. The main function of the nose includes conditioning (warming and humidifying) of inhaled air and protection (filtering, trigeminal function and olfaction) of the respiratory part of the lower airway against inhalation of exogenous particles and airborne irritants [1–4]. The submucosal blood vessels, the muco-ciliary transport system and the local immune mechanisms are the main tissue components involved in these functions. The function of these elements is influenced, at least in part, by biologically active agents released from both sensory and efferent sympathetic and parasympathetic autonomic nerves [5, 6]. A very dense sympathetic innervation is present around both resistance and capacitance vessels of the nasal mucosa [7, 8]. In contrast, parasympathetic fibres are very dense around nasal mucus glands [9, 10]. Noradrenaline (NA) is considered as the classical transmitter released from postganglionic sympathetic nerves [11]. Neuropeptide Y (NPY) is a 36 amino acid peptide first discovered in the porcine brain [12]. The co-localization of NPY with NA has been observed in many peripheral sympathetic nerve fibres. In the nasal mucosa, co-existence of NA and NPY immunoreactivity was observed in sympathetic nerves present around both resistance and capacitance vessels of most species including mankind [13–15]. Presence of NPY was also demonstrated in adrenergic nerves found in both lower airway and vascular smooth muscle of the respiratory tract in several different species [16, 17]. Somatostatin was also observed in association with NA and NPY in nasal sympathetic nerves [18]. Further, NPY has also been found in parasympathetic nerves in the airways in association with vasoactive intestinal polypeptide (VIP) and peptide histidine isoleucine (PHI) in humans as well as in various other species [14, 19–21].

The NPY Family of Peptides in Immune Disorders, Inflammation, Angiogenesis and Cancer,
edited by Zofia Zukowska and Giora Z. Feuerstein

Functional studies

The first studies regarding the physiological and pharmacological role of NPY in the upper airway mucosa were performed in animals. In the pig nasal mucosa, local intra-arterial injections of NPY as well as sympathetic nerve stimulation (SNS) were shown to produce long lasting vasoconstriction resistant to adrenergic blockers of both resistance and capacitance vessels [22]. The vasoconstrictive effect of NPY was also observed in the rabbit maxillary sinus mucosa [23]. In humans, exogenous NPY is a powerful vasoconstrictor of the nasal mucosa vascular bed in both control subjects and patients with perennial rhinitis [24–26]. A decrease in nasal nitric oxide production was also observed after endonasal application of *exogenous* NPY [25]. Several studies have shown that exogenous NPY as well as *endogenous* NPY, which is released after SNS, both elicit an increase in tracheobronchial vascular resistance which is slow in onset but long lasting [17, 27–31].

NPY effect on airway mucus secretions

Exogenous NPY does not reduce glandular secretions in the human nasal mucus [26, 32]. However, a randomized double-blind, three way, crossover placebo-controlled study has shown that pretreatment with exogenous NPY significantly reduces both nasal obstruction and mucus secretion induced by allergen challenge [33].

NPY receptors

The multiple biological effects of NPY are exerted through G-protein-coupled receptors. The NPY receptor family includes the Y1 receptor first characterized as a post-synaptic receptor, the Y2 receptor, known as a presynaptic receptor and the Y3 receptor considered as the NPY-preferring receptor. The Y4 receptor was characterized as a pancreatic polypeptide receptor. The Y5 is involved in feeding behavior and the Y6 receptor was recently cloned but its function remains unknown [34]. The predominant NPY receptor type in nasal mucosa blood vessels is of the Y1 type [25]. However, the presence of Y2 receptors is strongly suggested by *in vivo* studies with Y2 agonists (see below) [35, 36].

Endogenous release of NPY

After SNS, NPY was detectable in the nasal venous effluent only after high frequency stimulation whereas NA was secreted already upon low frequency stimulation [13]. Co-release of NPY and catecholamines has been demonstrated in man during intense physical exercise [37]. Interestingly, physical exercise has been shown to induce significant reduction of subjective and objective nasal airway resistance, most likely secondary to vasoconstriction of nasal capacitance vessels, in both control and rhinitis patients [38, 39]. Variations of plasma NPY concentrations over time correlated better with post-exercise nasal vasoconstriction than NA levels [40].

Modulator effects of NPY

There is evidence suggesting that NPY released from adrenergic nerves modulates the effector response to transmitters originating from several neuronal pathways (Fig. 1). In the lower airways, parasympathetic nerve activity can be inhibited by NPY [41, 42]. In anaesthetized cats, sympathetic nerve stimulation induced significant and prolonged attenuation of the vasodilator response to subsequent parasympathetic stimulation. Exogenous NPY mimicked the effect of sympathetic stimulation in attenuating subsequent parasympathetic evoked vasodilation [43]. Similar observations were made in anaesthetized dogs. Attenuation of parasympathetic-evoked vasodilation and mucus secretion could be mimicked by the NPY analog N-acetyl (Leu^{28}, Leu^{31}) NPY 24–36 [44], a Y2-receptor agonist [45]. In contrast, the NPY analog (Leu^{31}, Pro^{34}), a potent Y1 receptor agonist [46] did not have any effect on subsequent vasodilation and mucus secretion induced by parasympathetic stimulation [47]. All together, these observations suggest that nasal sympathetic nerve stimulation attenuate parasympathetic vasodilation via NPY release acting on prejunctional Y2 receptors. Observations obtained from *in vitro* experiments using electrical field stimulation on guinea pig tracheal smooth muscle strongly suggest that NPY modulates cholinergic neurotransmission via prejunctional mechanisms [41]. Further studies have shown that NPY has an inhibitory action on tachykinin release such as substance P (SP) from peripheral endings of capsaicin-sensitive airway sensory nerves [48].

In vitro, exogenous NPY modulates the contractile response of guinea pig tracheal smooth muscle induced by VIP, NA, SP and 5-hydroxytryptamine at both the pre- and postjunctional level [49].

Intranasal or intrabronchial pretreatment with TASP-V, a potent NPY Y2 receptor agonist [50], reduces both nasal obstruction and bronchoconstriction produced by histamine challenge in the pig. In healthy human volunteers TASP-V significantly lessens the nasal airway resistance increase induced by local application of histamine [51]. In the airways, blood flow, tracheal and bronchial smooth muscle tone, mucus and fluid production can be influenced by NPY. These functional effects can be direct or indirect via modulation of other neurotransmitters. The neuromodulatory role of NPY has been well documented on both parasympathetic activity and

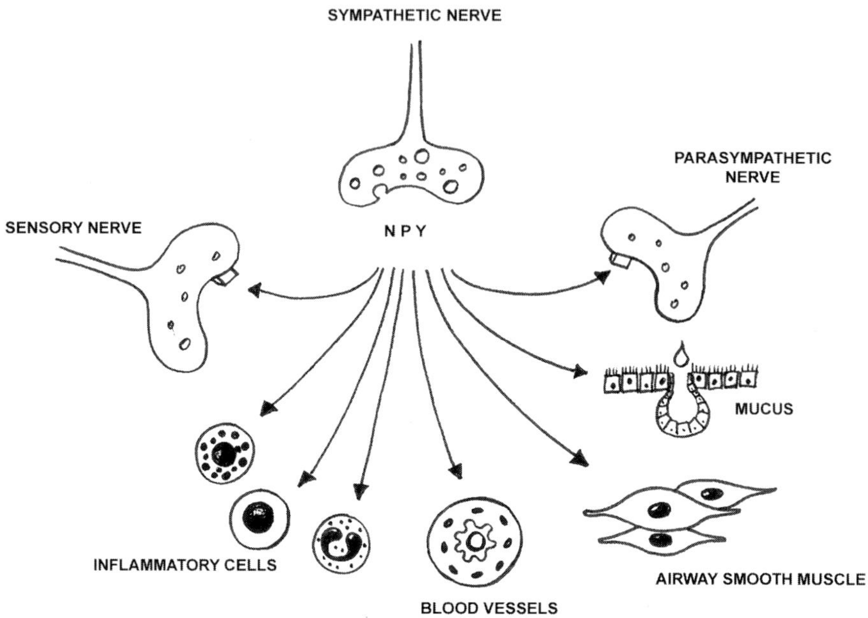

Figure 1
NPY released from sympathetic nerves, modulates airway homeostasis via several targets including parasympathetic and sensory nerves, mucus production, tracheobronchial smooth muscles, blood vessels and inflammatory cells such as mast cells, T lymphocytes, neutrophils and NK cells.

neurogenic inflammation, i.e., plasma extravasation elicited by sensory C-fibres stimulation. Most experiments suggest that NPY modulates neurotransmitter release by prejunctional mechanisms [48, 52].

NPY metabolism

Some proteases have been shown to be involved in the metabolism of biologically active peptides in various tissues [53]. The enzymes neutral endopeptidase (NEP) and dipeptidylpeptidase IV (DPPIV) are involved in NPY degradation. The activity of NEP can be inhibited by phosphoramidon. The inhibitory effect of NPY on the contraction of human bronchial segments *in vitro* could be enhanced in the presence of phosphoramidon. This observation suggests that NEP modulates the effect of NPY in the human airway by an inactivation mechanism [54]. Similar observations were made regarding DPPIV. In patients suffering from chronic rhinosinusitis or bronchitis [36], the activity of both NEP and DPPIV is significantly reduced. This suggests that the catabolism of NPY and other neuropeptides could also be involved in upper airway homeostasis [55]. Neutral endopeptidase activity and concentration of sensory neuropeptides in the human nasal mucosa varies proportionally to patient symptoms [55, 56]. These observations could explain the significant increase of NPY as well as SP and VIP in airway diseases.

NPY and the immune system

Both cellular and humoral functions of the immune system are modulated directly or indirectly by NPY. In this respect, NPY was shown to influence differentiation of T helper cells, activation of NK cells, and release of inflammatory mediators.

One important mechanism in the development of allergic rhinitis and asthma is the differentiation of Th1 and Th2 cells [57, 58]. NPY can induce IL-4 synthesis and reduce interferon-γ (IFN-γ) production [58]. By suppressing Th1 cell differentiation by IFN-γ and stimulating Th2 cell production via IL4, NPY induces *in vitro* a Th2 shift and an inhibition of the synthesis of anti-gene specific IgG2a [59]. At the same time, the production of IL-4 by Th1 cells and IFN-γ by Th2 cell are increased in the presence of NPY [60, 61]. The role of mast cells in allergic reactions is well documented [62, 63]. Rat peritoneal and human skin mast cell release of histamine can be induced by NPY as well as its C-terminal fragments [64–68]. These observations probably indicate direct activation of G proteins [64, 65].

Airway inflammation is associated with the presence of cytokine and oxygen radicals [69–72]. In the presence of NPY, the release of oxygen radicals from murine monocytes is increased (*in vitro*) [73]. Further evidence suggests that NPY is involved in the modulation of airway inflammatory response both in the periphery and when applied intracerebrally [74–78]. Indeed, NPY has important modulator effects on the release of both cytokine and oxidative agents [79].

NPY in airway diseases

No difference in the density of NPY-immunoreactivity has been found in bronchial mucosal nerve of asthmatic patients compared to healthy controls [80]. In contrast, NPY-immunoreactivity was found to be significantly reduced in bronchial smooth muscle nerves from both asthmatic and patients suffering from chronic bronchitis [81]. Elevation of NPY and NA in plasma has also been observed during acute asthma [82].

Conclusion

There is an abundant literature available regarding the involvement of NPY in airway homeostasis. However, there is no NPY-related therapeutic agent yet available or under clinical development for the treatment of airway diseases. Several aspects regarding the role of NPY in the physiological control of airway function and immunological mechanisms suggest that the development of selective agonists and antagonist should have potential therapeutic applications.

Acknowledgements
This work was supported by a grant from the Swiss National Fund for Scientific Research FNSRS (n° 3100A0-100621-1) to JSL.

References

1 Proetz AW (1953) Respiratory air currents and their clinical aspects. *J Laryngol Otol* 67(1): 1–27
2 Cole P (1954) Respiratory mucosal vascular responses, air conditioning and thermo regulation. *J Laryngol Otol* 68(9): 613–622
3 Cole P (1954) Recordings of respiratory air temperature. *J Laryngol Otol* 68(5): 295–307
4 Williams HL (1973) Nasal physiology. In: WB Saunders (ed): *Otolaryngology*. Philadelphia, p. 330
5 Barnes PJ, Baraniuk JN, Belvisi MG (1991) Neuropeptides in the respiratory tract. Part I. *Am Rev Respir Dis* 144: 1187–1198
6 Barnes PJ, Baraniuk JN, Belvisi MG (1991) Neuropeptides in the respiratory tract. Part II. *Am Rev Respir Dis* 144(6): 1391–1399
7 Dahlström A, Fuxe K (1965) The adrenergic innervation of the nasal mucosa in certain mammals. *Acta Otolaryngol* 59: 65–72
8 Anggard A, Densert O (1974) Adrenergic innervation of the nasal mucosa in cat. A histological and physiological study. *Acta Otolaryngol* 78(3–4): 232–241
9 Cauna N (1970) The fine structure of the arteriovenous anastomosis and its nerve supply in the human nasal respiratory mucosa. *Anat Rec* 168(1): 9–21
10 Cauna N (1970) Electron microscopy of the nasal vascular bed and its nerve supply. *Ann Otol Rhinol Laryngol* 79(3): 443–450
11 von Euler US (1948) Identification of the sympathomimetic ergone in adrenergic nerves of cattle (Sympathetin N) with laevonoradrenaline. *Acta Physiol Scand* 16: 63–74
12 Tatemoto K, Carlquist M, Mutt V (1982) Neuropeptide Y – a novel brain peptide with structural similarities to peptide YY and pancreatic polypeptide. *Nature* 296(5858): 659–660

13 Lacroix JS, Stjarne P, Anggard A, Lundberg JM (1989) Sympathetic vascular control of the pig nasal mucosa (III): Co-release of noradrenaline and neuropeptide Y. *Acta Physiol Scand* 135(1): 17–28

14 Lacroix JS, Anggard A, Hokfelt T, O'Hare MM, Fahrenkrug J, Lundberg JM (1990) Neuropeptide Y: presence in sympathetic and parasympathetic innervation of the nasal mucosa. *Cell Tissue Res* 259(1): 119–128

15 Baraniuk JN, Castellino S, Lundgren JD, Goff J, Mullol J, Merida M, Shelhamer JH, Kaliner MA (1990) Neuropeptide Y (NPY) in human nasal mucosa. *Am J Respir Cell Mol Biol* 3(2): 165–173

16 Sheppard MN, Polak JM, Allen JM, Bloom SR (1984) Neuropeptide tyrosine (NPY): a newly discovered peptide is present in the mammalian respiratory tract. *Thorax* 39(5): 326–330

17 Widdicombe JG (1991) Neural control of airway vasculature and edema. *Am Rev Respir Dis* 143(3 Pt 2): S18–S21

18 Lacroix JS, Auberson S, Morel DR, Theodorsson E, Hokfelt T, Lundberg JM (1992) Vascular control of the pig nasal mucosa: distribution and effect of somatostatin in relation to noradrenaline and neuropeptide Y. *Regul Pept* 40(3): 373–387

19 Heppt W, Peiser C, Cryer A, Dinh QT, Zweng M, Witt C, Fischer A, Groneberg DA (2002) Innervation of human nasal mucosa in environmentally triggered hyperreflectoric rhinitis. *J Occup Environ Med* 44(10): 924–929

20 Groneberg DA, Heppt W, Cryer A, Wussow A, Peiser C, Zweng M, Dinh QT, Witt C, Fischer A (2003) Toxic rhinitis-induced changes of human nasal mucosa innervation. *Toxicol Pathol* 31(3): 326–331

21 Groneberg DA, Heppt W, Welker P, Peiser C, Dinh QT, Cryer A, Zweng M, Witt C, Fischer A (2003) Aspirin-sensitive rhinitis-associated changes in upper airway innervation. *Eur Respir J* 22(6): 986–991

22 Lacroix JS, Stjarne P, Anggard A, Lundberg JM (1988) Sympathetic vascular control of the pig nasal mucosa (2): Reserpine-resistant, non-adrenergic nervous responses in relation to neuropeptide Y and ATP. *Acta Physiol Scand* 133(2): 183–197

23 Cervin A (1992) Neuropeptide Y 16–36 inhibits mucociliary activity but does not affect blood flow in the rabbit maxillary sinus *in vivo*. *Regul Pept* 39(2–3): 237–246

24 Lacroix JS, Ricchetti AP, Morel D, Mossimann B, Waeber B, Grouzmann E (1996) Intranasal administration of neuropeptide Y in man: systemic absorption and functional effects. *Br J Pharmacol* 118(8): 2079–2084

25 Cervin A, Onnerfalt J, Edvinsson L, Grundemar L (1999) Functional effects of neuropeptide Y receptors on blood flow and nitric oxide levels in the human nose. *Am J Respir Crit Care Med* 160(5 Pt 1): 1724–1728

26 Baraniuk JN, Silver PB, Kaliner MA, Barnes PJ (1992) Neuropcptide Y is a vasoconstrictor in human nasal mucosa. *J Appl Physiol* 73(5): 1867–1872

27 Widdicombe JG (1990) The NANC system and airway vasculature. Arch Int Pharmacodyn Ther 303: 83–99

28 Laitinen LA, Laitinen A, Salonen RO, Widdicombe JG (1987) Vascular actions of airway neuropeptides. *Am Rev Respir Dis* 136(6 Pt 2): S59–S64
29 Salonen RO, Webber SE, Widdicombe JG (1988) Effects of neuropeptides and capsaicin on the canine tracheal vasculature *in vivo*. *Br J Pharmacol* 95(4): 1262–1270
30 Matran R (1991) Neural control of lower airway vasculature. Involvement of classical transmitters and neuropeptides. *Acta Physiol Scand* (Suppl) 601: 1–54
31 Franco-Cereceda A, Matran R, Alving K, Lundberg JM (1995) Sympathetic vascular control of the laryngeo-tracheal, bronchial and pulmonary circulation in the pig: evidence for non-adrenergic mechanisms involving neuropeptide Y. *Acta Physiol Scand* 155(2): 193–204
32 Mullol J, Rieves RD, Baraniuk JN, Lundgren JD, Merida M, Hausfeld JH, Shelhamer JH, Kaliner MA (1992) The effects of neuropeptides on mucous glycoprotein secretion from human nasal mucosa *in vitro*. *Neuropeptides* 21(4): 231–238
33 Lacroix JS, Mosimann BL (1996) Attenuation of allergen-evoked nasal responses by local pretreatment with exogenous neuropeptide Y in atopic patients. *J Allergy Clin Immunol* 98(3): 611–616
34 Michel MC, Beck-Sickinger A, Cox H, Doods HN, Herzog H, Larhammar D, Quirion R, Schwartz T, Westfall T (1998) XVI. International Union of Pharmacology recommendations for the nomenclature of neuropeptide Y, peptide YY, and pancreatic polypeptide receptors. *Pharmacol Rev* 50(1): 143–150
35 Grundemar L (1997) Characterization of the receptor response for the neuropeptide Y-evoked suppression of parasympathetically-mediated contractions in the guinea pig trachea. *Regul Pept* 71(2): 97–101
36 Landis BN, Grouzman E, Lacroix JS (2005) Modulation of histamine induced bronchoconstriction after DPP IV pretreatment. *Regul Pept; in press*
37 Pernow J, Lundberg JM, Kaijser L, Hjemdahl P, Theodorsson-Norheim E, Martinsson A, Pernow B (1986) Plasma neuropeptide Y-like immunoreactivity and catecholamines during various degrees of sympathetic activation in man. *Clin Physiol* 6(6): 561–578
38 Richerson HB, Seebohm PM (1968) Nasal airway response to exercise. *J Allergy* 41(5): 269–284
39 Dallimore NS, Eccles R (1977) Changes in human nasal resistance associated with exercise, hyperventilation and rebreathing. *Acta Otolaryngol* 84(5–6): 416–421
40 Lacroix JS, Correia F, Fathi M, Grouzmann E (1997) Post-exercise nasal vasoconstriction and hyporeactivity: possible involvement of neuropeptide Y. *Acta Otolaryngol* 117(4): 609–613
41 Stretton CD, Barnes PJ (1988) Modulation of cholinergic neurotransmission in guinea-pig trachea by neuropeptide Y. *Br J Pharmacol* 93(3): 672–678
42 Matran R, Martling CR, Lundberg JM (1989) Inhibition of cholinergic and non-adrenergic, non-cholinergic bronchoconstriction in the guinea pig mediated by neuropeptide Y and alpha 2-adrenoceptors and opiate receptors. *Eur J Pharmacol* 163(1): 15–23
43 Lacroix JS, Ulman LG, Potter EK (1994) Sympathetic and parasympathetic interaction

in vascular control of the nasal mucosa in anaesthetized cats. *J Physiol* 480(Pt 2): 325–331

44 Potter EK, Barden JA, McCloskey MJ, Selbie LA, Tseng A, Herzog H, Shine J (1994) A novel neuropeptide Y analog, N-acetyl [Leu28,Leu31]neuropeptide Y-(24-36), with functional specificity for the presynaptic (Y2) receptor. *Eur J Pharmacol* 267(3): 253–262

45 Lacroix JS, Ulman LG, Potter EK (1994) Modulation by neuropeptide Y of parasympathetic nerve-evoked nasal vasodilatation via Y2 prejunctional receptor. *Br J Pharmacol* 113(2): 479–484

46 Herzog H, Hort YJ, Ball HJ, Hayes G, Shine J, Selbie LA (1992) Cloned human neuropeptide Y receptor couples to two different second messenger systems. *Proc Natl Acad Sci USA* 89(13): 5794–5798

47 Revington M, Lacroix JS, Potter EK (1997) Sympathetic and parasympathetic interaction in vascular and secretory control of the nasal mucosa in anaesthetized dogs. *J Physiol* 505(Pt 3): 823–831

48 Stretton CD, Belvisi MG, Barnes PJ (1990) Neuropeptide Y modulates non-adrenergic, non-cholinergic neural bronchoconstriction *in vivo* and *in vitro*. *Neuropeptides* 17(4): 163–170

49 Benchekroun TM, Fournier A, St-Pierre S, Cadieux A (1992) Inhibitory action of neuropeptide Y on agonist-induced responses in isolated guinea pig trachea. *Eur J Pharmacol* 216(3): 421–428

50 Grouzmann E, Buclin T, Martire M, Cannizzaro C, Dorner B, Razaname A, Mutter M (1997) Characterization of a selective antagonist of neuropeptide Y at the Y2 receptor. Synthesis and pharmacological evaluation of a Y2 antagonist. *J Biol Chem* 272(12): 7699–7706

51 Malis DD, Grouzmann E, Morel DR, Mutter M, Lacroix LS (1999) Influence of TASP-V, a novel neuropeptide Y (NPY) Y2 agonist, on nasal and bronchial responses evoked by histamine in anaesthetized pigs and in humans. *Br J Pharmacol* 126(4): 989–996

52 Takahashi T, Ichinose M, Yamauchi H, Miura M, Nakajima N, Ishikawa J, Inoue H, Takishima T, Shirato K (1993) Neuropeptide Y inhibits neurogenic inflammation in guinea pig airways. *J Appl Physiol* 75(1): 103–107

53 Nadelhaft I, Vera PL (1991) Neurons labelled after the application of tracer to the distal stump of the transected hypogastric nerve in the rat. *J Auton Nerv Syst* 36(2): 87–96

54 Fujiwara H, Kurihara N, Hirata K, Ohta K, Kanazawa H, Takeda T (1993) Effect of neuropeptide Y on human bronchus and its modulation of neutral endopeptidase. *J Allergy Clin Immunol* 92(1 Pt 1): 89–94

55 Lacroix JS, Kurt AM, Pochon N, Bretton C, Lundberg JM, Deshusses J (1995) Neutral endopeptidase activity and concentration of sensory neuropeptide in the human nasal mucosa. *Eur Arch Otorhinolaryngol* 252(8): 465–468

56 Grouzmann E, Monod M, Landis B, Wilk S, Brakch N, Nicoucar K, Giger R, Malis D, Szalay-Quinodoz I, Cavadas C, et al. (2002) Loss of dipeptidylpeptidase IV activity in

chronic rhinosinusitis contributes to the neurogenic inflammation induced by substance P in the nasal mucosa. *FASEB J* 16(9): 1132–1134

57 Chung F (2001) Anti-inflammatory cytokines in asthma and allergy: interleukin-10, interleukin-12, interferon-gamma. *Mediators Inflamm* 10(2): 51–59

58 Kawamura N, Tamura H, Obana S, Wenner M, Ishikawa T, Nakata M, Yamamoto H (1998) Differential effects of neuropeptides on cytokine production by mouse helper T cell subsets. *Neuroimmunomodulation* 5(1–2): 9–15

59 Bedoui S, Kawamura N, Straub RH, Pabst R, Yamamura T, von Horsten S (2003) Relevance of neuropeptide Y for the neuroimmune crosstalk. *J Neuroimmunol* 134(1–2): 1–11

60 Levite M (1998) Neuropeptides, by direct interaction with T cells, induce cytokine secretion and break the commitment to a distinct T helper phenotype. *Proc Natl Acad Sci USA* 95(21): 12544–12549

61 Levite M (2000) Nerve-driven immunity. The direct effects of neurotransmitters on T-cell function. *Ann NY Acad Sci* 917: 307–321

62 Bradding P (2003) The role of the mast cell in asthma: a reassessment. *Curr Opin Allergy Clin Immunol* 3(1): 45–50

63 Groneberg DA, Welker P, Fischer TC, Dinh QT, Grutzkau A, Peiser C, Wahn U, Henz BM, Fischer A (2003) Down-regulation of vasoactive intestinal polypeptide receptor expression in atopic dermatitis. *J Allergy Clin Immunol* 111(5): 1099–1105

64 Mousli M, Trifilieff A, Pelton JT, Gies JP, Landry Y (1995) Structural requirements for neuropeptide Y in mast cell and G protein activation. *Eur J Pharmacol* 289(1): 125–133

65 Emadi-Khiav B, Mousli M, Bronner C, Landry Y (1995) Human and rat cutaneous mast cells: involvement of a G protein in the response to peptidergic stimuli. *Eur J Pharmacol* 272(1): 97–102

66 Shen GH, Grundemar L, Zukowska-Grojec Z, Hakanson R, Wahlestedt C (1991) C-terminal neuropeptide Y fragments are mast cell-dependent vasodepressor agents. *Eur J Pharmacol* 204(3): 249–256

67 Cross LJ, Beck-Sickinger AG, Bienert M, Gaida W, Jung G, Krause E, Ennis M (1996) Structure activity studies of mast cell activation and hypotension induced by neuropeptide Y (NPY), centrally truncated and C-terminal NPY analogues. *Br J Pharmacol* 117(2): 325–332

68 Grundemar L, Krstenansky JL, Hakanson R (1994) Neuropeptide Y and truncated neuropeptide Y analogs evoke histamine release from rat peritoneal mast cells. A direct effect on G proteins? *Eur J Pharmacol* 258(1–2): 163–166

69 Barnes PJ (2001) Cytokine modulators as novel therapies for airway disease. *Eur Respir J* Suppl 34: 67s–77s

70 Barnes PJ (2001) Cytokine modulators for allergic diseases. *Curr Opin Allergy Clin Immunol* 1(6): 555–560

71 Bowler RP, Crapo JD (2002) Oxidative stress in airways: is there a role for extracellular superoxide dismutase? *Am J Respir Crit Care Med* 166(12 Pt 2): S38–S43

72 Henricks PA, Nijkamp FP (2001) Reactive oxygen species as mediators in asthma. *Pulm Pharmacol Ther* 14(6): 409–420

73 De la Fuente M, Del Rio M, Medina S (2001) Changes with aging in the modulation by neuropeptide Y of murine peritoneal macrophage functions. *J Neuroimmunol* 116(2): 156–167

74 Dimitrijevic M, Stanojevic S, Vujic V, Kovacevic-Jovanovic V, Beck-Sickinger A, Demuth H, von Horsten S (2002) Effect of neuropeptide Y on inflammatory paw edema in the rat: involvement of peripheral NPY Y1 and Y5 receptors and interaction with dipeptidyl-peptidase IV (CD26). *J Neuroimmunol* 129(1–2): 35–42

75 von Horsten S, Ballof J, Helfritz F, Nave H, Meyer D, Schmidt RE, Stalp M, Klemm A, Tschernig T, Pabst R (1998) Modulation of innate immune functions by intracerebroventricularly applied neuropeptide Y: dose and time dependent effects. *Life Sci* 63(11): 909–922

76 von Horsten S, Exton NG, Exton MS, Helfritz F, Nave H, Ballof J, Stalp M, Pabst R (1998) Brain NPY Y1 receptors rapidly mediate the behavioral response to novelty and a compartment-specific modulation of granulocyte function in blood and spleen. *Brain Res* 806(2): 282–286

77 von Horsten S, Nave H, Ballof J, Helfritz F, Meyer D, Schmidt RE, Stalp M, Exton NG, Exton MS, Straub RH, et al. (1998) Centrally applied NPY mimics immunoactivation induced by non-analgesic doses of met-enkephalin. *Neuroreport* 9(17): 3881–3885

78 Song C, Leonard BE (1998) Comparison between the effects of sigma receptor ligand JO 1784 and neuropeptide Y on immune functions. *Eur J Pharmacol* 345(1): 79–87

79 Straub RH, Schaller T, Miller LE, von Horsten S, Jessop DS, Falk W, Scholmerich J (2000) Neuropeptide Y cotransmission with norepinephrine in the sympathetic nerve-macrophage interplay. *J Neurochem* 75(6): 2464–2471

80 Howarth PH, Springall DR, Redington AE, Djukanovic R, Holgate ST, Polak JM (1995) Neuropeptide-containing nerves in endobronchial biopsies from asthmatic and nonasthmatic subjects. *Am J Respir Cell Mol Biol* 13(3): 288–296

81 Chanez P, Springall D, Vignola AM, Moradoghi-Hattvani A, Polak JM, Godard P, Bousquet J (1998) Bronchial mucosal immunoreactivity of sensory neuropeptides in severe airway diseases. *Am J Respir Crit Care Med* 158(3): 985–990

82 Dahlof C, Dahlof P, Lundberg JM, Strombom U (1988) Elevated plasma concentration of neuropeptide Y and low level of circulating adrenaline in elderly asthmatics during rest and acute severe asthma. *Pulm Pharmacol* 1(1): 3–6

NPY analgesia: moving from acute to chronic pain

Bradley K. Taylor

Department of Pharmacology, SL83, Tulane University Health Sciences Center, New Orleans, LA 70112, USA

Previous reviews have described the behavioral effects of NPY in models of acute pain [1], as well as changes in NPY and NPY receptor gene expression in pain pathways following tissue or nerve injury [2]. Particularly striking is the massive upregulation of NPY in the cell bodies and central terminals of sensory neurons. This chapter focuses on the functional relevance of the NPY system to the development and modulation of chronic pain. Emphasis is given to recent studies demonstrating that agonists and antagonists at NPY receptors modulate the behavioral, biochemical, and molecular markers of inflammatory and neuropathic pain. The results indicate that the development of novel NPY agonists should yield a powerful new pharmacotherapy for the treatment of chronic pain.

Introduction

Normal physiological pain is an early warning system that helps to prevent or minimize tissue damage in the face of potential bodily harm. Also referred to as acute or transient pain, this signaling system elicits a coordinated response constellation, including behavioral withdrawal from the threatening stimulus [3]. An arsenal of analgesic drugs for the treatment of this type of pain is readily available, and includes aspirin, acetaminophen, anti-inflammatory drugs such as ibuprofen, and opioids such as morphine. Much more difficult, however, is the treatment of pathological (chronic) pain. Thus, severe pain that is chronic (lasting more than 6 months) represents a major complication associated with tissue or neuronal injury, and affects an enormous percentage of our population. Clinical examples include migraine (~10% of the population), severe back pain (~10%), chronic widespread pain (~13%), fibromyalgia (~15%), and neuropathic pain (~3%) [4]. Chronic pain reduces the patient's ability to work, walk, or sleep [5], and is associated with many deleterious physiological effects. Over the past two decades, the development of animal models of arthritic and neuropathic pain that closely mimic their respective human conditions has led to an explosion of research on the pathophysiological neural changes underlying chronic pain [6]. Thus, it has become clear that tissue or

The NPY Family of Peptides in Immune Disorders, Inflammation, Angiogenesis and Cancer,
edited by Zofia Zukowska and Giora Z. Feuerstein

nerve injury produce dramatic alterations in the gene expression of peptidergic neurotransmitters and receptors at the level of the primary afferent neuron, spinal cord dorsal horn, and brain. Such plastic changes are thought to take the subject from a state of acute pain to one of chronic pain [27]. As reviewed previously, a profound example of plasticity involves neuropeptide Y and the NPY Y1 receptor [2, 7]. Both are highly expressed at key sites of pain transmission, including lamina II of the spinal cord dorsal horn [8–10], and tissue or nerve injury dramatically alters their expression profiles. These anatomical findings position the NPY system as a key player in the development of pathological pain. This chapter begins with a review of the behavioral pharmacology of NPY in models of acute pain, and then discusses more recent findings in models of inflammatory and neuropathic pain.

NPY in animal models of acute pain

Intrathecal NPY receptor agonists inhibit acute nociception

The primary afferent neuron relays information from the periphery to the CNS. Noxious somatosensory stimuli activate such neurons, leading to the synaptic release of pronociceptive neurotransmitters such as glutamate, substance P (SP) and calcitonin gene-related peptide (CGRP), in the outer layers (laminae I and II) of the dorsal horn of the spinal cord. To target this site in awake behaving rodents, pharmacological studies typically use the intrathecal route of administration [11]. In recent years, we and others have repeatedly shown that intrathecal administration of NPY receptor agonists reduces behavioral responsiveness to noxious heat (antinociception), an effect that is attenuated in Y1 deletion mutant mice [12–14]. A single injection can produce antinociceptive actions for 3 h, suggesting that NPY is either resistant to peptidase cleavage in the cerebrospinal fluid, is cleaved into smaller fragments that also inhibit spinal nociceptive transmission, or that NPY itself induces long-lasting neuronal effects [12, 14]. These effects of NPY are not likely secondary to non-specific behavioral effects or to changes in local thermal regulation, since even the highest dose of NPY (30 μg) affects neither motor coordination nor paw skin temperature in the rat [14]. In anesthetized rats, intrathecal NPY decreases nociception as assessed with a flexor reflex, but only at higher doses [15, 16]. The paradoxical observation that lower doses of NPY *increase* the flexor reflex requires further study. Albeit, the majority of behavioral studies in awake rodents indicate that agonists at Y1 and/or Y2 receptors inhibit the spinal processing of noxious stimuli.

The formalin test is a key model system of acute ongoing pain that is frequently used to initially validate novel analgesic systems. This is because it evokes a constellation of very robust behavioral, neural and physiological responses that are easy to detect and quantify [17, 18]. Our laboratory determined that intrathecal administration of NPY dose-dependently inhibited the flinching, licking, pressor, and tachycardia responses associated with the intraplantar injection of dilute formalin [19]. This is accompanied by a reduction in formalin-evoked expression of Fos immunoreactivity in the dorsal horn [20]. Furthermore, Y1 deletion mutant mice exhibit exaggerated behavioral signs of nociception following the administration of noxious formalin, acetic acid, or magnesium sulfate [13]. Taken together, these studies suggest that NPY acts at Y1 receptors to inhibit spinal nociceptive neurons, leading to analgesic actions.

Supraspinal NPY produces mixed effects on acute nociception

NPY and the NPY Y1 receptor are expressed in multiple brain regions thought to modulate pain [21–23]. Early studies using intracerebroventricular administration of NPY reported contradictory findings: some described antinociception, while others described pronociception or no effect [1]. More recent studies have begun to resolve these findings with the use of intracranial injection into discreet brain regions. For example, microinjection of NPY into the rostral ventral medulla, periaqueductal gray, or nucleus accumbens *decreases* withdrawal reflexes to noxious heat or mechanical stimuli [24–26]. Conversely, NPY into the nucleus gracilis *increases* behavioral responsiveness to low-intensity mechanical stimulation, possibly reflective of a pronociceptive action [27]. Such opposing actions of NPY at different brain regions provide one explanation as to why intracerebroventricular administration, which delivers NPY to multiple brain regions, exerts inconsistent actions. Further studies using selective antagonists are required to address an alternative explanation based on receptor subtype.

NPY in animal models of chronic inflammatory pain

Newer pain models include the intraplantar administration of inflammogens such as carrageenan or complete Freund's adjuvant (CFA). These injections lead not only to tissue injury and inflammation, but also to the development of allodynia (hypersensitivity to non-noxious stimuli) and hyperalgesia (increased response to noxious stimuli) that lasts days to weeks [3]. Such hypersensitivity is typically manifested as an increase in responsiveness to thermal or mechanical stimuli, and is also observed in the setting of clinical pain.

NPY inhibits behavioral signs of inflammatory pain

To test the hypothesis that intrathecal administration of NPY inhibits inflammatory hyperalgesia, our laboratory evaluated the effects of NPY on thermal and mechanical hypersensitivity in the carrageenan and CFA models. We found that intrathecal NPY but not saline produced a long-lasting and dose-dependent inhibition of thermal and mechanical hypersensitivity [14, 28]. This data supports the idea that inflammation enhances sensitivity to the inhibitory effects of NPY [14]. Furthermore, because BIBO 3304, a Y1 receptor antagonist, almost completely reversed the inhibitory effect of NPY in the complete Freund's adjuvant model, we concluded that spinal Y1 receptors contribute to the inhibitory effects of NPY. Similarly, intraplantar administration of the Y1 agonist, Leu^{31}-Pro^{36}-NPY, inhibited the mechanical hypersensitivity evoked by locally-administered capsaicin, the pungent substance of hot chili peppers [Gibbs and Hargreaves, in preparation]. Thus, exogenous administration of NPY, either peripherally or spinally, reduces behavioral signs of inflammatory pain.

We also tested the hypothesis that *endogenous* activation of the Y1 receptor tonically inhibits inflammatory hyperalgesia, and reported that intrathecal administration of BIBO 3304 (in the absence of NPY) increased thermal hypersensitivity to CFA [14]. Furthermore, Y1 deletion mutant mice displayed exaggerated behavioral responses to the intraplantar administration of carrageenan [13]. Taken together, these data indicate that *endogenous* Y1 receptor activation contributes to inflammatory hyperalgesia. The primary afferent neuron is not a plausible source of endogenously-released NPY, because immunoreactivity and mRNA is virtually absent in DRG neurons, even during inflammation [8, 9, 29, 30]. Furthermore, the spontaneous release of NPY is not significantly altered by electrical stimulation of peripheral nerves [10]. A more likely source is the NPY-containing cell bodies of interneurons in lamina I–III [8, 31]. Indeed, many of these interneurons also express GABA and form synaptic contacts with pain-related neurons in the dorsal horn [32]. With respect to a possible source of NPY from supraspinal sites, it is not known whether the sparse NPY-containing projection from bulbospinal neurons might increase and become more predominant following injury [33]. But irrespective of the source of spinal NPY, an important overall hypothesis is that injury-induced upregulation of NPY in the spinal cord serves a vital purpose: to inhibit pronociceptive neurotransmission, and therefore limit the progression of chronic pain.

Inflammation upregulates NPY and the Y1 receptor

Groundbreaking studies by Hokfelt and colleagues have clearly demonstrated that inflammation is associated with increased NPY and Y1 receptor expression in the dorsal horn (Fig. 1). First, using *in situ* hybridization and immunohistochemistry, they demonstrated strong expression of Y1 mRNA and protein in deep lamina II neurons, indicating the presence of Y1 receptors in areas of the spinal cord related to nociceptive transmission [34]. Second, they found that unilateral injection of CFA markedly increased Y1 mRNA in the ipsilateral spinal dorsal horn laminae II and

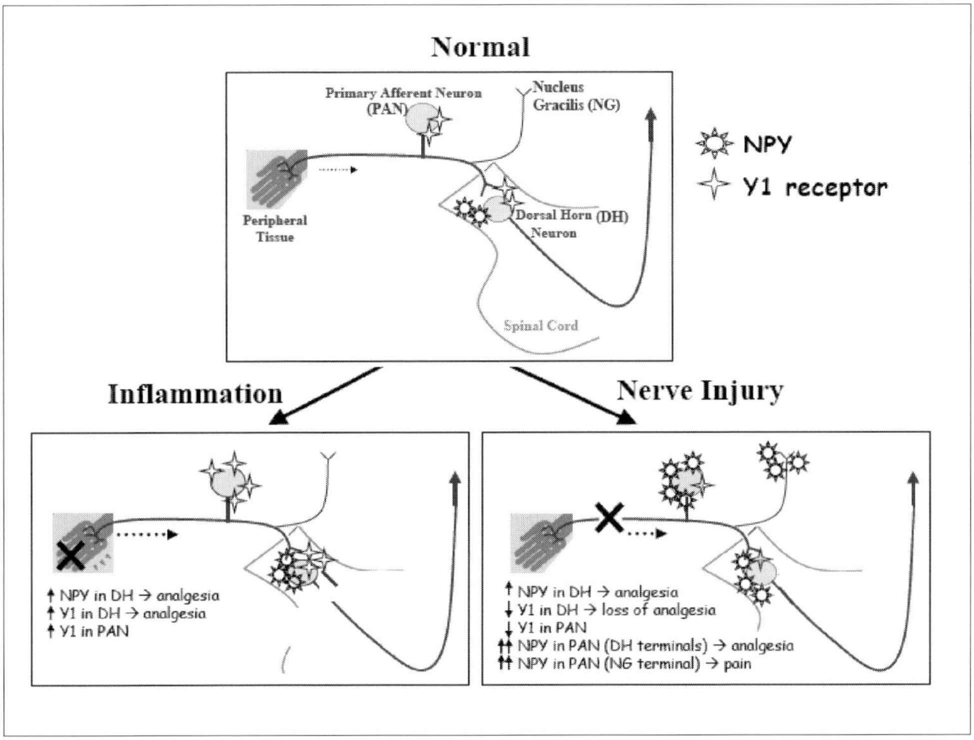

Figure 1

Expression of NPY and Y1 receptors in the somatosensory system after inflammation or nerve injury. Inflammation increases the intensity of NPY and Y1 immunostaining in the dorsal horn. Because spinal NPY-Y1 transmission is predominantly inhibitory, these changes likely lead to feedback inhibition of inflammatory pain (analgesia). Similarly, peripheral nerve injury robustly increases NPY staining in the primary afferent neurons in the dorsal root ganglia (DRG), their central terminals within the dorsal horn, and probably in spinal interneurons as well. These changes likely lead to feedback inhibition of neuropathic pain. The loss of Y1 staining in the dorsal horn may represent loss of feedback inhibition, and ultimately exaggerated neuropathic pain. Nerve injury-induced up-regulation of NPY in the DRG is associated with the axon terminal de novo *expression of NPY in the nucleus gracilis, which, opposite to the dorsal horn, likely facilitates the behavioral signs of neuropathic pain. The functional relevance of Y1 receptor plasticity in the dorsal root ganglia to pain control is not at all clear.*

III, indicating a greater number of targets for NPY in the presence of inflammation [9]. Third, they found that CFA increased NPY mRNA and NPY immunoreactivity in lamina II of ipsilateral dorsal horn neurons [9, 34, but see 29]. The greater number of Y1 receptor targets on the inflamed side could result in the enhanced behavioral sensitivity to the inhibitory effects of NPY [9]. Taken together, behavioral and immunohistochemical studies support the idea that inflammation strengthens inhibitory NPY tone. Further studies are necessary to determine whether enhanced release of NPY and Y1-mediated inhibition of spinal nociceptive transmission ultimately results in a compensatory, adaptive inhibition of pain in the setting of inflammation [14].

NPY in animal models of neuropathic pain

The quality and pattern of altered sensitivity in neuropathic pain clearly differs from transient or inflammatory pain. For example, a cold stimulus such as ice may reduce inflammatory pain in the normal person, but produce excruciating pain in the neuropathic pain patient. These qualitative differences in sensation suggest that nerve injury leads to a *reorganization* of sensory transmission pathways that persist long after the normal healing period. Recent data indicates that alterations in the neural NPY systems contribute to such reorganization. These alterations include biochemical, anatomical, and physiological changes in the somatosensory system at the level of the primary afferent neuron, spinal cord dorsal horn, and brain. Intense research is determining the relative contributions of these reactive changes to neuropathic pain.

Mixed effects of NPY on behavioral signs of neuropathic pain

Y1 receptor deletion mutant mice display exaggerated behavioral signs of allodynia after nerve injury, suggesting that endogenous NPY acts at Y1 receptors to prevent the development of neuropathic pain [13]. Although these conclusions should be viewed with caution because the differences were small and possibly confounded by altered baseline sensitivity to thermal stimuli, they are supported by our findings that intrathecal administration of NPY dose-dependently inhibited allodynia and hyperalgesia [35]. Furthermore, microinjection of NPY to pain modulatory centers such as the rostral ventral medulla (RVM) reduces signs of allodynia and hyperalgesia following nerve injury [36]. On the other hand, NPY appears to mediate the development of *pro*-nociceptive actions at the dorsal column nuclei of the somatosensory system [27]. Thus, lumbar peripheral nerve injury massively upregulates the expression of NPY in the nucleus gracilis (Fig. 1). Importantly, intracranial administration of anti-NPY antiserum or BIBO 3304 into the nucleus gracilis reverses nerve injury-induced mechanical allodynia. Whether these opposing actions of NPY can be overcome as NPY agonists are developed for the treatment of chronic pain is an important consideration.

Nerve injury dramatically alters the expression of NPY and NPY receptors

The dorsal column/medial lemniscus pathway normally mediates the transmission of non-noxious stimuli, but this system undergoes profound plasticity after peripheral nerve injury. As illustrated in Figure 1, most striking is the massive upregulation of NPY in the cell bodies of large primary afferent neurons of the DRG and trigeminal ganglion [37–42]. These changes are accompanied by increased NPY expression in the central terminals of these neurons, at laminae III–IV [9, 41–43] and dorsal column nuclei [27, 44]. Because NPY levels at these sites are very low or absent in the normal animal (or in animals with inflammation), such plasticity is viewed as a switch in "neurochemical signature" [3]. A key unresolved question relates to the significance of this plasticity to pain modulation: is there a temporal relationship between these changes in gene expression and the magnitude of allodynia or hyperalgesia? Some studies support such a correlation [45–47], while others do not [48, 49].

Similar questions with regards to plasticity of NPY receptors after nerve injury also remain unanswered. Early competition studies with Y1 and Y2 agonists suggested that both receptors contribute binding sites to lamina I–IV of the dorsal horn, and the number of these sites *increase* after axotomy [50]. More definitive studies using immunohistochemical techniques, however, indicate that the dorsal horn expression of Y1 protein actually *decreases* after partial or complete sciatic nerve injury [47]. The functional significance of NPY upregulation in the face of decreased Y1 receptor immunoreactivity requires a greater understanding of the effect of nerve injury on Y1 receptor transport, downregulation and internalization.

In contrast to the spinal cord, the function of nerve-injury induced upregulation of NPY in the dorsal column nuclei is less ambiguous. A particularly intriguing idea is that upregulation of NPY in the nucleus gracilis after lumbar spinal nerve damage selectively mediates tactile allodynia [27]. After nerve injury, intracranial administration of NPY antiserum or the Y1 receptor antagonist BIBO3304 reversed behavioral signs of tactile allodynia. While future studies are needed to demonstrate NPY release and the presence of Y1 receptors in the nucleus gracilis, these results raise the possibility that a Y1 receptor *antagonist* will reduce clinical neuropathic pain. This contrasts with the bulk of the data in the spinal cord indicating that receptor *agonists* will reduce clinical inflammatory pain (but see [36]).

Presynaptic mechanisms of the spinal analgesic actions of NPY

Intrathecal administration of NPY inhibits the activity of spinal nociresponsive neurons, but by what neurochemical mechanism? One possibility is that NPY acts postsynaptically to inhibit the activity of spinal pain transmission neurons. In support of this, dorsal horn neurons expressing a key marker of nociceptive neurons, the neurokinin 1 receptor ("SP receptor"), receive synaptic contacts from terminals containing NPY [51]. However, as illustrated in Figure 2, a more substantial literature indicates that NPY acts presynaptically to inhibit the release of nociceptive transmitters from primary afferent neurons. First, NPY binding sites decorate the cell bodies of small diameter primary afferent neurons that contain nociceptive neuropeptides such as SP and CGRP [52, 53]. These binding sites appear to be transported centrifugally from the dorsal root ganglia to the central terminals of primary afferent neurons, since isolation of these terminals from the DRG by dorsal rhizotomy decreases the number of dorsal horn binding sites [53, 54]. Second, in dorsal root ganglion neurons, NPY reduces an essential requirement for neurotransmitter release: Ca^{2+} conductance [55]. Third, NPY acts via presynaptic Y2 receptors to attenuate excitatory postsynaptic currents (EPSC) in spinal cord slices [56]. Fourth, NPY inhibits the potassium- or capsaicin-evoked release of immunoreactive SP and CGRP from primary afferent neurons [55, 57, 58]. Importantly, NPY inhibits the *in vivo* release of immunoreactive SP in anesthetized rats [59]. Our laboratory has recently confirmed these results with in vivo microdialysis of spinal substance P and functional anatomical studies assessing internalization of the NK1 receptor [28].

At which receptor subtype does NPY exert its presynaptic effects in the dorsal horn? Of the four NPY receptors cloned in the rat, Y1, Y2, Y4, and Y5 [2, 60], most attention has been paid to those expressed in ascending pain pathways: Y1 and Y2. These receptors have been generally regarded to serve postsynaptic and presynaptic functions, respectively [61], and therefore one would expect the presynaptic Y2 receptor to be the primary target of NPY in the dorsal horn. In support of this idea, Y2 mRNA is found on the cell bodies of DRG neurons [62], and therefore might be transported to the central terminals of primary afferents. Furthermore, electrophysiological studies indicate that Y2 but not Y1 receptor agonists inhibit EPSCs in spinal cord slices, indicating a predominant role for the Y2 receptor in mediating the analgesic effects of NPY [56].

At the spinal cord, however, the current dogma indicating an exclusive role for the Y2 receptor in the presynaptic effects of NPY must be reconsidered. In particular, the majority of the behavioral studies described above indicate that it is the Y1 receptor that mediates the antiallodynic and antihyperalgesic actions of NPY, likely by inhibiting neurotransmitter release [13, 14, 28, 59]. Furthermore, the Y1 receptor is largely co-localized with selective markers of primary afferent neurons, CGRP and TRPV1, and so is poised to modulate the release of pronociceptive neurotransmitters [53, 58]. Indeed, the Y1 antagonist, BIBP3226, reverses the inhibitory effect of NPY on CGRP release in spinal cord slices [58]. Similar experiments with newly-developed Y2 receptor antibodies, siRNA, deletion mutant mice, and receptor antagonists will undoubtedly contribute to a better understanding of the contribu-

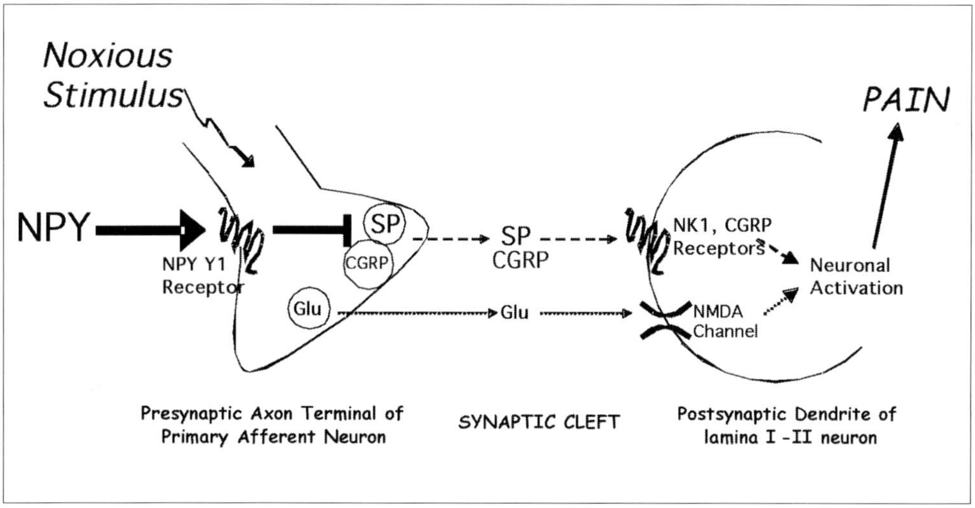

Figure 2
Proposed mechanism for the analgesic actions of NPY. A noxious somatosensory stimulus (wavy line) depolarizes the presynaptic axon terminal of unmyelinated or thinly myelinated primary afferent neurons. This evokes the release of pronociceptive peptides (e.g., substance P and calcitonin gene related peptide, dashed lines) and amino acids (e.g., glutamate, dotted lines) into the synaptic cleft. Upon reaching their binding sites (e.g., neurokinin-1 and CGRP receptors and NMDA channels) on nociresponsive neurons in the superficial lamina of the dorsal horn, various intracellular events lead to activation and depolarization, ultimately triggering the transmission of nociceptive signals to the brain. As illustrated by the bold arrow and bold line, recent data indicates that exogenously applied NPY acts at presynaptic NPY Y1 receptors to inhibit peptide release, decrease the activation of dorsal horn neurons, and thus reduce the behavioral signs of allodynia and hyperalgesia.

tion of the Y2 receptor to the antinociceptive and antiallodynic effects of NPY. These and other studies will test the important overall hypothesis that injury-induced downregulation of NPY receptors in the spinal cord represent a loss of spinal nociceptive inhibition, and therefore contribute to the progression of chronic pain. Exogenous administration of NPY receptor agonists may overcome these deficiencies, and thus become an important class of pharmacotherapeutics for the management of chronic pain.

Acknowledgements
This work was supported by grant NS45954 to BKT.

References

1 Wettstein JG, Earley B, Junien JL (1995) Central nervous system pharmacology of neuropeptide Y. *Pharmacol Ther* 65(3): 397–414

2 Hokfelt T, Zhang X, Xu ZQ, Ji RR, Shi T, Corness J, Kerekes N, Landry M, Holmberg K, Broberger C (1997) Phenotypic regulation in dorsal root ganglion neurons after nerve injury: focus on peptides and their receptors. In: D Borsook (ed): *Molecular Neurobiology of Pain*, Vol. 9: 115–143. IASP Press, Seattle, WA

3 Taylor BK (2001) Pathophysiologic mechanisms of neuropathic pain. *Curr Pain Headache Rep* 5(2): 151–61

4 Crombie IK, Croft PR, Linton SJ, LeResche L, van Korff M (1999) *Epidemiology of Pain*. IASP Press, Seattle, WA

5 Wall PD, Melzack R (1999) *Textbook of Pain*. Churchill Livingstone, New York, NY

6 Woolf CJ, Salter MW (2000) Neuronal plasticity: increasing the gain in pain. *Science* 288(5472): 1765–1769

7 Hokfelt T, Zhang X, Wiesenfeld-Hallin Z (1994) Messenger plasticity in primary sensory neurons following axotomy and its functional implications. *Trends Neurosci* 17(1): 22–30

8 Gibson SJ, Polak JM, Allen JM, Adrian TE, Kelly JS, Bloom SR (1984) The distribution and origin of a novel brain peptide, neuropeptide Y, in the spinal cord of several mammals. *J Comparative Neurol* 227(1): 78–91

9 Ji RR, Zhang X, Wiesenfeld-Hallin Z, Hokfelt T (1994) Expression of neuropeptide Y and neuropeptide Y (Y1) receptor mRNA in rat spinal cord and dorsal root ganglia following peripheral tissue inflammation. *J Neurosci* 14(11 Pt 1): 6423–6434

10 Mark MA, Jarrott B, Colvin LA, MacMillan SJ, Duggan AW (1997) The release of immunoreactive neuropeptide Y in the spinal cord of the anaesthetized rat and cat. *Brain Res* 754(1–2): 195–203

11 Yaksh TL, Rudy TA (1976) Chronic catheterization of the spinal subarachnoid space. *Physiol Behav* 17: 1031–1036

12 Hua XY, Boublik JH, Spicer MA, Rivier JE, Brown MR, Yaksh TL (1991) The antinociceptive effects of spinally administered neuropeptide Y in the rat: systematic studies on structure-activity relationship. *J Pharmacol Exp Ther* 258(1): 243–248

13 Naveilhan P, Hassani H, Lucas G, Blakeman KH, Hao JX, Xu XJ, Wiesenfeld-Hallin Z, Thoren P, Ernfors P (2001) Reduced antinociception and plasma extravasation in mice lacking a neuropeptide Y receptor. *Nature* 409(6819): 513–517

14 Taiwo OB, Taylor BK (2002) Antihyperalgesic effects of intrathecal neuropeptide Y during inflammation are mediated by Y1 receptors. *Pain* 96(3): 353–363

15 Xu IS, Luo L, Ji RR, Hokfelt T, Xu XJ, Wiesenfeld-Hallin Z (1998) The effect of intrathecal neuropeptide Y on the flexor reflex in rats after carrageenan-induced inflammation. *Neuropeptides* 32(5): 447–452

16 Xu XJ, Hao JX, Hokfelt T, Wiesenfeld-Hallin Z (1994) The effects of intrathecal neu-

ropeptide Y on the spinal nociceptive flexor reflex in rats with intact sciatic nerves and after peripheral axotomy. *Neuroscience* 63(3): 817–826
17 Taylor BK, Akana SF, Peterson MA, Dallman MF, Basbaum AI (1998) Pituitary-adrenocortical responses to persistent noxious stimuli in the awake rat: endogenous corticosterone does not reduce nociception in the formalin test. *Endocrinology* 139(5): 2407–2413
18 Taylor BK, Peterson MA, Basbaum AI (1995) Persistent cardiovascular and behavioral nociceptive responses to subcutaneous formalin require peripheral nerve input. *J Neurosci* 15(11): 7575–7584
19 Mahinda TB, Taylor BK (2004) Intrathecal neuropeptide Y inhibits behavioral and cardiovascular responses to noxious inflammatory stimuli in awake rats. *Physiol Behav* 80(5): 703–711
20 Intondi AB, Dahlgren MN, Taylor BK (2004) Intrathecal neuropeptide Y reduces noxious stimulus-evoked Fos expression in the dorsal horn. In: *Society for Neuroscience, Washington, D.C.*, 2004, Program No. 864.10. Society for Neuroscience
21 Chronwall BM (1985) Anatomy and physiology of the neuroendocrine arcuate nucleus. *Peptides* 6 (Suppl 2): 1–11
22 Kopp J, Xu ZQ, Zhang X, Pedrazzini T, Herzog H, Kresse A, Wong H, Walsh JH, Hokfelt T (2002) Expression of the neuropeptide Y Y1 receptor in the CNS of rat and of wild-type and Y1 receptor knock-out mice. Focus on immunohistochemical localization. *Neuroscience* 111(3): 443–532
23 Wolak ML, DeJoseph MR, Cator AD, Mokashi AS, Brownfield MS, Urban JH (2003) Comparative distribution of neuropeptide Y Y1 and Y5 receptors in the rat brain by using immunohistochemistry. *J Comp Neurol* 464(3): 285–311
24 Li Y, Li JJ, Yu LC (2002) Anti-nociceptive effect of neuropeptide Y in the nucleus accumbens of rats: an involvement of opioid receptors in the effect. *Brain Res* 940(1–2): 69–78
25 Wang JZ, Lundeberg T, Yu L (2000) Antinociceptive effects induced by intra-periaqueductal grey administration of neuropeptide Y in rats. *Brain Res* 859(2): 361–363
26 Zhang Y-X, Lundeberg T, Yu L-C (2000) Involvement of neuropeptide Y and Y1 receptor in antinociception in nucleus raphe magnus of rats. *Regul Peptides* 95: 109–113
27 Ossipov MH, Zhang ET, Carvajal C, Gardell L, Quirion R, Dumont Y, Lai J, Porreca F (2002) Selective mediation of nerve injury-induced tactile hypersensitivity by neuropeptide Y. *J Neurosci* 22(22): 9858–9867
28 Taylor BK, Robinson CM, Afrah AW, Kuphal KE, Stiller C-O (2003) NPY reduces noxious stimulus-induced substance P release in the dorsal horn *in vivo*. *Society for Neuroscience* Abstracts 29
29 Wakisaka S, Kajander KC, Bennett GJ (1992) Effects of peripheral nerve injuries and tissue inflammation on the levels of neuropeptide Y-like immunoreactivity in rat primary afferent neurons. *Brain Res* 598(1–2): 349–352
30 Zhang X, Nicholas AP, Hokfelt T (1993) Ultrastructural studies on peptides in the dor-

sal horn of the spinal cord – I. Co-existence of galanin with other peptides in primary afferents in normal rats. *Neuroscience* 57(2): 365–384

31 Sasek CA, Elde RP (1985) Distribution of neuropeptide Y-like immunoreactivity and its relationship to FMRF-amide-like immunoreactivity in the sixth lumbar and first sacral spinal cord segments of the rat. *J Neurosci* 5(7): 1729–1739

32 Polgar E, Shehab SA, Watt C, Todd AJ (1999) GABAergic neurons that contain neuropeptide Y selectively target cells with the neurokinin 1 receptor in laminae III and IV of the rat spinal cord. *J Neurosci* 19(7): 2637–2646

33 Holets VR, Hokfelt T, Rokaeus A, Terenius L, Goldstein M (1988) Locus coeruleus neurons in the rat containing neuropeptide Y, tyrosine hydroxylase or galanin and their efferent projections to the spinal cord, cerebral cortex and hypothalamus. *Neuroscience* 24(3): 893–906

34 Zhang X, Bao L, Xu ZQ, Kopp J, Arvidsson U, Elde R, Hokfelt T (1994) Localization of neuropeptide Y Y1 receptors in the rat nervous system with special reference to somatic receptors on small dorsal root ganglion neurons. *Proc of Nat Acad Sci USA* 91(24): 11738–11742

35 Kuphal K, Pedrazzini T, Taylor BK (2001) Increased acute and inflammatory nociception and reduced spinal NPY inhibition of neuropathic nociception in neuropeptide Y (NPY) Y1 receptor knockout mice. In: *Society for Neurosciences 2001*, p. 1895. Society for Neurosciences

36 Abhyankar SS, Taylor BK (2002) NPY acts at Y1 receptors in the RVM to inhibit neuropathic pain. In: *Society for Neuroscience, Orlando, 2002*. Society for Neuroscience.

37 Benoliel R, Eliav E, Iadarola MJ (2001) Neuropeptide Y in the trigeminal ganglion following chronic constriction injury of the rat infraorbital nerve: is there correlation to somatosensory parameters? *Pain* 91: 111–121

38 Fukuoka T, Tokunaga A, Kondo E, Miki K, Tachibana T, Noguchi K (1998) Change in mRNAs for neuropeptides and the GABA(A) receptor in dorsal root ganglion neurons in a rat experimental neuropathic pain model. *Pain* 78(1): 13–26

39 Nahin RL, Ren K, De Leon M, Ruda M (1994) Primary sensory neurons exhibit altered gene expression in a rat model of neuropathic pain. *Pain* 58(1): 95–108

40 Noguchi K, De Leon M, Nahin RL, Senba E, Ruda MA (1993) Quantification of axotomy-induced alteration of neuropeptide mRNAs in dorsal root ganglion neurons with special reference to neuropeptide Y mRNA and the effects of neonatal capsaicin treatment. *J Neurosci Res* 35(1): 54–66

41 Shi TJ, Cui JG, Meyerson BA, Linderoth B, Hokfelt T (1999) Regulation of galanin and neuropeptide Y in dorsal root ganglia and dorsal horn in rat mononeuropathic models: possible relation to tactile hypersensitivity. *Neuroscience* 93(2): 741–757

42 Wakisaka S, Kajander KC, Bennett GJ (1991) Increased neuropeptide Y (NPY)-like immunoreactivity in rat sensory neurons following peripheral axotomy. *Neurosci Lett* 124(2): 200–203

43 Stone LS, Vulchanova L, Riedl MS, Wang J, Williams FG, Wilcox GL, Elde R (1999)

Effects of peripheral nerve injury on alpha-2A and alpha-2C adrenergic receptor immunoreactivity in the rat spinal cord. *Neuroscience* 93(4): 1399–1407

44 Zhang X, Meister B, Elde R, Verge VM, Hokfelt T (1993) Large calibre primary afferent neurons projecting to the gracile nucleus express neuropeptide Y after sciatic nerve lesions: an immunohistochemical and in situ hybridization study in rats. *Eur J Neurosci* 5(11): 1510–1519

45 Benoliel R, Eliav E, Iadarola MJ (2001) Neuropeptide Y in trigeminal ganglion following chronic constriction injury of the rat infraorbital nerve: is there correlation to somatosensory parameters? *Pain* 91(1–2): 111–121

46 Munglani R, Bond A, Smith GD, Harrison SM, Elliot PJ, Birch PJ, Hunt SP (1995) Changes in neuronal markers in a mononeuropathic rat model relationship between neuropeptide Y, pre-emptive drug treatment and long-term mechanical hyperalgesia. *Pain* 63(1): 21–31

47 Brumovsky PR, Bergman E, Liu HX, Hokfelt T, Villar MJ (2004) Effect of a graded single constriction of the rat sciatic nerve on pain behavior and expression of immunoreactive NPY and NPY Y1 receptor in DRG neurons and spinal cord. *Brain Res* 1006(1): 87–99

48 Munglani R, Harrison SM, Smith GD, Bountra C, Birch PJ, Elliot PJ, Hunt SP (1996) Neuropeptide changes persist in spinal cord despite resolving hyperalgesia in a rat model of mononeuropathy. *Brain Res* 743(1–2): 102–108.

49 Kim HJ, Back SK, Kim J, Sung B, Hong SK, Na HS (2003) Increases in spinal vasoactive intestinal polypeptide and neuropeptide Y are not sufficient for the genesis of neuropathic pain in rats. *Neurosci Lett* 342(1–2): 109–113

50 Zhang X, Ji RR, Nilsson S, Villar M, Ubink R, Ju G, Wiesenfeld-Hallin Z, Hokfelt T (1995) Neuropeptide Y and galanin binding sites in rat and monkey lumbar dorsal root ganglia and spinal cord and effect of peripheral axotomy. *Eur J Neurosci* 7(3): 367–380

51 Polgar E, Fowler JH, McGill MM, Todd AJ (1999) The types of neuron which contain protein kinase C gamma in rat spinal cord. *Brain Res* 833(1): 71–80

52 Mantyh PW, Allen CJ, Rogers S, DeMaster E, Ghilardi JR, Mosconi T, Kruger L, Mannon PJ, Taylor IL, Vigna SR (1994) Some sensory neurons express neuropeptide Y receptors: potential paracrine inhibition of primary afferent nociceptors following peripheral nerve injury. *J Neurosci* 14(6): 3958–3968

53 Brumovsky PR, Shi TJ, Matsuda H, Kopp J, Villar MJ, Hokfelt T (2002) NPY Y1 receptors are present in axonal processes of DRG neurons. *Exp Neurol* 174(1): 1–10

54 Kar S, Quirion R (1992) Quantitative autoradiographic localization of [125I]neuropeptide Y receptor binding sites in rat spinal cord and the effects of neonatal capsaicin, dorsal rhizotomy and peripheral axotomy. *Brain Res* 574(1–2): 333–337

55 Walker MW, Ewald DA, Perney TM, Miller RJ (1988) Neuropeptide Y modulates neurotransmitter release and Ca^{2+} currents in rat sensory neurons. *J Neurosci* 8(7): 2438–2446

56 Moran TD, Colmers WF, Smith PA (2004) Opioid-like actions of neuropeptide Y in rat

substantia gelatinosa: Y1 suppression of inhibition and Y2 suppression of excitation. *J Neurophysiol* 92(6): 3266–3275

57 Bleakman D, Colmers WF, Fournier A, Miller RJ (1991) Neuropeptide Y inhibits Ca^{2+} influx into cultured dorsal root ganglion neurones of the rat via a Y2 receptor. *Br J Pharmacol* 103(3): 1781–1789

58 Gibbs J, Flores CM, Hargreaves KM (2004) Neuropeptide Y inhibits capsaicin-sensitive nociceptors via a Y1-receptor-mediated mechanism. *Neuroscience* 125(3): 703–709

59 Duggan AW, Hope PJ, Lang CW (1991) Microinjection of neuropeptide Y into the superficial dorsal horn reduces stimulus-evoked release of immunoreactive substance P in the anaesthetized cat. *Neuroscience* 44(3): 733–740

60 Wahlestedt C, Reis DJ (1993) Neuropeptide Y-related peptides and their receptors – are the receptors potential therapeutic drug targets? *Annu Rev Pharmacol Toxicol* 33: 309–352

61 King PJ, Williams G, Doods H, Widdowson PS (2000) Effect of a selective neuropeptide Y Y(2) receptor antagonist, BIIE0246 on neuropeptide Y release. *Eur J Pharmacol* 396(1): R1–R3

62 Zhang X, Shi T, Holmberg K, Landry M, Huang W, Xiao H, Ju G, Hokfelt T (1997) Expression and regulation of the neuropeptide Y Y2 receptor in sensory and autonomic ganglia. *Proc Nat Acad Sci USA* 94(2): 729–734

The NPY family of peptides and angiogenesis

NPY as a mediator of neurogenic angiogenesis and revascularization of ischemic tissues

Edward W. Lee, Jason U. Tilan, Lydia E. Kuo and Zofia Zukowska

Georgetown University Medical Center, Department of Physiology and Biophysics, 3900 Reservoir Rd NW, Washington, DC 20057, USA

Introduction

Tissue development, growth and regeneration all require adequate nourishment and blood supply. During embryogenesis, vessels are formed from precursor cells (vasculogenesis) and subsequently expanded by sprouting from arteries (arteriogenesis) or venules/capillaries (capillary angiogenesis) [1]; the latter is often referred to as simply, "angiogenesis". In the adult organism, the tissues' angiogenic activity is normally low, involves primarily (although not exclusively) capillary angiogenesis, and is limited to certain organs, such as the uterus during the menstrual cycle [2] or skeletal muscle undergoing remodeling during training [3]. It can be reactivated, however, in some diseases, for example retinopathies and cancer, where it contributes to pathological changes (see chapters by Bjorndahl et al., and Kitlinska). Angiogenesis/arteriogenesis is also activated in ischemic tissues as the body's compensatory mechanism to restore impaired organ blood flow. In recent years, interest and research in this area have exploded, triggered by the therapeutic potential of using angiogenic growth factors as a new way to revascularize ischemic tissues by stimulating the formation of collateral vessels [4, 5]. However, despite many new angiogenic factors being discovered, and some of them already being tested in clinical trials in patients with ischemic vascular disease of the limbs [6–8] and the heart [9–12], we are still far from understanding how to make normal, fully functional collateral arteries. While factors such as vascular endothelial growth factor (VEGF) which has been studied the most are clearly effective in producing capillary angiogenesis, they are insufficient or unable to reconstruct normal arteriogenesis. Hence, the quest for new mediators and an understanding of how multiple factors interplay in the regulation of angiogenesis *versus* arteriogenesis still continues. Many excellent state-of-the-art reviews have been written on this subject [13, 14], therefore here, we will only briefly present the basics of angio/arteriogenesis, and focus our overview on a new angiogenic factor which our group has discovered – neuropeptide Y (NPY), a neurotransmitter derived from the sympathetic nerves [15]. We will

The NPY Family of Peptides in Immune Disorders, Inflammation, Angiogenesis and Cancer, edited by Zofia Zukowska and Giora Z. Feuerstein

discuss its role as a mediator of neurogenic angiogenesis in the growth of many organs, from skeletal muscles to tumors and adipose tissue, focusing on the peptide's contribution to the revascularization of ischemic limbs. An in-depth review of NPY-mediated angiogenesis in tumors, retinopathy and wound healing can be found in chapters by Kitlinska and Bjorndahl et al, respectively.

Complexity of angiogenesis and arteriogenesis: Can a single growth factor do it?

Angio- and arterio-genesis are complex multi-step processes. They start with the discontinuation and/or dysfunction of the endothelium, followed by the adhesion of endothelial cells to the basement membrane, release of proteases, degradation of matrix, and subsequent cell migration, proliferation and differentiation into a capillary network [14]. To form arteries and veins, vascular smooth muscle cells (VSMC) migrate, proliferate, join newly formed capillary tubes, and lay down new extracellular matrix. Numerous angiogenic growth factors have been identified, most notably basic fibroblast growth factor (bFGF) and VEGF [1, 13, 14]. Most of these factors are derived from mesenchymal cells, residing in the vascular wall or blood-borne, or released by endothelial progenitor cells in the bone marrow and locally in the target tissues [14]. Their role in ischemic angiogenesis and therapeutic potential for treatment of ischemic vascular diseases has been suggested by many studies [11, 13, 14]. However, the therapy with these angiogenic factors have also revealed their side effects, such as the stimulation of tumor angiogenesis and growth for fibroblast growth factors (FGFs) [14], and new vessels of poor quality and increased permeability, leading to the formation of hemangiomas for VEGF [16]. A recent study using the combination of an angiogenic factor, VEGF, with angiopoetin, which stabilizes the vessel wall, appears to have overcome the vessel permeability problem [17] – indicating that multiple growth factors may be necessary to recapitulate normal physiological angio/arteriogenesis.

The alternative approach has been proposed by the Epstein's group [18], who suggested the use of bone marrow-derived endothelial progenitor cells as "factories" of angiogenic growth factors, alone or transfected with a "master switch" gene, which would upregulate the expression of many factors activated by tissue ischemia. Such a gene is hypoxia-inducible factor-1α (HIF-1α) [19], a transcription factor induced in target hypoxic cells, where it upregulates the expression of multiple factors, whose genes carry a HIF-1α-binding domain [19]. These include VEGF, PlGF and angiopoetin-2 [20]. Theoretically, transfection of human endothelial progenitor cells with a HIF-1α adenoviral vector may upregulate expression of these growth factors and enhance the angiogenic response in ischemic tissues. This approach could provide a promising new way for the treatment with a *combination* of angiogenic factors that are collectively upregulated by HIF-1α. Such a therapy, however, would require genetic manipulation of endothelial progenitor cells which in humans is still in its infancy.

Sympathetic nerves are activated at angiogenic sites

Yet another approach for an integrated revascularization therapy would be to find a small molecular mediator as the primary responder to tissue ischemia. Therefore, we asked these questions: What starts the tissue response to hypoxia? Which vascular components are activated first? Which molecules are released first and what do they do in terms of angio/arteriogenesis? These questions led us to suspect that peripheral vascular nerves and their small molecular neurotransmitters may be the answer.

All vessels of the body, except the thoracic aorta, are innervated by sensory and sympathetic nerves. These nerves are the first line of defense, alerting the body of vascular injury or tissue ischemia. Sympathetic nerves are continuously active, but their activity increases in conditions of ischemia [21, 22]. They release norepinephrine (NE) and its co-transmitters, NPY and purines [15, 23], which cooperate in providing normal vasoconstrictor tone. Due to their vasoconstrictive effects, the sympathetic nerves were believed to be deleterious for ischemic tissues and further impair tissue perfusion. For these reasons, local sympathectomy was used to alleviate ischemic pain in peripheral vascular insufficiency [24]. While this procedure is moderately effective in relieving short-term ischemic pain [24], chronically, it produces mixed results [24, 25], and therefore has become less frequently used in recent years. This raises the possibility that, in addition to detrimental vasoconstrictive activities, the sympathetic nerves possess other, beneficial effects for tissues they innervate. It has been often forgotten that the sympathetic nerves used to be referred to as trophic, i.e., nourishing the growth of the innervated tissues [26].

NPY is an angiogenic neurotransmitter

Sympathetic trophic activity was believed to be of the adrenergic origin. However, neither NE nor epinephrine is known to exert any direct or potent vascular growth-promoting or angiogenic actions [27]. We therefore hypothesized that it is NPY that is the sympathetic trophic and angiogenic factor. Several lines of evidence led us to this hypothesis.

NPY is a 36 amino acid amidated peptide [28] and a neurotransmitter in all sympathetic nerves innervating blood vessels. Human coronary [29] and cerebral [30] arteries are richly innervated by NPY-positive fibers. Blood vessels also possess NPY receptors, of which the Y1 receptor is the predominant type, mediating vasoconstriction [15, 31] (see chapters by Edvinsson et al., Toro et al., and Westfall) and proliferation of vascular smooth muscle cells [32]. NPY is released from the nerves during their activation. Some of the strongest stimuli for NPY release are exposure to cold [31], strenuous exercise [33], particularly when combined with hypoxia [34] and myocardial infarction [35]; but NPY responsiveness declines with age [36]. The same conditions are also known to affect angiogenesis. Capillary growth is augmented in exercising skeletal muscles [3] and ischemic myocardium [11], and this angiogenic activity also decreases with age [37]. Moreover, both angiogenesis and NPY levels are known to be the highest early in development [38], coinciding with organogenesis and growth.

NPY stimulates angiogenesis *in vitro* and *in vivo*

In addition to being a potent mitogen for VSMC [32, 39], NPY stimulates endothelial cell adhesion [40, 41], migration and proliferation [40]. In both cell types, NPY-induced growth is bimodal ranging from sub-pM to nM concentrations (see chapter by Kitlinska for details). In VSMCs, NPY-induced mitogenesis is mediated by Y1 and Y5 receptors [32], in the endothelial cells, this effect appears to be due to the activation of Y2 and Y5 receptors [42].

Human endothelium constitutively expresses a complete NPY autocrine system: Y1 and Y2 receptors, the peptide itself and its processing enzyme, dipeptidyl peptidase IV (DPPIV) which converts NPY1-36 to NPY3-36 [40]. This conversion reduces NPY's affinity for the Y1 receptor, but forms a Y2/Y5-selective agonist, NPY3-36 [43]. All NPY receptors [40] and DPPIV [44] are markedly upregulated during endothelial cell migration and differentiation. NPY itself, bFGF and nitric oxide (NO) are also positive regulators, and in addition, induce Y5 receptor expression [45]. DPPIV activity and the conversion of NPY1-36 to NPY3-36 appears to be a prerequisite for NPY's angiogenic activity, as the neutralization of the enzyme by an antibody markedly impairs endothelial cell migration in response to NPY1-36, but not NPY3-36, in the wound healing assay [44].

The potency and efficacy of NPY in stimulating angiogenesis *in vitro* is similar to that of bFGF and VEGF [40], and is in part mediated via the release of these factors [42]. Also, as with VEGF activity, NPY-mediated angiogenesis (aortic sprouting) is dependent on NO [42], which by itself is angiogenic [46]. The NO-dependent pathway in NPY's actions was also observed by others [47] in cerebral vessels, which respond with vasodilation following the activation of Y2 receptors. Importantly, mice deficient in endothelial nitric oxide synthase (eNOS) are completely unresponsive to NPY-induced aortic sprouting in a manner that is similar to that of VEGF [42]. Thus, it appears that endothelial NO is essential for NPY's angiogenic signaling.

Which NPY receptor is the major angiogenic receptor has been a subject of many investigations (also see chapters by Bjorndal et al and Kitlinska). Studies of NPY-induced angiogenesis in wound healing [48] and retinopathy [48, 49] all suggest the importance of the Y2 receptor. In aging, Kitlinska et al. [50] have found that NPY's angiogenic activity is reduced in parallel with the loss of the Y2 receptor and DPPIV expression. The role of this receptor as NPY's main angiogenic receptor has been recently confirmed in Y2 knockout mice, in which NPY-induced aortic sprouting, capillary formation on Matrigel [51], retinal angiogenesis [48] and oxygen-induced retinopathy [49] were all markedly reduced.

Hindlimb ischemia activates NPY-Y2 receptor system

Given NPY's angiogenic and vascular growth-promoting activity, we hypothesized that the peptide contributes to revascularization of the ischemic tissues. To test this hypothesis, we used rat/murine hindlimb models where femoral artery occlusion produces calf muscle ischemia (capillary angiogenesis zone), which in turn stimulates ischemic angiogenesis. In contrast, shear stress in the large arteries above the occlusion activates arteriogenesis in an attempt to develop collaterals (collateralization zone) (Fig. 1).

In the ischemic leg, as compared to the contralateral non-ischemic one, venous plasma NPY levels increased while femoral arterial content of the peptide decreased, suggesting the peptide's release from vascular stores [42]. This response was, at least in part, neurogenic since ipsilateral lumbar sympathectomy reduced it [42]. The remaining sympathectomy-resistant NPY component could be either due to the incompleteness of the one-sided denervation, or to the NPY released from non-neurogenic stores such as platelets [52], or immune [53] and endothelial cells [40]; due to both de novo peptide synthesis, as well as uptake of sympathetically-derived NPY [40].

Limb ischemia also induced changes in NPY receptor expression. As in other ischemic tissue models, congestive heart failure [54] and retinopathy [49], during leg ischemia, a major shift occurred in the NPY receptor system towards the induction/upregulation of Y2 receptors and DPPIV. The Y2 receptor and DPPIV mRNA became upregulated rapidly within the first 48 h and remained elevated throughout the next 2–4 weeks [55]. Interestingly, the upregulation of the Y2 receptor and DPPIV expression occurred only in the ischemic *gastrocnemius* muscle, i.e., in the capillary angiogenesis zone (Fig. 1B). In contrast, in the *adductor* muscle, corresponding to the area of the femoral artery occlusion and collateralization zone, the Y2 receptor was only slightly elevated, while the Y1 receptor and NPY mRNA were markedly upregulated (Fig. 1A).

The differential patterns of NPY receptor upregulation suggest that different receptors and intermediary factors may be involved in capillary angiogenesis in the

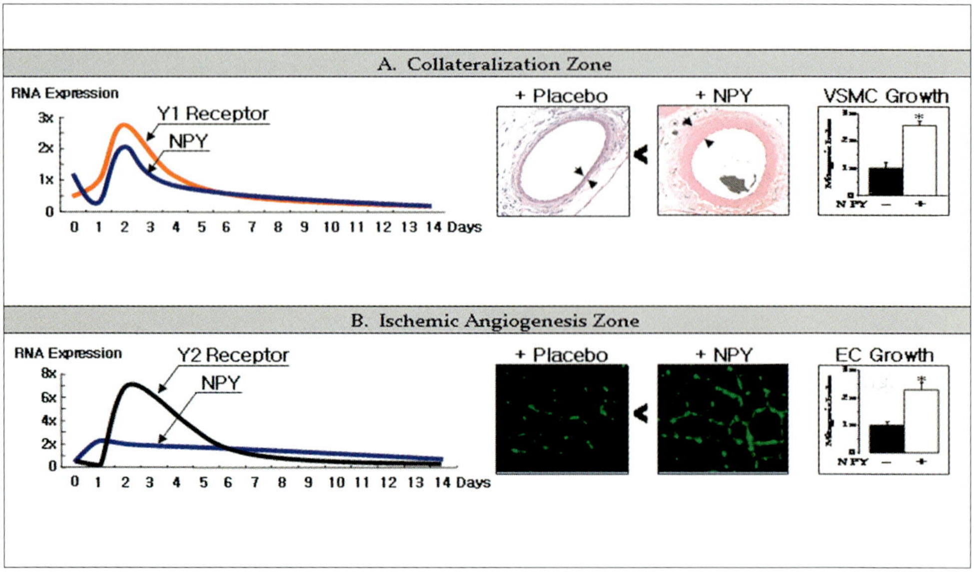

Figure 1
NPY and revascularization of rat ischemic hindlimb (A) In the collateralization zone (adductor muscle) NPY and Y1 receptor expression (based on Real Time RT-PCR data) is rapidly upregulated. Two weeks later, formation of collateral artery is observed, further augmented by NPY (1 µg/pellet/14 days, into the popliteal fossa). NPY's arteriogenic activity is corroborated by peptide's ability to stimulate VSMC proliferation via Y1 receptors. (B) In the ischemic angiogenesis zone (gastrocnemius muscle), ischemia rapidly upregulates mostly Y2 receptor, and to a lesser degree NPY mRNAs, and this is associated with increase in density of cd31-positive capillaries. The latter reflects peptide's ability to stimulate endothelial cell (EC) proliferation by activating Y2 and Y5 receptors.

ischemic tissues (Y2/DPPIV), as opposed to arteriogenesis/collateralization in the area of shear stress (Y1/NPY) [55]. These receptor changes may correspond to NPY's growth promoting activity on endothelial cells, which require Y2/Y5/DPPIV, and on VSMCs, which require Y1/Y5. Since arteriogenesis includes all processes: endothelial and VSMC proliferation and differentiation, it is likely that all NPY receptors would participate in it. Interestingly, spontaneous collateralization of the rat ischemic limb does not fully restore its blood flow and, at the same time, it is also rare that the Y5 receptor is induced by ischemia or shear stress alone [42]. However, its expression is markedly upregulated by NPY and other angiogenic factors such as bFGF (see chapter by Kitlinska). Whether or not the Y5 receptor acts as an amplifier of angiogenesis and collateralization, and could potentially be a therapeutic target, is an intriguing possibility.

NPY stimulates revascularization of ischemic hindlimb

To further test the hypothesis that NPY is a mediator of ischemic tissue revascularization, the peptide was administered as a slow-release pellet (1 µg/14 days) into the popliteal fossa below the femoral artery occlusion in the rat hindlimb ischemia model. NPY stimulated formation of large muscular arteries in the *adductor* muscle at the level of arterial occlusion, and induced capillary angiogenesis in the *gastrocnemius* muscle and the footpad (Fig. 1) [42]. These actions led to almost complete restoration of blood flow, as well as a marked improvement in contractile performance of ischemic muscles [42]. This occurred at physiological plasma NPY concentrations (< 10 nM) similar to those induced by stress or exercise in rats [15, 31]. Although these NPY levels may be vasoconstrictive, a Y1-selective receptor antagonist did not alter any of the aspects of the NPY-mediated ischemic angiogenesis or improvements of blood flow. NPY's ability to revascularize ischemic tissue was, however, markedly impaired in Y2 knockout mice, suggesting that the Y2 receptor is critical for this process [51] (Fig. 2B).

Such a powerful effect of restoring ischemic tissue blood flow and function is probably related to NPY's ability to stimulate growth of not only capillaries in the area of ischemia, but also the formation of collaterals [32, 56]. In this respect, NPY resembles bFGF rather than VEGF, which was particularly striking when these growth factors were compared in the aortic sprouting assay [40]. While vessel sprouting stimulated by VEGF resembled abnormal, leaky, and engorged vessels characteristic of pathological angiogenesis, for example in tumors [14] or retinopathies [57], NPY stimulated the formation of long, normal-looking aortic sprouts [58].

The notion that NPY is a mediator of physiological angiogenesis, as occurring in ischemic or exercising muscles [59] was further supported in NPY transgenic rats, over-expressing the peptide under its natural promoter [60]. These rats have elevated plasma and tissue NPY levels, but have no apparent pathologies [60]. However, their gastrocnemius muscles were hyper-vascularized, with a two- to three-fold increase in capillary per fiber ratio, and a greater muscle fiber area than their wild type counterparts [42] – suggesting that NPY plays a role in angiogenesis during normal development (Fig. 2A). Again, vascularization of the non-ischemic and ischemic tissues was markedly reduced in $Y2^{-/-}$ mice (Fig. 2B) [51]. In addition, spontaneous sprouting of the aortic rings was greatly delayed and impaired in mice null for the NPY gene (Fig. 2C) [42].

Taken together, these observations indicate that NPY is an important mediator of physiological angio- and arteriogenesis in ischemic, as well as non-ischemic tissues. The primary angiogenic receptors are Y2 and Y5, but Y1 or Y5 receptor activation may be necessary for arteriogenesis. Given the angiogenic potency of NPY and its ability to form normal looking, smooth muscle containing vessels – Y2/Y5 agonists may be very attractive targets for drugs aimed at the

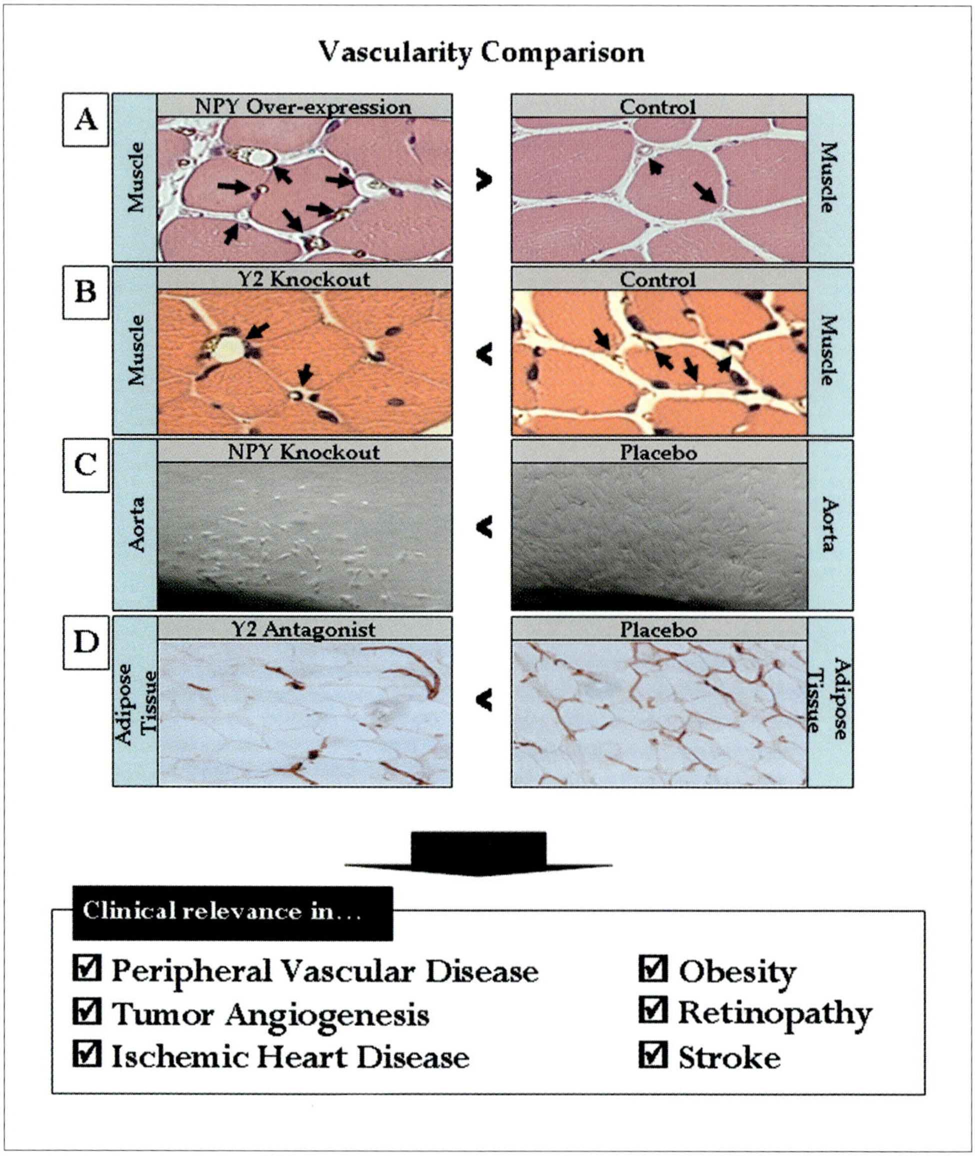

Figure 2
Elevated tissue levels of NPY in the skeletal muscles of NPY over-expressing rats are associated with increased capillary density per muscle fiber (A). Elimination or blockade of NPY-Y2 receptor signaling markedly reduces capillary density per skeletal muscle fiber of NPY $Y2^{-/-}$ *mice (B), aorting sprouting in* $NPY^{-/-}$ *mice (C) and capillary density per adipose tissue in obese mice (D)*

revascularization of ischemic limbs and hearts, in peripheral vascular and heart diseases.

NPY-mediated angiogenesis in tissue growth and regeneration

Ischemic angiogenesis not only occurs in pathological conditions such as following arterial occlusion, but is a general phenomenon that underlies the growth of any tissue. Under-perfusion of a developing tumor leads to its ischemia, which in turn stimulates angiogenesis, restores blood flow, and promotes further cancer expansion. The relationship between tumor ischemia, angiogenesis and growth became clearer when anti-angiogenic therapies for cancer did not necessarily result in tumor suppression [61]. In one such study, inhibition of VEGF-mediated angiogenesis was shown, quite unsuspectedly, to promote tumor growth by inducing its ischemia, which led to a greater expansion of malignant cells on one hand and the development of more normal looking vessels which improved cancer blood flow on the other [62]. Studies by Kitlinska [63], from our laboratory, indicated that the benefits of antagonizing NPY-mediated angiogenesis for cancer may be due to such therapy being bi-directional: it inhibits the peptide's ability to form normal vessels, as well as its proliferative actions on tumor cells (see chapter by Kitlinska).

Similar mechanisms of ischemia-driven angiogenesis may play a role in the development of a highly vascularized organ such as adipose tissue [64]. It is also a tissue that is well innervated by sympathetic nerves; hence, NPY may potentially play an angiogenic and growth-promoting role. Using $Lep^{ob/ob}$ mice, which are genetically obese, we have recently shown that a NPY slow release pellet (1 μg/14 days, the same as those used in the ischemia studies), placed subcutaneously in the abdominal area, stimulates adipose tissue growth and vascularization [65]. In contrast, daily injections of a specific NPY-Y2 receptor antagonist (which, *in vitro*, inhibits endothelial cell differentiation into capillary tubes [66]) reduced adipose tissue vascularization and fat deposits of $Lep^{ob/ob}$ mice by 40% within 2 weeks [65] (Fig. 2D). Similar, albeit smaller, changes were also observed in the wild type non-obese mice [65]. Thus, NPY and its Y2 receptors are an endogenous angiogenic system active in the adipose tissue, where it may contribute to the development of obesity. Hence, Y2 receptor antagonists, by virtue of their anti-angiogenic properties, could be a new treatment against excessive adiposity.

Since sympathetic nerves innervate all of the tissues of the body, its neurotransmitter, NPY can potentially exert angiogenic activity and promote growth and regeneration of any tissue. One example yet to be explored is the liver, which is highly vascularized, innervated and receives splanchnic blood flow rich in NPY-like peptides (such as peptide YY and pancreatic polypeptide) from the gut. What role these peptides play in liver angiogenesis and regeneration is an interesting question to be studied.

Last, but not least, there is the central and peripheral nervous system itself, which is critically dependent on blood supply and is also highly abundant in NPY [67]. In neuroblastomas, which are cancers derived from the sympathetic neurons, NPY is released and acts as an autocrine tumor growth factor, but it also stimulates angiogenesis by activating Y2/Y5 receptors [63]. In the hippocampal neurons, as shown by Howell et al. [68] (see chapter by Gray and Scharfman in this book), NPY increases neurogenesis by activating Y1 receptors, but its indirect actions via the stimulation of angiogenesis have not been studied yet.

We propose that NPY is a universal mediator of neurogenic angiogenesis, activated by tissue ischemia and/or injury, and contributes to tissue growth and regeneration. It is a part of a larger trophic cycle, by which nerves communicate with vascular cells and target tissues, involving a multiplicity of mediators released from these cells. Some of the potential upstream regulators of NPY expression are neurotrophins, and its downstream mediators include VEGF and bFGF. The central role of NPY would rely on its function as a neurotransmitter, thus possibly being the molecule that starts the cascade of events leading to the coordinated activation of multiple pathways and tissue response to ischemia (Fig. 3).

Summary and conclusions

NPY is a sympathetic neurotransmitter and a mediator of neurogenic angiogenesis in many tissues, such as skeletal muscles, tumors, retina, skin, adipose tissue and possibly the nervous system itself. It stimulates endothelial cell adhesion, migration, proliferation and differentiation leading to the formation of capillaries by activating the Y2 and Y5 receptors. NPY also stimulates vascular smooth muscle cell proliferation via the Y1 and Y5 receptors, and thus is able to form collateral arteries. Ischemia upregulates the NPY-Y2 receptor system and the peptide improves revascularization and ischemic skeletal muscle function. We propose that NPY is a mediator of a sympathetic trophic cycle, acting to adjust vessel development to that of nerves and *vice versa*, and thus promoting tissue growth. As a neurotransmitter, NPY is positioned at the beginning of the cascade that activates downstream mediators leading to angiogenesis, arteriogenesis and neurogenesis in a tissue-specific manner. These potent and multi-faceted activities of NPY make the peptide's Y2/Y5 receptor agonists valuable new targets for revascularization therapy, and its Y2/Y5 antagonists potential drugs to inhibit excessive growth of organs such as adipose tissue or tumors.

Acknowledgments
This study was supported by grants from the National Institutes of Health: HL67357 to Dr. Zofia Zukowska.

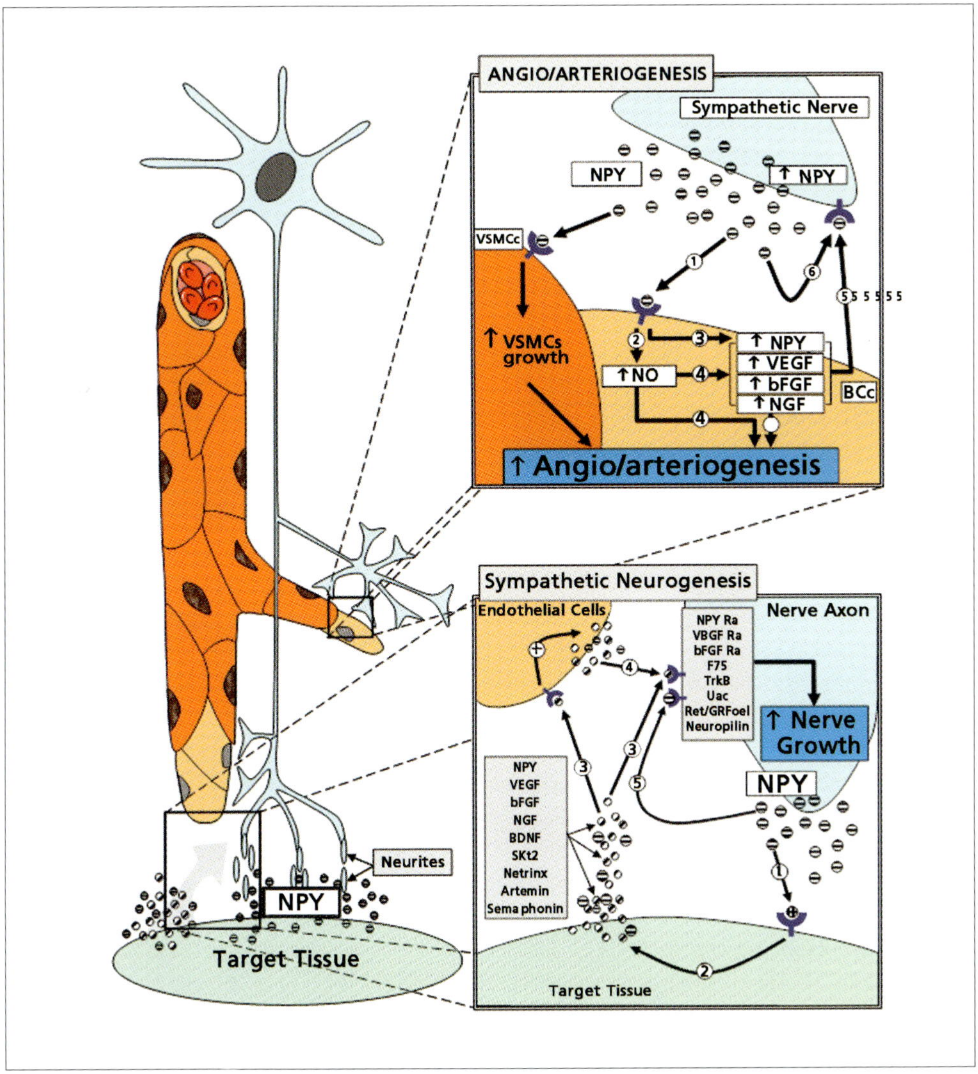

Figure 3

Proposed model of NPY's role in neurogenic angiogenesis and a cross talk with neurogenesis. Sympathetic nerve-released NPY stimulates growth of VSMC (Y1R) and endothelial cells (EC) and leads to angio- and arteriogenesis via Y2 and Y5 receptors. NPY also stimulates release of secondary mediators from VSMCs, ECs and target tissue, which are angiogenic and also upregulate neuronal expression of NPY. The reciprocal cross talk also exists between vascular cells and target tissue which release their own NPY and/or NPY's downstream angiogenic mediators (listed in inserts) and stimulate neurogenesis, adjusting vessels-to-nerve ratios.

References

1 Folkman J (1995) Angiogenesis in cancer, vascular, rheumatoid and other disease. *Nat Med* 1: 27–31

2 Gargett CE, Rogers PA (2001) Human endometrial angiogenesis. *Reproduction* 121: 181–186

3 Hudlicka O, Brown MD (2000) Modulators of angiogenesis. In: GM Rubanyi (ed): *Angiogenesis in Health and Disease*. Marcel Dekker, New York, Basel, 215–244

4 Sellke FW, Simons M (1999) Angiogenesis in cardiovascular disease: current status and therapeutic potential. *Drugs* 58: 391–396

5 Folkman J (1998) Therapeutic angiogenesis in ischemic limbs. *Circulation* 97: 1108–1110

6 Kim HJ, Jang SY, Park JI, Byun J, Kim DI, Do YS, Kim JM, Kim S, Kim BM, Kim WB et al (2004) Vascular endothelial growth factor-induced angiogenic gene therapy in patients with peripheral artery disease. *Exp Mol Med* 36: 336–344

7 Rajagopalan S, Olin JW, Young S, Erikson M, Grossman PM, Mendelsohn FO, Regensteiner JG, Hiatt WR, Annex BH (2004) Design of the Del-1 for therapeutic angiogenesis trial (DELTA-1), a phase II multicenter, double-blind, placebo-controlled trial of VLTS-589 in subjects with intermittent claudication secondary to peripheral arterial disease. *Hum Gene Ther* 15: 619–624

8 Lederman RJ, Mendelsohn FO, Anderson RD, Saucedo JF, Tenaglia AN, Hermiller JB, Hillegass WB, Rocha-Singh K, Moon TE, Whitehouse MJ et al (2002) Therapeutic angiogenesis with recombinant fibroblast growth factor-2 for intermittent claudication (the TRAFFIC study): a randomised trial. *Lancet* 359: 2053–2058

9 Vale PR, Losordo DW, Milliken CE, McDonald MC, Gravelin LM, Curry CM, Esakof DD, Maysky M, Symes JF, Isner JM (2001) Randomized, single-blind, placebo-controlled pilot study of catheter-based myocardial gene transfer for therapeutic angiogenesis using left ventricular electromechanical mapping in patients with chronic myocardial ischemia. *Circulation* 103: 2138–2143

10 Hedman M, Hartikainen J, Syvanne M, Stjernvall J, Hedman A, Kivela A, Vanninen E, Mussalo H, Kauppila E, Simula S et al. (2003) Safety and feasibility of catheter-based local intracoronary vascular endothelial growth factor gene transfer in the prevention of postangioplasty and in-stent restenosis and in the treatment of chronic myocardial ischemia: phase II results of the Kuopio Angiogenesis Trial (KAT). *Circulation* 107: 2677–2683

11 Rosinberg A, Khan TA, Sellke FW, Laham RJ (2004) Therapeutic angiogenesis for myocardial ischemia. *Expert Rev Cardiovasc Ther* 2: 271–283

12 Henry TD, Annex BH, McKendall GR, Azrin MA, Lopez JJ, Giordano FJ, Shah PK, Willerson JT, Benza RL, Berman DS et al (2003) The VIVA trial: Vascular endothelial growth factor in ischemia for vascular angiogenesis. *Circulation* 107: 1359–1365

13 Jain RK (2005) Normalization of tumor vasculature: an emerging concept in antiangiogenic therapy. *Science* 307: 58–62

14 Carmeliet P (2003) Angiogenesis in health and disease. *Nat Med* 9: 653–660
15 Zukowska-Grojec Z, Wahlestedt C (1993) Origin and actions of neuropeptide Y in the cardiovascular system. In: WF Colmers, C Wahlestedt (eds): *The Biology of Neuropeptide Y and Related Peptides*. Human Press, Totowa, New Jersey, 315–388
16 Ozawa CR, Banfi A, Glazer NL, Thurston G, Springer ML, Kraft PE, McDonald DM, Blau HM (2004) Microenvironmental VEGF concentration, not total dose, determines a threshold between normal and aberrant angiogenesis. *J Clin Invest* 113: 516–527
17 Shyu KG, Chang H, Isner JM (2003) Synergistic effect of angiopoietin-1 and vascular endothelial growth factor on neoangiogenesis in hypercholesterolemic rabbit model with acute hindlimb ischemia. *Life Sci* 73: 563–579
18 Kinnaird T, Stabile E, Burnett MS, Epstein SE (2004) Bone-marrow-derived cells for enhancing collateral development: mechanisms, animal data, and initial clinical experiences. *Circ Res* 95: 354–363
19 Pugh CW, Ratcliffe PJ (2003) Regulation of angiogenesis by hypoxia: role of the HIF system. *Nat Med* 9: 677–684
20 Kelly BD, Hackett SF, Hirota K, Oshima Y, Cai Z, Berg-Dixon S, Rowan A, Yan Z, Campochiaro PA, Semenza GL (2003) Cell type-specific regulation of angiogenic growth factor gene expression and induction of angiogenesis in nonischemic tissue by a constitutively active form of hypoxia-inducible factor 1. *Circ Res* 93: 1074–1081
21 Ganguly PK, Dhalla KS, Shao Q, Beamish RE, Dhalla NS (1997) Differential changes in sympathetic activity in left and right ventricles in congestive heart failure after myocardial infarction. *Am Heart J* 133: 340–345
22 Takatsu H, Duncker CM, Arai M, Becker LC (1997) Cardiac sympathetic nerve function assessed by [131I]metaiodobenzylguanidine after ischemia and reperfusion in anesthetized dogs. *J Nucl Cardiol* 4: 35–41
23 Burnstock G (1987) Local control of blood pressure by purines. *Blood Vessels* 24: 156–160
24 Alexander JP (1994) Chemical lumbar sympathectomy in patients with severe lower limb ischaemia. *Ulster Med J* 63: 137–143
25 Perez-Burkhardt JL, Gonzalez-Fajardo JA, Martin JF, Carpintero Mediavilla LA, Mateo Gutierrez AM (1999) Lumbar sympathectomy as isolated technique for the treatment of lower limbs chronic ischemia. *J Cardiovasc Surg (Torino)* 40: 7–13
26 Bevan RD (1984) Trophic effects of peripheral adrenergic nerves on vascular structure. *Hypertension* 6: III19–III26
27 Blaes N, Boissel JP (1983) Growth-stimulating effect of catecholamines on rat aortic smooth muscle cells in culture. *J Cell Physiol* 116: 167–172
28 Tatemoto K, Carlquist M, Mutt V (1982) Neuropeptide Y – a novel brain peptide with structural similarities to peptide YY and pancreatic polypeptide. *Nature* 296: 659–660
29 Allen YS, Bloom SR, Polak JM (1986) The neuropeptide Y-immunoreactive neuronal system: discovery, anatomy and involvement in neurodegenerative disease. *Hum Neurobiol* 5: 227–234

30 Edvinsson L, Emson P, McCulloch J, Tatemoto K, Uddman R (1983) Neuropeptide Y: cerebrovascular innervation and vasomotor effects in the cat. *Neurosci Lett* 43: 79–84
31 Zukowska-Grojec Z, Dayao EK, Karwatowska-Prokopczuk E, Hauser GJ, Doods HN (1996) Stress-induced mesenteric vasoconstriction in rats is mediated by neuropeptide Y Y1 receptors. *Am J Physiol* 270: H796–800
32 Pons J, Kitlinska J, Ji H, Lee EW, Zukowska Z (2003) Mitogenic actions of neuropeptide Y in vascular smooth muscle cells: synergetic interactions with the beta-adrenergic system. *Can J Physiol Pharmacol* 81: 177–185
33 Lundberg JM, Martinsson A, Hemsen A, Theodorsson-Norheim E, Svedenhag J, Ekblom B, Hjemdahl P (1985) Co-release of neuropeptide Y and catecholamines during physical exercise in man. *Biochem Biophys Res Commun* 133: 30–36
34 Kaijser L, Pernow J, Berglund B, Grubbstrom J, Lundberg JM (1994) Neuropeptide Y release from human heart is enhanced during prolonged exercise in hypoxia. *J Appl Physiol* 76: 1346–1349
35 Kaijser L, Pernow J, Berglund B, Lundberg JM (1990) Neuropeptide Y is released together with noradrenaline from the human heart during exercise and hypoxia. *Clin Physiol* 10: 179–188
36 Jensen EW, Espersen K, Kanstrup IL, Christensen NJ (1994) Exercise-induced changes in plasma catecholamines and neuropeptide Y: relation to age and sampling times. *J Appl Physiol* 76: 1269–1273
37 Rivard A, Fabre JE, Silver M, Chen D, Murohara T, Kearney M, Magner M, Asahara T, Isner JM (1999) Age-dependent impairment of angiogenesis. *Circulation* 99: 111–120
38 Gordon L, Polak JM, Moscoso GJ, Smith A, Kuhn DM, Wharton J (1993) Development of the peptidergic innervation of human heart. *J Anat* 183: 131–140
39 Zukowska-Grojec Z, Karwatowska-Prokopczuk E, Fisher TA, Ji H (1998) Mechanisms of vascular growth-promoting effects of neuropeptide Y: role of its inducible receptors. *Regul Pept* 75–76: 231–238
40 Zukowska-Grojec Z, Karwatowska-Prokopczuk E, Rose W, Rone J, Movafagh S, Ji H, Yeh Y, Chen WT, Kleinman HK, Grouzmann E et al (1998) Neuropeptide Y: a novel angiogenic factor from the sympathetic nerves and endothelium. *Circ Res* 83: 187–195
41 Sung CP, Arleth AJ, Feuerstein GZ (1991) Neuropeptide Y upregulates the adhesiveness of human endothelial cells for leukocytes. *Circ Res* 68: 314–318
42 Lee EW, Michalkiewicz M, Kitlinska J, Kalezic I, Switalska H, Yoo P, Sangkharat A, Ji H, Li L, Michalkiewicz T et al. (2003) Neuropeptide Y induces ischemic angiogenesis and restores function of ischemic skeletal muscles. *J Clin Invest* 111: 1853–1862
43 Grandt D, Schimiczek M, Rascher W, Feth F, Shively J, Lee TD, Davis MT, Reeve JR Jr, Michel MC (1996) Neuropeptide Y 3-36 is an endogenous ligand selective for Y2 receptors. *Regul Pept* 67: 33–37
44 Ghersi G, Chen W, Lee EW, Zukowska Z (2001) Critical role of dipeptidyl peptidase IV in neuropeptide Y-mediated endothelial cell migration in response to wounding. *Peptides* 22: 453–458
45 Lee EW, Movafagh S, Eckhardt L, Karwatowska E, Colton C, Grant DS, Zukowska Z.

(2001) Neuropeptide Y (NPY) stimulates angiogenesis via nitric oxide (NO)-dependent pathways. *Circulation* 104: 592

46 Ziche M, Morbidelli L (2000) Nitric oxide and angiogenesis. J Neurooncol 50: 139–148

47 You J, Edvinsson L, Bryan RM Jr (2001) Neuropeptide Y-mediated constriction and dilation in rat middle cerebral arteries. *J Cereb Blood Flow Metab* 21: 77–84

48 Ekstrand AJ, Cao R, Bjorndahl M, Nystrom S, Jonsson-Rylander AC, Hassani H, Hallberg B, Nordlander M, Cao Y (2003) Deletion of neuropeptide Y (NPY) 2 receptor in mice results in blockage of NPY-induced angiogenesis and delayed wound healing. *Proc Natl Acad Sci USA* 100: 6033–6038

49 Koulu M, Movafagh S, Tuohimaa J, Jaakkola U, Kallio J, Pesonen U, Geng Y, Karvonen MK, Vainio-Jylha E, Pollonen M et al. (2004) Neuropeptide Y and Y2-receptor are involved in development of diabetic retinopathy and retinal neovascularization. *Ann Med* 36: 232–240

50 Kitlinska J, Lee EW, Movafagh S, Pons J, Zukowska Z (2002) Neuropeptide Y-induced angiogenesis in aging. *Peptides* 23: 71–77

51 Lee EW, Grant DS, Movafagh S, Zukowska Z (2003) Impaired angiogenesis in neuropeptide Y (NPY)-Y2 receptor knockout mice. *Peptides* 24: 99–106

52 Myers AK, Farhat MY, Shen GH, Debinski W, Wahlestedt C, Zukowska-Grojec Z (1990) Platelets as a source and site of action for neuropeptide Y. *Ann NY Acd Sci* 611: 408–411

53 von Horsten S, Exton NG, Exton MS, Helfritz F, Nave H, Ballof J, Stalp M, Pabst R (1998) Brain NPY Y1 receptors rapidly mediate the behavioral response to novelty and a compartment-specific modulation of granulocyte function in blood and spleen. *Brain Res* 806: 282–286

54 Lee EW, Callanan EY, Winaver J, Haramati A, Zukowska Z (2005) A role of neuropeptide Y (NPY) and NPY receptors in congestive heart failure. *Am J Physiol Heart Circ Physiol; in revision*

55 Tilan JU, Karlsson J, Abe K, Kuo L, Lee EW, Everhart L, Li L, Kitlinska J, Fricke S, Thornell L et al (2005) Rapid activation of the neuropeptide (NPY) receptor system during ischemic angiogenesis in rat and human limb muscles. *Faseb J* 19: A1663.1931.1616

56 Zukowska-Grojec Z, Karwatowska-Prokopczuk E, Fisher TA, Ji H (1998) Mechanisms of vascular growth-promoting effects of neuropeptide Y: role of its inducible receptors. *Regul Pept* 75–76: 231–238

57 Alon T, Hemo I, Itin A, Pe'er J, Stone J, Keshet E (1995) Vascular endothelial growth factor acts as a survival factor for newly formed retinal vessels and has implications for retinopathy of prematurity. *Nat Med* 1: 1024–1028

58 Zukowska Z, Grant DS, Lee EW (2003) Neuropeptide Y: a novel mechanism for ischemic angiogenesis. *Trends Cardiovasc Med* 13: 86–92

59 Brown MD, Hudlicka O (2003) Modulation of physiological angiogenesis in skeletal muscle by mechanical forces: involvement of VEGF and metalloproteinases. *Angiogenesis* 6: 1–14

60 Michalkiewicz M, Michalkiewicz T (2000) Developing transgenic neuropeptide Y rats. *Methods Mol Biol* 153– 73–89

61 Heymach JV, Desai J, Manola J, Davis DW, McConkey DJ, Harmon D, Ryan DP, Goss G, Quigley T, Van den Abbeele AD et al (2004) Phase II study of the antiangiogenic agent SU5416 in patients with advanced soft tissue sarcomas. *Clin Cancer Res* 10: 5732–5740

62 Marx J (2003) Angiogenesis. A boost for tumor starvation. *Science* 301: 452–454

63 Kitlinska J, Abe K, Kuo L, Pons J, Yu M, Li L, Tilan J, Everhart L, Lee EW, Zukowska Z et al. (2005) Differential effects of neuropeptide Y on the growth and vascularization of neural crest-derived tumors. *Cancer Res* 65: 1719–1728

64 Rupnick MA, Panigrahy D, Zhang CY, Dallabrida SM, Lowell BB, Langer R, Folkman MJ (2002) Adipose tissue mass can be regulated through the vasculature. *Proc Natl Acad Sci USA* 99: 10730–10735

65 Kuo LE, Lee EW, Zukowska Z (2003) Angiogenic and growth-promoting actions of neuropeptide Y in adipose tissue: a novel link in the pathogenesis of central obesity. *Circulation* 108: IV-1–IV-1050

66 Movafagh S, Kitlinska J, Zukowska Z (2005) Multiple NPY receptors mediate growth promoting effect of NPY on human endothelial cells. *Faseb J; in revision*

67 Hokfelt T, Broberger C, Zhang X, Diez M, Kopp J, Xu Z, Landry M, Bao L, Schalling M, Koistinaho J et al (1998) Neuropeptide Y: some viewpoints on a multifaceted peptide in the normal and diseased nervous system. *Brain Res Rev* 26: 154–166

68 Howell OW, Doyle K, Goodman JH, Scharfman HE, Herzog H, Pringle A, Beck-Sickinger AG, Gray WP (2005) Neuropeptide Y stimulates neuronal precursor proliferation in the post-natal and adult dentate gyrus. *J Neurochem* 93: 560–570

NPY-induced angiogenesis in retinopathy and wound healing

Meit Björndahl[1], Renhai Cao[1], Luxun Xue[2] and Yihai Cao[1]

[1]Microbiology and Tumor Biology Center, Karolinska Institutet, 171 77 Stockholm, Sweden;
[2]Zhenzhou University Cancer Center, Zhenzhou, Henan, China

Angiogenic activity of NPY

Neuropeptide Y (NPY) is a member of a large family that includes additional polypeptides; peptide YY, gut-derived peptide, and pancreatic polypeptides [1]. NPY, one of the most abundant peptides in the brain and heart, has pleiotropic activities ranging from stimulation of food intake and obesity, regulation of cardiovascular function, and neuroendocrine function [1–3]. The first clue that NPY might stimulate angiogenesis came from *in vitro* studies, in which NPY stimulated human endothelial cell migration, attachment, and DNA synthesis [4–6]. In a Matrigel assay using human umbilical vein endothelial cells, NPY induces capillary tube formation. It also induces sprouting from aorta rings and formation of new blood vessels in an *in vivo* model [4, 6].

The *in vivo* angiogenic activity of NPY has been further validated using a mouse corneal micropocket assay. When various doses of NYP together with a slow-release polymer were implanted into the mouse cornea, NYP induced corneal neovascularization in a dose dependent manner [7]. Because of the avascular nature of the cornea tissue, the NPY-induced neovascularization can be compared with vascular structures induced by other angiogenic factors, such as vascular endothelial growth factor A (VEGF-A) and fibroblast growth factor-2 (FGF-2). Interestingly, the angiogenic response induced by NPY is structurally similar to the FGF-2-induced vascular networks, in which the microvessels are well separated from each other and grow toward the implanted pellets [7, 8]. This type of vasculature is completely different from VEGF-A-induced vessels, which appeared as vascular plexuses due to fusion of microvessels/capillaries into large luminal networks. These results indicate that NPY is a direct angiogenic stimulator that induces vascular "tree"-like structures. In the chick chorioallantoic membrane model, NPY stimulates vascular sprouting from pre-existing blood vessels in developing embryos [7]. This sprouting effect of NPY appears in a dose-dependent manner.

The NPY Family of Peptides in Immune Disorders, Inflammation, Angiogenesis and Cancer,
edited by Zofia Zukowska and Giora Z. Feuerstein

NPY receptors and angiogenesis

NPY binds and activates six receptors, Y1–Y6, all of which belong to a family of $G_{i/o}$-heptahelical receptors [1, 9]. Several of these receptors predominantly act on the vascular system [1, 5–7, 10, 11]. For example, the Y1 receptor mediates vasoconstriction via regulation of contraction and proliferation of vascular smooth muscle cells (VSMCs) [5, 12–15]. The Y2 receptor is constitutively expressed on vascular endothelial cells [6]. Similar to Y1 receptor, Y5 is also expressed on VSMCs and transduces mitotic signals [11, 15]. The distribution of NPY receptors on vascular cells suggests that NPY is an important neurotransmitter controlling blood flow and vessel growth.

In addition to tissue-specific regulation of expression of receptor subtypes, enzymatic processing may further regulate the biological activity of NPY and influence its affinity for different receptor subtypes. Intact N- and C-terminal sequences are required for interaction with the Y1 receptor [16–20]. In HUVECs, NPY is co-localized with dipeptidyl peptidase IV (DPPIV), an endothelial serine protease which cleaves Tyr(1)-Pro(2) from $NPY_{1\text{-}36}$ to form $NPY_{3\text{-}36}$ [21]. This cleavage results in the formation of a potent angiogenic C-terminal fragment, NPY_{3-36}, which is unable to activate the Y1 receptor but activates non-Y1 receptors, such as Y2, Y3, and Y5 in the central and peripheral nervous system. Thus, DPPIV terminates the Y1 activity of NPY and converts it to a potent Y2 agonist [22].

The Y2 receptor is not expressed, or expressed at very low levels, in intact vessels or non-growing endothelial cells and VSMC [6, 11]. However, injury of these cells or ischemia upregulates the expression of Y2, as well as Y1 and Y5 [4, 23]. Localization of the Y2 receptor on endothelial cells suggests that this receptor might be directly involved in mediating NPY-induced angiogenesis [6, 7]. The fact that $NPY_{3\text{-}36}$ was as competent as native NPY in promoting capillary tube formation *in vitro* suggests that the Y2 receptor, possibly in cooperation with the Y5 receptor, mediates endothelial cell activity [6]. Similarly in the mouse corneal model, $NPY_{3\text{-}36}$ also exhibits an angiogenic activity indistinguishable from the unprocessed NPY [7]. In contrast, [LeuPro]NPY, a potent activator for Y1 and Y5 receptors lacking affinity for the Y2 receptor, is unable to induce endothelial cell activity *in vitro* and angiogenesis *in vivo* [5–7]. These studies further suggest that the Y2 receptor is responsible for mediation of NPY-induced angiogenesis.

The conclusion that the Y2 receptor is an angiogenic receptor for NPY was obtained from studies using Y2 receptor deficient mice. Two independent studies show that elimination of the Y2 receptor in mice impairs NPY-induced neovascularization [7, 10]. In the aortic ring assay, NPY-induced endothelial cell sprouting is almost completely abolished in $Y2^{-/-}$ mice. A significant reduction of *in vivo* Matrigel capillary formation is also observed in $Y2^{-/-}$ mice [10]. In a hindlimb ischemia model, NPY induces collateral growth in the ischemic region of skeletal muscle [23]. In $Y2^{-/-}$ mice, NPY-induced neovascularization in the ischemic hindlimb model is significantly decreased [10]. In the mouse corneal assay, NPY is unable to induce corneal neovascularization in $Y2^{-/-}$ mice although other angiogenic factors including FGF-2 and VEGF-A-induced angiogenesis is not affected in the Y2 knockout mice [7].

NPY and wound healing

The angiogenic activity of NPY suggests that this neurotransmitter may play a physiological role in regulation of angiogenesis-dependent processes, such as wound healing [7]. It has been shown that wounding of endothelial cells results in upregulated expression of DPPIV in migrating cells, and that it causes the enzyme to be distributed to the edge of the wound. The proteolytic activity of DPPIV has been found be essential for the ability of NPY to promote cell migration and wound closure through generation of NPY_{3-36} [21]. In mice, NPY significantly accelerates the healing of excisional full skin wounds. In contrast, a significantly delayed wound repair is found in $Y2^{-/-}$ mice, suggesting a physiological role of NPY in skin wound healing. FGF-2, a known potent wound healing factor, can correct the delayed wound healing phenotype in the $Y2^{-/-}$ mice [7]. However, exogenous NPY is unable to correct the wound healing defect in $Y2^{-/-}$ mice. These data further validate the finding that Y2 is the critical receptor for NPY-induced angiogenesis and wound healing. Consistent with these findings, NPY-treated wounds in wild type mice contain a high density of capillaries/microvessels whereas this wound promoting activity is impaired in $Y2^{-/-}$ mice, which demonstrate a reduced vascularity in the wound. It seems that inflammation may not play a role in NPY-induced wound healing as a similar numbers of inflammatory cells, such as macrophages, are present in both wild type and $Y2^{-/-}$ mice.

NPY and diabetic retinopathy

Diabetic retinopathy is one of the most common causes of blindness. This pathological condition is characterized by increased proliferation of blood vessels, vascular occlusion, angiogenesis, microaneurysms, hemorrhages, and infarction affecting the retina of the eye. In diabetic patients, hyperglycemia is a primary factor that causes diabetic retinopathy [24, 25]. However, there appear to be a difference in incidences of retinopathy between patients with type I or type II diabetes, suggesting that additional factors are involved [24]. The development of retinopathy is associated with progressing ischemia in the retina. Due to the ischemic and hypoxic environment in the retina, VEGF is produced at high levels by retinal pigment epithelial cells, as well as other cell types, in order to compensate the low levels of oxygen. This overexpression of VEGF stimulates new blood vessel growth [24, 26,

27]. However, VEGF-induced blood vessels usually have functional defects in their structures, which contain disorganized and leaky vessels [7]. As a result, neovascularization in the diabetic retina lead to tissue edema [27].

A common polymorphism in the human NPY gene was recently identified with the Leucine^7 to Proline^7 substitution (Leu^{7}Pro^7) in the signal peptide part of the NPY. The Leu^{7}Pro^7 polymorphism in the prepro-NPY gene exists in 6–13% of Caucasian populations [28]. The biological consequences of the Leu^{7}Pro^7 polymorphism include increased levels of blood lipids, decreased levels of insulin, accelerated atherosclerosis, and diabetitc retinopathy [28–31]. The underlying cellular mechanisms by which the Leu^{7}Pro^7 polymorphism causes these changes is not fully understood, but may involve an altered processing and secretion of NPY [32]. Low levels of insulin and high levels of lipids in the blood often lead to the development of type II diabetes among Leu^{7}Pro^7 polymorphism individuals. Most type II diabetic patients will develop certain degrees of retinopathy during their life spans, and diabetic retinopathy is considered the number one cause of blindness in diabetic patients. Positive correlations in clinical patients have been reported between the presence of the Leu^{7}Pro^7 polymorphism and development of diabetic retinopathy in patients with type II diabetes [30].

The angiogenic activity of NPY may contribute to the development of diabetes and retinopathy via two mechanisms: (1) Obesity is a prevalent factor leading to development of type 2 diabetes [33]. Although stimulation of food intake by the central nervous system is involved in NPY-induced obesity [3], elevated levels of circulating NPY in Leu^{7}Pro^7 polymorphism individuals may also play peripheral roles in development of obesity. Like any other healthy or pathological tissue growth, expansion of adipose tissue is also dependent on angiogenesis [34–36]. The high levels of plasma NPY in Leu^{7}Pro^7 polymorphism individuals may switch on the angiogenic phenotype in the adipose tissue and thus directly promote adipose tissue growth. The development and progression of obesity could further lead to severity of diabetes. (2) Direct role of NPY on retinal neovascularization. Deletion of the Y2 receptor, a critical receptor for NPY-mediated angiogenesis, results in prevention of retinal neovascularization in a hyperoxia mouse model. Further, Y2 receptor antagonists also inhibit hyproxia-induced neovascularization in a rat model [30]. These findings demonstrate that NPY is a critical factor for development of new blood vessels in the retina and that the Y2 receptor is responsible for transducing the angiogenic signals.

Conclusions

Accumulating evidence shows that NPY is a direct pro-angiogenic factor that promotes blood vessel growth in several tissues and organs. The mechanisms underlying its angiogenic effect requires activation of the Y2 receptor, which is expressed

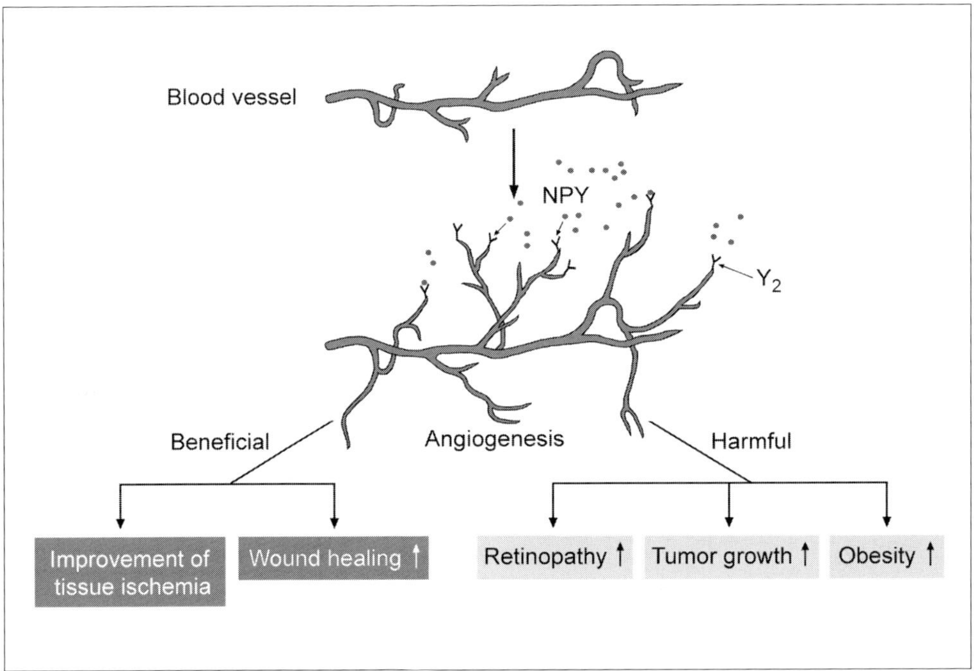

Figure 1
Schematic diagram of roles of NPY-induced angiogenesis in contribution to both physiological and pathological angiogenesis-related processes. The pro-angiogenic activity of NPY may be used in the treatment of disorders such as ischemic skeletal and cardiac muscles and wound healing. Paradoxically, NYP antagonists may instead be used for the treatment of cancer, diabetic retinopathy, and obesity.

on endothelial cells, although other receptors including Y1 and Y5 may also be involved. Due to its angiogenic activity, NPY exhibits a broader spectrum of physiological and pathological functions including wound healing, revascularization of ischemic tissue, development of diabetic retinopathy, progression of tumor growth and obesity (Fig. 1). Like most other pro-angiogenic factors, NPY may have paradoxical implications in the treatment of angiogenesis-related disorders. Due to its angiogenic effect, NPY might be potential as a therapeutic agent in the treatment of ischemic diseases, such as ischemic myocardium and skeletal muscle tissues. Blockage of the angiogenic function of NPY, on the other hand, may be used in the treatment and prevention of diabetic retinopathy. Thus, understanding the angiogenic functions and underlying molecular mechanisms of NPY may become an important target for disease therapy.

References

1 Blomqvist AG, Herzog H (1997) Y-receptor subtypes – how many more? *Trends Neurosci* 20: 294–298
2 Duhault J, Boulanger M, Chamorro S, Boutin JA, Della Zuana O, Douillet E, Fauchere JL, Feletou M, Germain M, Husson B et al (2000) Food intake regulation in rodents: Y5 or Y1 NPY receptors or both? *Can J Physiol Pharmacol* 78: 173–185
3 Gerald C, Walker MW, Criscione L, Gustafson EL, Batzl-Hartmann C, Smith KE, Vaysse P, Durkin MM, Laz TM, Linemeyer DL et al (1996) A receptor subtype involved in neuropeptide-Y-induced food intake. *Nature* 382: 168–171
4 Zukowska Z, Grant DS, Lee EW (2003) Neuropeptide Y: a novel mechanism for ischemic angiogenesis. *Trends Cardiovasc Med* 13: 86–92
5 Zukowska-Grojec Z, Karwatowska-Prokopczuk E, Fisher TA, Ji H (1998) Mechanisms of vascular growth-promoting effects of neuropeptide Y: role of its inducible receptors. *Regul Pept* 75–76: 231–238
6 Zukowska-Grojec Z, Karwatowska-Prokopczuk E, Rose W, Rone J, Movafagh S, Ji H, Yeh Y, Chen WT, Kleinman HK, Grouzmann E et al (1998) Neuropeptide Y: a novel angiogenic factor from the sympathetic nerves and endothelium. *Circ Res* 83: 187–195
7 Ekstrand AJ, Cao R, Bjorndahl M, Nystrom S, Jonsson-Rylander AC, Hassani H, Hallberg B, Nordlander M, Cao Y (2003) Deletion of neuropeptide Y (NPY) 2 receptor in mice results in blockage of NPY-induced angiogenesis and delayed wound healing. *Proc Natl Acad Sci USA* 100: 6033–6038
8 Cao R, Bjorndahl MA, Religa P, Clasper S, Garvin S, Galter D, Meister B, Ikomi F, Tritsaris K, Dissing S et al. (2004) PDGF-BB induces intratumoral lymphangiogenesis and promotes lymphatic metastasis. *Cancer Cell* 6: 333–345
9 Michel MC, Beck-Sickinger A, Cox H, Doods HN, Herzog H, Larhammar D, Quirion R, Schwartz T, Westfall T (1998) XVI. International Union of Pharmacology recommendations for the nomenclature of neuropeptide Y, peptide YY, and pancreatic polypeptide receptors. *Pharmacol Rev* 50: 143–150
10 Lee EW, Grant DS, Movafagh S, Zukowska Z (2003) Impaired angiogenesis in neuropeptide Y (NPY)-Y2 receptor knockout mice. *Peptides* 24: 99–106
11 Pons J, Kitlinska J, Ji H, Lee EW, Zukowska Z (2003) Mitogenic actions of neuropeptide Y in vascular smooth muscle cells: synergetic interactions with the beta-adrenergic system. *Can J Physiol Pharmacol* 81: 177–185
12 Erlinge D, Brunkwall J, Edvinsson L (1994) Neuropeptide Y stimulates proliferation of human vascular smooth muscle cells: cooperation with noradrenaline and ATP. *Regul Pept* 50: 259–265
13 Erlinge D, Yoo H, Edvinsson L, Reis DJ, Wahlestedt C (1993) Mitogenic effects of ATP on vascular smooth muscle cells versus other growth factors and sympathetic cotransmitters. *Am J Physiol* 265: 1089–1097
14 Shigeri Y, Nakajima S, Fujimoto M (1995) Neuropeptide YY1 receptors-mediated

increase in intracellular Ca^{2+} concentration via phospholipase C-dependent pathway in porcine aortic smooth muscle cells. *J Biochem* 118: 515–520

15 Zukowska-Grojec Z, Pruszczyk P, Colton C, Yao J, Shen GH, Myers AK, Wahlestedt C (1993) Mitogenic effect of neuropeptide Y in rat vascular smooth muscle cells. *Peptides* 14: 263–268

16 Fuhlendorff J, Gether U, Aakerlund L, Langeland-Johansen N, Thogersen H, Melberg SG, Olsen UB, Thastrup O, Schwartz TW (1990) [Leu31, Pro34]neuropeptide Y: a specific Y1 receptor agonist. *Proc Natl Acad Sci USA* 87: 182–186

17 Grundemar L, Jonas SE, Morner N, Hogestatt ED, Wahlestedt C, Hakanson R (1992) Characterization of vascular neuropeptide Y receptors. *Br J Pharmacol* 105: 45–50

18 Grundemar L, Krstenansky JL, Hakanson R (1993) Activation of neuropeptide Y1 and neuropeptide Y2 receptors by substituted and truncated neuropeptide Y analogs: identification of signal epitopes. *Eur J Pharmacol* 232: 271–278

19 Schwartz TW, Fuhlendorff J, Kjems LL, Kristensen MS, Vervelde M, O'Hare M, Krstenansky JL, Bjornholm B (1990) Signal epitopes in the three-dimensional structure of neuropeptide Y. Interaction with Y1, Y2, and pancreatic polypeptide receptors. *Ann NY Acad Sci* 611: 35–47

20 Wahlestedt C, Grundemar L, Hakanson R, Heilig M, Shen GH, Zukowska-Grojec Z, Reis DJ (1990) Neuropeptide Y receptor subtypes, Y1 and Y2. *Ann NY Acad Sci* 611: 7–26

21 Ghersi G, Chen W, Lee EW, Zukowska Z (2001) Critical role of dipeptidyl peptidase IV in neuropeptide Y-mediated endothelial cell migration in response to wounding. *Peptides* 22: 453–458

22 Mentlein R, Dahms P, Grandt D, Kruger R (1993) Proteolytic processing of neuropeptide Y and peptide YY by dipeptidyl peptidase IV. *Regul Pept* 49: 133–144

23 Lee EW, Michalkiewicz M, Kitlinska J, Kalezic I, Switalska H, Yoo P, Sangkharat A, Ji H, Li L, Michalkiewicz T et al. (2003) Neuropeptide Y induces ischemic angiogenesis and restores function of ischemic skeletal muscles. *J Clin Invest* 111: 1853–1862

24 Porta M, Bandello F (2002) Diabetic retinopathy A clinical update. *Diabetologia* 45: 1617–1634

25 Stitt AW (2003) The role of advanced glycation in the pathogenesis of diabetic retinopathy. *Exp Mol Pathol* 75: 95–108

26 Duh E, Aiello LP (1999) Vascular endothelial growth factor and diabetes: the agonist versus antagonist paradox. *Diabetes* 48: 1899–1906

27 Yamagishi S, Inagaki Y, Okamoto T, Amano S, Koga K, Takeuchi M, Makita Z (2002) Advanced glycation end product-induced apoptosis and overexpression of vascular endothelial growth factor and monocyte chemoattractant protein-1 in human-cultured mesangial cells. *J Biol Chem* 277: 20309–20315

28 Karvonen MK, Pesonen U, Koulu M, Niskanen L, Laakso M, Rissanen A, Dekker JM, Hart LM, Valve R, Uusitupa MI (1998) Association of a leucine(7)-to-proline(7) polymorphism in the signal peptide of neuropeptide Y with high serum cholesterol and LDL cholesterol levels. *Nat Med* 4: 1434–1437

29 Niskanen L, Voutilainen-Kaunisto R, Terasvirta M, Karvonen MK, Valve R, Pesonen U, Laakso M, Uusitupa MI, Koulu M (2000) Leucine 7 to proline 7 polymorphism in the neuropeptide y gene is associated with retinopathy in type 2 diabetes. *Exp Clin Endocrinol Diabetes* 108: 235–236

30 Koulu M, Movafagh S, Tuohimaa J, Jaakkola U, Kallio J, Pesonen U, Geng Y, Karvonen MK, Vainio-Jylha E, Pollonen M et al. (2004) Neuropeptide Y and Y2-receptor are involved in development of diabetic retinopathy and retinal neovascularization. *Ann Med* 36: 232–240

31 Karvonen MK, Valkonen VP, Lakka TA, Salonen R, Koulu M, Pesonen U, Tuomainen TP, Kauhanen J, Nyyssonen K, Lakka H et al. (2001) Leucine7 to proline7 polymorphism in the preproneuropeptide Y is associated with the progression of carotid atherosclerosis, blood pressure and serum lipids in Finnish men. *Atherosclerosis* 159: 145–151

32 Kallio J, Pesonen U, Kaipio K, Karvonen MK, Jaakkola U, Heinonen OJ, Uusitupa MI, Koulu M (2001) Altered intracellular processing and release of neuropeptide Y due to leucine 7 to proline 7 polymorphism in the signal peptide of preproneuropeptide Y in humans. *Faseb J* 15: 1242–1244

33 Hibbert-Jones E, Regan G, Bramwell J (2004) What do we know about... diabetes and obesity in adults and children? *J Fam Health Care* 14: 95–98

34 Rupnick MA, Panigrahy D, Zhang CY, Dallabrida SM, Lowell BB, Langer R, Folkman MJ (2002) Adipose tissue mass can be regulated through the vasculature. *Proc Natl Acad Sci USA* 99: 10730–10735

35 Cao R, Brakenhielm E, Wahlestedt C, Thyberg J, Cao Y (2001) Leptin induces vascular permeability and synergistically stimulates angiogenesis with FGF-2 and VEGF. *Proc Natl Acad Sci USA* 98: 6390–6395

36 Brakenhielm E, Cao R, Gao B, Angelin B, Cannon B, Parini P, Cao Y (2004) Angiogenesis inhibitor, TNP-470, prevents diet-induced and genetic obesity in mice. *Circ Res* 94: 1579–1588

NPY in vascular inflammation, remodeling and atherosclerosis

Lijun Li, Ken Abe, Edward W. Lee and Zofia Zukowska

Department of Physiology and Biophysics, Georgetown University Medical Center, 3900 Reservoir Rd NW, Washington, DC 20057, USA

Introduction

Previous chapters have described in details how neuropeptide Y (NPY) and related peptides modulate function of immune, inflammatory, vascular smooth muscle and endothelial cells, as well as sensory neurons. The accumulated evidence suggests that NPY is a potent vascular smooth muscle mitogen and angiogenic factor, which can also be pro- or anti-inflammatory, anti- or pro-nociceptive – in a receptor- and tissue-specific manner. For example, NPY stimulation of Y1 receptors in macrophages and leukocytes increases their adhesion, migration and release of proinflammatory cytokines such as IL-6 and IL-1β (see chapter by Bedoui and von Hoersten). In contrast, NPY's effect on lymphocytes may vary: in arthritic joints, peptide suppresses inflammation via Y1 receptors but in systemic lupus erythematosus (SLE), elevated peptide levels are associated with development of the disease (Bedoui and von Hoersten).

Recent experimental and clinical data suggests that atherosclerosis is also an inflammatory disease [1]. Monocyte transmigration into the intima and formation of activated macrophages, uptake of oxidized lipids and formation of foam cells are considered the hallmark and an initial step of the atherosclerotic process [1–3]. Mice deficient in macrophage scavenger receptor, cd36, have attenuated plaque formation [4] and monocyte deficiency *per se* is also protective against atherosclerosis in ApoE knockout mice [4], an animal model of the disease [5]. The role of T lymphocytes in development of atherosclerotic lesions has been less recognized but also gained attention in recent years [6]. A particularly pronounced immune component is seen in vascular remodeling that occurs after transplantation [7].

While immune cells are known to be regulated by cytokines and the endocrine system, and whether or not nerves, sympathetic or sensory, play any role in triggering or modulating immune responses in the vascular wall remains poorly understood. We hypothesized that neuroimmunomodulation occurs within the vessel wall following surgical interventions such as balloon angioplasty – and that NPY is a

The NPY Family of Peptides in Immune Disorders, Inflammation, Angiogenesis and Cancer, edited by Zofia Zukowska and Giora Z. Feuerstein

major mediator of this cross talk. Using rodent models of vascular injury, we have shown that neuroinflammation, which is elicited following carotid artery angioplasty by local perivascular administration of NPY, is mimicked by chronic stress, and both are completely prevented by specific Y1 receptor antagonist [8]. In addition, strains of mice which have elevated plasma NPY levels, due to stress or overexpression in platelets, have exaggerated angioplasty-induced vascular neointimal and inflammatory responses, while these lesions are markedly attenuated in NPY$^{-/-}$ mice (Abe et al., unpublished).

NPY is a vascular mitogen

All arteries of the body, except the thoracic aorta, in humans and rodents, are richly innervated by the sympathetic nerves, which contain both neurotransmitters norepinephrine (NE) and NPY. While NE is released with the slightest sympathetic activation, the release of NPY is usually delayed and occurs following more prolonged and intense nerve stimulation [9] (see [35]). Once released, NPY acts on its vascular Y1 receptors, which are the predominant type of NPY receptors in blood vessels, and causes vasoconstriction [10]. NPY also potentiates contractile effects of NE [11] and both actions are responsible for a major role that the peptide plays in control of vascular tone during stress [10].

The same Y1 receptors mediate NPY's proliferative effects on vascular smooth muscle cells (VSMC) [12, 13] (see chapter by Kitlinska et al.). However, during VSMC growth, another NPY's receptor, the Y5, is induced, and also contributes to peptide's mitogenic action. Interactions between Y1 and Y5, possibly involving their heterodimerization (see chapters by Kitlinska and [36]), amplify NPY's proliferative activity and are responsible for bimodality of the NPY-growth response curve. As a result, the peptide exerts high affinity mitogenic effects already at sub-picomolar concentrations, in addition to having activity of lower affinity, at nanomolar concentrations [12]. Both growth responses are inhibited by Y1 or Y5 receptor antagonist, but their complete blockade requires a combination of both inhibitors [12].

NPY induces medial hypertrophy and stimulates neointima formation

To determine if NPY plays a role in vascular remodeling *in vivo*, we used an established model of balloon angioplasty of the rat carotid artery. Normally, the balloon injury-induced denudation of the endothelium stimulates VSMC proliferation and migration from the advential-medial layer into the space of former intima, where they form *neo*intima. This process does not appear to lead to any significant inflammatory reactions or thrombus formation unless vascular integrity is damaged, and vessels remain patent, albeit narrowed. Many growth factors such as angiotensin II, endothelin and basic fibroblast growth factors have been shown to augment neointima formation in this model [14].

Angioplasty of the rat carotid artery activated vascular NPY system. Two weeks after surgery, NPY and its receptors, Y1 and Y5, were markedly upregulated (at mRNA and protein levels) in VSMCs of the neointima, while in the media, primarily Y1 receptor immunostaining was increased [8]. Administration of NPY as a slow release pellet at low physiological concentration (1–10 µg/pellet/14 days) caused even greater, massive, upregulation of Y1 and Y5 receptor immunostaining in neointimal cells as well as in macrophages, which normally, after angioplasty alone, are very sparse in the vessel wall (Li et al., unpublished). Similar upregulation of its own receptor expression by NPY was observed *in vitro* in VSMC [12].

Fourteen days after NPY (1–10 µg/14 days) was placed at the side of the balloon-injured rat carotid artery, media thickness increased by up to 50% (Pons, 2003 #8357). Remarkably, this increase was completely prevented by continuous infusion of a Y1 receptor antagonist (H409/22 acetate, 0.02 µmol/kg/min/14 days i.v.) but not by the Y5 receptor antagonist [12]. Since NPY had no effect when placed by the side of the intact carotid artery, the ability of the peptide to induce medial hypertrophy appears to require other mediators, present only in the injured vessel. This observation suggests that NPY-Y1 receptors may contribute to medial hypertrophy in conditions when blood vessels are “activated” by other stimuli, for example, by endothelial denudation. Since endothelial damage is postulated to be involved in many cardiovascular diseases, including hypertension, amplification of VSMC growth provided by NPY and its Y1 receptors may be a mediator of vascular medial hypertrophy in these states. The contribution of this pathway might be particularly important in certain forms of essential hypertension, where sympathetic activity is increased. Since Y1 antagonist was effective in blocking NPY’s effect on medial hypertrophy, drugs inhibiting Y1 receptors could be useful in therapy of hypertension, similarly to how angiotensin converting enzyme inhibitors have been, reducing vascular tone not only by antagonizing vasoconstrictive but also growth-promoting activities of these peptides.

NPY stimulates thrombosis, vascular inflammation and atherosclerosis

In placebo-treated rats, angioplasty stimulated formation of a concentric neointima, with no inflammatory reaction [8]. In contrast, neointima which formed in the NPY-treated vessels was highly and dose-dependently augmented, and in addition, was non-laminar and, at the highest dose (10 µg/14 days), occlusive [8]. Also, the NPY-induced neointimal lesion, unlike the one produced by angioplasty alone, contained numerous microvessels, thrombus and deposits of matrix, macrophages and lipids (Fig. 1). Co-localization of macrophages and lipids suggested that these cells may have become foam cells, the cells that are the hallmark of atherosclerosis. Thus, the

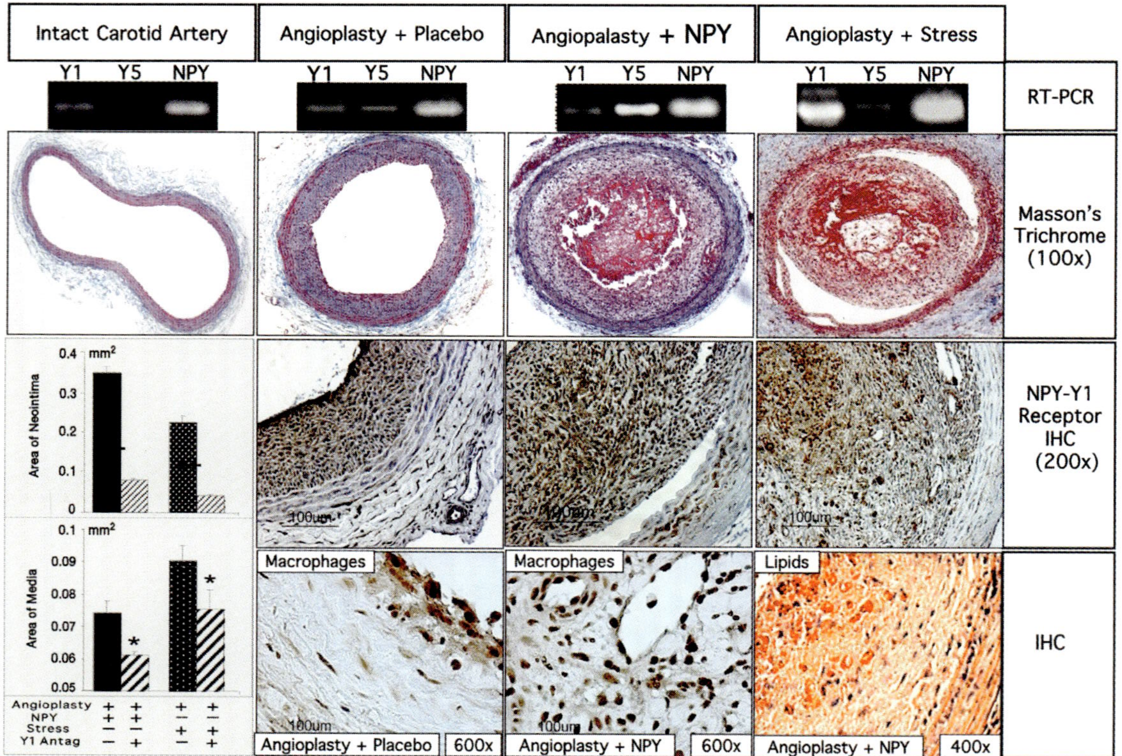

Figure 1. NPY- and stress-induced changes in the rat carotid arteries following balloon angioplasty, both prevented by Y1 receptor antagonist. The NPY-Y1-mediated vascular remodeling involves up-regulation of Y1 receptors in the vessel wall, inflammation (cd68+ macrophages), thrombosis, neointima formation, neovascularization and lipid deposition (Red Oil O+), leading to vessel occlusion within 2 weeks.

Figure 1

NPY- and stress-induced changes in the rat carotid arteries following balloon angioplasty, both prevented by Y1 receptor antagonist. The NPY-Y1-mediated vascular remodeling involves inflammation, thrombosis, neointima formation, neovascularization and lipid deposition, leading to vessel occlusion within 2 weeks.

NPY-induced lesion resembled an advanced atherosclerotic plaque, although it had formed within only 2 weeks and in rats which were devoid of any genetic or diet-induced alteration in lipid metabolism.

How NPY could elicit such pronounced pro-atherosclerotic effects in such a short period of time is puzzling. Perhaps one hypothesis that could explain this is that NPY, being a neurogenic mediator, is positioned at the top of the cascade of events, which then leads to activation of multiple pathways. NPY has indeed pleiotropic activities, some of which are direct and other via activation of other mediators. NPY, in addition to being a VSMC mitogen, stimulates chemotaxis and proliferation of endothelial cells [15] and monocytes [16] – and this activity may be pivotal in its ability to cause macrophage transmigration into the neointima and angiogenesis of the lesion. In addition, the peptide is known for its metabolic activity to stimulate appetite, activate lipoprotein lipase, promote lipid storage and increase plasma lipids [17]. While these activities have been believed to be only of central origin, where NPY is a potent orexigenic peptide, recent studies from our lab suggest that NPY is adipogenic and lipogenic by peripheral actions on the adipose tissue itself. Furthermore, NPY may stimulate vascular inflammation and shift the profile of T-lymphocytes from anti-inflammatory Th1 to proinflammatory Th2 [18], similarly to what occurs during formation of atheroma. Finally, the peptide is potently angiogenic [15] and can stimulate formation of new vasa vasorum, feeding the neointimal lesion and promoting its growth. This action is in part indirect and involves release of vascular endothelial growth factor and nitric oxide [19].

Interestingly, NPY signal peptide gene polymorphism, Leu^7Pro^7, is associated with similar derangements as those evoked by NPY in rats subjected to angioplasty. The mutation, which is rather frequent in Finns and other Northern Europeans (6–8%) [20], is functional and leads to higher stress-induced plasma NPY levels [21], and associates with hyperlipidemia and accelerated atherosclerosis of human carotid arteries [22] (see chapter by Pesonen). Thus, it appears that high NPY may be a risk factor for atherosclerosis. What conditions predispose to a "high NPY state"? First and foremost, stress, as NPY is a neurotransmitter released during sympathetic nerve activation.

Stress mimics NPY-induced lesions by activating Y1 receptors

To determine if endogenous NPY, released during stress, exerts similar pro-atherosclerotic effects after angioplasty, the experiment was repeated in stressed versus non-stressed rats. Stress consisted of rats standing in 1 cm of ice-cold water for 2 h, starting with the period just before angioplasty, and continued daily for 2 weeks (Li et al., unpublished). We have shown previously that such a stress is a powerful neurogenic stimulus for NPY release [23], which then activates Y1 receptors and causes vasoconstriction [10]. NPY-immunoreactivity increased in the platelet-poor- and platelet-rich-plasmas, and in the latter, it attained the levels similar to those induced by angioplasty alone [8].

Two weeks after angioplasty, balloon-injured carotid arteries of stressed rats resembled the NPY-treated vessels (Fig. 1). While in unstressed rats, a concentric laminar neointima with sparse monocytes was formed, in the stressed animals, vessels were occluded with a lesion containing neointima, thrombus, microvessels and a massive invasion of the neointima with immune cells, such as macrophages, which co-localized with deposition of lipids (Fig. 1). Unlike NPY-treated vessels [8], however, stress induced greater adventitial thickening and invasion of immune and smooth muscle cells from the adventitia into the media and neointima, sometimes with dissolution of the external lamina. These changes suggested a development of a neurogenic-type inflammation, with concentration of the immune infiltrate at the adventitial-medial border, the place of entry of the sympathetic and sensory nerves into the vessel wall.

Remarkably, *all* of the stress-induced changes were *prevented* by a continuous administration of Y1 receptor antagonist (H409/22, 0.02 mmol/kg/min/14 days, i.v.), at the dose which blocked the effects of the exogenous NPY (Fig. 1). Thus, stress, by releasing NPY, stimulated, via activation of Y1 receptors, a cascade of inflammatory events leading to vessel remodeling with an occlusive atherosclerotic-like lesion. Which cells expressing these Y1 receptors are primary responders remains to be determined but the likely candidates are monocytes/macrophages. As others have shown, macrophages are attracted to adrenergic nerves, and inflammation or injury upregulates their Y1 receptors [24]. Exogenous NPY, when combined with β-adrenergic agonists, increases monocyte release of cytokines such as IL-1β and IL-6 and also stimulates macrophage proliferation and migration [25]. Interestingly, these effects are also inhibited by the same type of the Y1 receptor antagonist as used in our study (see chapter by Bedoui and von Hoersten). Taken together, these observations suggest that triggering nerve-macrophage interplay by stress-induced sympathetic nerve activation upregulates their NPY-Y1 system, which then accelerates neurogenic inflammation and vascular remodeling after angioplasty. Hence, Y1 receptor antagonists may be useful as therapy against restenosis.

Platelet NPY as a risk factor for exaggerated vascular remodeling

In addition to exerting local actions in the vascular wall, including upregulation of the NPY system, balloon angioplasty *per se* had significant systemic effects. In rats 2 weeks after angioplasty, platelet NPY levels doubled, although plasma NPY levels remained unchanged [8]. Rat megakaryocytes, like those of some mice (see below), synthesize NPY, which is then packaged into platelet vesicles and released during aggregation [26]. Thus, it appears that vessel injury by angioplasty activated some humoral or neurogenic mechanism by which either platelet uptake, or more likely megakaryocyte synthesis of NPY, increased. This increase was prevented by the Y1 receptor antagonist suggesting that it's not due to blockade of peptide uptake by platelets since Y1 receptors are known to participate in peptide's internalization [27] and opposite effects would have been expected from their blockade.

The role of platelet NPY in angioplasty-induced vascular lesions is completely unknown. Few previous studies which dealt with the platelet NPY, including that of Ericsson et al. [28], Ogawa et al. [29] and ours [30], have established that some strains of mice (like rats but not normal humans or pigs) synthesize authentic NPY in megakaryocytes and package it into platelets, and these mice often have some autoimmune abnormalities, such as SLE-like. Since NPY appears to contribute to pathogenesis of this disease, which involves many vascular abnormalities such as perivasculitis nodosa, we speculated that similar association may occur with neointimal lesions, which develop after angioplasty. The mechanism(s) of this process is poorly understood but recent data suggest that platelet aggregation may be critical for generation of proinflammatory cytokines and triggering leukocyte/monocyte activation, which then activates inflammatory cascade [31]. What effects NPY has on platelet aggregation is controversial as both stimulatory effect in rat platelets and inhibitory in human platelets have been reported [30].

The hypothesis that platelet NPY provides an additional source of the peptide, which is activated by angioplasty and amplifies vascular responses to injury, was then tested in two murine strains known to genetically vary in their megakaryocyte/platelet NPY content (Abe et al., unpublished). In two models, the ligation of the carotid artery and wire-induced angioplasty of the femoral artery, C57/BL mice, which do *not* express platelet NPY, were resistant, while Sv129/J, which synthesize NPY in megakaryocytes, were sensitive to vascular injury and formed significant neointima (Abe et al., unpublished) (Fig. 2). Of interest is that C57/BL mice, which were completely unable to produce any neointima in response to vascular interventions, are often used in atherosclerosis studies, as a strain prone to metabolic abnormalities and a wild type control for $ApoE^{-/-}$, a standard model of the disease. This implies that mechanisms of angioplasty-induced vascular remodeling and atherosclerosis differ, as ample evidence indicates [32], but also suggests that platelet NPY plays an amplifying role in formation of neointimal lesions following vascular interventions.

Neuronal-megakaryocyte interaction contributes to vascular remodeling

The remarkable association of the ability of the mouse strain to respond to vascular injury with the expression of platelet NPY and its upregulation following angioplasty, suggested a possibility of neuronal or humoral control of megakaryocyte NPY production (since platelets lack ability for synthesis and peptide uptake is unlike, see above). C57/BL mice, which are devoid of platelet NPY, had low level of

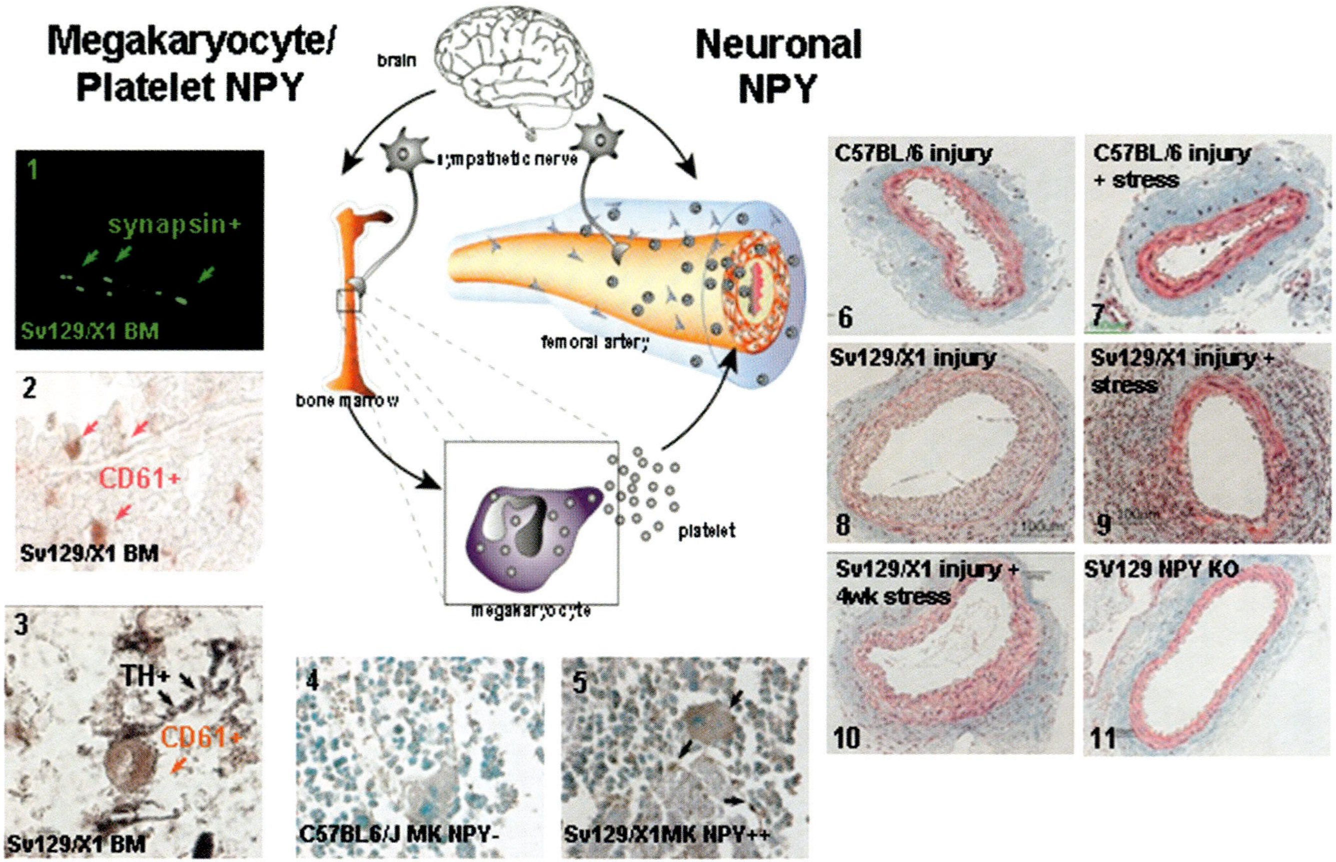

Figure 2
Postulated interplay between an injured vascular wall, sympathetic nerves, bone marrow-derived megakaryocytes and platelets in the development of vascular remodeling after angioplasty and stress

expression of NPY mRNA in the bone marrow and none in megakaryocytes, while Sv129/J expressed it abundantly in both (Fig. 2). In SV129/J but not C57/BL mice, stress markedly upregulated bone marrow NPY mRNA, while increasing platelet NPY levels. In the former, this was associated with marked increase in neointima formation and an induction of a thrombotic and inflammatory lesion in the femoral artery injured by wire angioplasty (Fig. 2). Latest studies on NPY-/-/Sv129/J mice suggest that angioplasty- and stress-induced vascular lesions are visibly inhibited in these animals. These observations indicate that endogenous NPY, platelet- and nerve-derived, is necessary for formation of neointima, thrombus and inflammatory vascular reactions after angioplasty, at least in rodents. While rats and mice obviously are not humans, whether or not some men are like rats in terms of their platelet NPY remains to be determined. Interestingly, Nilsson et al. [33] have recently reported higher platelet but not plasma NPY levels in patients with depression.

Summary and conclusions

There are no acceptable rodent models of restenosis which would develop vascular lesions similar to human. Models of atherosclerosis in rats and mice exist but they require genetic modification of lipid handling systems, for example ApoE knockouts, which are rarely the sole basis of human disease. Moreover, vascular remodeling developing in those models lacks resemblance to human lesions since they are not thrombotic or inflammatory. While stress has been implicated in pathogenesis of cardiovascular diseases, its role in atherosclerosis has been thought to be indirect and poorly defined. Our studies have shown for the first time that repeated stress is a potent amplifier of vascular inflammation, and thrombus and neointima formation after angioplasty in both rats and mice. NPY, the sympathetic vasoconstrictor neurotransmitter and a vascular growth factor, elicits similar effects via its Y1 receptors. Stress releases NPY from the sympathetic nerves and upregulates megakaryocyte/platelet NPY system, and both appear to play a role in development of an occlusive vascular lesion. In stressed rodents or NPY-treated rats, post-angioplasty vascular lesions resemble atherosclerotic-like plaques by containing thrombus, matrix, microvessels, macrophages and lipid deposition. Platelet NPY appears to be an additional amplifier of these lesions, as mice which express it are susceptible, while those that do not are resistant, to angioplasty- and stress-induced vascular effects. Thus, stress and NPY are mediators of an inflammatory, lipid-rich and thrombotic vascular remodeling, which can lead to rapid vessel occlusion after angioplasty. Since both effects were completely prevented by a specific Y1 receptor antagonist, drugs directed at inhibiting these receptors may become a useful therapy against restenosis in humans. At particular risk would be patients with NPY-Leu^7Pro^7 signal peptide polymorphism, which has been associated with accelerated atherosclerosis and higher stress-induced NPY levels, men (as opposed to women), whose NPY system is upregulated by androgens [34], and population under severe or chronic stress, which increases NPY release.

Acknowledgements
This study was supported in part by grants from the National Institutes of Health: HL55310 to Dr. Zofia Zukowska.

References

1 Libby P (2002) Atherosclerosis: the new view. *Sci Am* 286: 46–55
2 Russell ME (1997) Graft arteriosclerosis: molecular features of macrophage activation. *Transplant Proc* 29: 2542–2543
3 Cipollone F, Marini M, Fazia M, Pini B, Iezzi A, Reale M, Paloscia L, Materazzo G, D'Annunzio E, Conti P et al. (2001) Elevated circulating levels of monocyte chemoattractant protein-1 in patients with restenosis after coronary angioplasty. *Arterioscler Thromb Vasc Biol* 21: 327–334
4 Nicholson AC, Han J, Febbraio M, Silversterin RL, Hajjar DP (2001) Role of CD36, the macrophage class B scavenger receptor, in atherosclerosis. *Ann NY Acad Sci* 947: 224–228
5 Plump AS, Smith JD, Hayek T, Aalto-Setala K, Walsh A, Verstuyft JG, Rubin EM, Breslow JL (1992) Severe hypercholesterolemia and atherosclerosis in apolipoprotein E-deficient mice created by homologous recombination in ES cells. *Cell* 71: 343–353
6 Buono C, Pang H, Uchida Y, Libby P, Sharpe AH, Lichtman AH (2004) B7-1/B7-2 costimulation regulates plaque antigen-specific T-cell responses and atherogenesis in low-density lipoprotein receptor-deficient mice. *Circulation* 109: 2009–2015
7 gano H, Libby P, Taylor MK, Hasegawa S, Stinn JL, Becker G, Tilney NL, Mitchell RN (1998) Coronary arteriosclerosis after T-cell-mediated injury in transplanted mouse hearts: role of interferon-gamma. *Am J Pathol* 152: 1187–1197
8 Li L, Lee EW, Ji H, Zukowska Z (2003) Neuropeptide Y-induced acceleration of postangioplasty occlusion of rat carotid artery. *Arterioscler Thromb Vasc Biol* 23: 1204–1210
9 Zukowska-Grojec Z, Konarska M, McCarty R (1988) Differential plasma catecholamine and neuropeptide Y responses to acute stress in rats. *Life Sci* 42: 1615–1624
10 Zukowska-Grojec Z, Dayao EK, Karwatowska-Prokopczuk E, Hauser GJ, Doods HN (1996) Stress-induced mesenteric vasoconstriction in rats is mediated by neuropeptide Y Y1 receptors. *Am J Physiol* 270: H796–H800
11 Wahlestedt C, Hakanson R, Vaz CA, Zukowska-Grojec Z (1990) Norepinephrine and neuropeptide Y: vasoconstrictor cooperation *in vivo* and *in vitro*. *Am J Physiol* 258: R736–R742
12 Pons J, Kitlinska J, Ji H, Lee EW, Zukowska Z (2003) Mitogenic actions of neuropep-

tide Y in vascular smooth muscle cells: synergetic interactions with the beta-adrenergic system. *Can J Physiol Pharmacol* 81: 177–185

13 Erlinge D, Brunkwall J, Edvinsson L (1994) Neuropeptide Y stimulates proliferation of human vascular smooth muscle cells: cooperation with noradrenaline and ATP. *Regul Pept* 50: 259–265

14 Laurent S, Vanhoutte P, Cavero I, Chabrier PE, Dupuis B, Elghozi JL, Hamon G, Janiak P, Juillet Y, Kher A et al (1996) The arterial wall: a new pharmacological and therapeutic target. *Fundam Clin Pharmacol* 10: 243–257

15 Zukowska-Grojec Z, Karwatowska-Prokopczuk E, Rose W, Rone J, Movafagh S, Ji H, Yeh Y, Chen WT, Kleinman HK, Grouzmann E et al (1998) Neuropeptide Y: a novel angiogenic factor from the sympathetic nerves and endothelium. *Circ Res* 83: 187–195

16 De la Fuente M, Bernaez I, Del Rio M, Hernanz A (1993) Stimulation of murine peritoneal macrophage functions by neuropeptide Y and peptide YY. Involvement of protein kinase C. *Immunology* 80: 259–265

17 Kalra SP, Clark JT, Sahu A, Dube MG, Kalra PS (1988) Control of feeding and sexual behaviors by neuropeptide Y: physiological implications. *Synapse* 2: 254–257

18 Bedoui S, Miyake S, Lin Y, Miyamoto K, Oki S, Kawamura N, Beck-Sickinger A, Von Horsten S, Yamamura T (2003) Neuropeptide Y (NPY) suppresses experimental autoimmune encephalomyelitis: NPY(1) receptor-specific inhibition of autoreactive Th1 responses *in vivo*. *J Immunol* 171: 3451–3458

19 Zukowska Z, Grant DS, Lee EW (2003) Neuropeptide y. A novel mechanism for ischemic angiogenesis. *Trends Cardiovasc Med* 13: 86–92

20 Karvonen MK, Pesonen U, Koulu M, Niskanen L, Laakso M, Rissanen A, Dekker JM, Hart LM, Valve R, Uusitupa MI (1998) Association of a leucine(7)-to-proline(7) polymorphism in the signal peptide of neuropeptide Y with high serum cholesterol and LDL cholesterol levels. *Nat Med* 4: 1434–1437

21 Kallio J, Pesonen U, Kaipio K, Karvonen MK, Jaakkola U, Heinonen OJ, Uusitupa MI, Koulu M (2001) Altered intracellular processing and release of neuropeptide Y due to leucine 7 to proline 7 polymorphism in the signal peptide of preproneuropeptide Y in humans. *Faseb J* 15: 1242–1244

22 Karvonen MK, Valkonen VP, Lakka TA, Salonen R, Koulu M, Pesonen U, Tuomainen TP, Kauhanen J, Nyyssonen K, Lakka HM et al. (2001) Leucine7 to proline7 polymorphism in the preproneuropeptide Y is associated with the progression of carotid atherosclerosis, blood pressure and serum lipids in Finnish men. *Atherosclerosis* 159: 145–151

23 Zukowska-Grojec Z (1995) Neuropeptide Y. A novel sympathetic stress hormone and more. *Ann NY Acad Sci* 771: 219–233

24 Bedoui S, Lechner S, Gebhardt T, Nave H, Beck-Sickinger AG, Straub RH, Pabst R, von Horsten S (2002) NPY modulates epinephrine-induced leukocytosis via Y-1 and Y-5 receptor activation *in vivo*: sympathetic co-transmission during leukocyte mobilization. J Neuroimmunol 132: 25–33

25 Medina S, Del Rio M, Ferrandez MD, Hernanz A, De la Fuente M (1998) Changes with

age in the modulation of natural killer activity of murine leukocytes by gastrin-releasing peptide, neuropeptide Y and sulfated cholecystokinin octapeptide. *Neuropeptides* 32: 549–555

26 Myers AK, Farhat MY, Vaz CA, Keiser HR, Zukowska-Grojec Z (1988) Release of immunoreactive-neuropeptide Y by rat platelets. *Biochem Biophys Res Commun* 155: 118–122

27 Berglund MM, Hipskind PA, Gehlert DR (2003) Recent developments in our understanding of the physiological role of PP-fold peptide receptor subtypes. *Exp Biol Med (Maywood)* 228: 217–244

28 Ericsson A, Schalling M, McIntyre KR, Lundberg JM, Larhammar D, Seroogy K, Hokfelt T, Persson H (1987) Detection of neuropeptide Y and its mRNA in megakaryocytes: enhanced levels in certain autoimmune mice. *Proc Natl Acad Sci USA* 84: 5585–5589

29 Ogawa T, Kitamura K, Kangawa K, Matsuo H, Eto T (1992) Platelet neuropeptide Y in spontaneously hypertensive rats. *J Hypertens* 10: 765–771

30 Myers AK, Torres Duarte AP, Zukowska-Grojec Z (1993) Immunoreactive neuropeptide Y (NPY) in plasma and platelets of rat and mouse strains and human volunteers. *Regul Pept* 47: 239–245

31 Wagner DD (2005) New links between inflammation and thrombosis. *Arterioscler Thromb Vasc Biol* 25: 1321–1324

32 Carmeliet P, Collen D (1997) Molecular genetics of the fibrinolytic and coagulation systems in haemostasis, thrombogenesis, restenosis and atherosclerosis. *Curr Opin Lipidol* 8: 118–125

33 Nilsson C, Karlsson G, Blennow K, Heilig M, Ekman R (1996) Differences in the neuropeptide Y-like immunoreactivity of the plasma and platelets of human volunteers and depressed patients. *Peptides* 17: 359–362

34 Lewandowski J, Pruszczyk P, Wocial B, Ignatowska-Switalska H, Romejko-Woniewicz E, Madej-Firek A, Januszewicz W, Zukowska-Grojec Z (1996) Sex hormone modulation of neuropeptide Y and cardiovascular responses to stress in humans. In: R McCarty, G Aguilera, E Sabban, R Kvetnansky (eds): *Stress: Molecular and Neurobiological Advances*. Gordon and Breach Science Publishers S.A., New York, USA, 569–578

35 Dutton MA, Lee EW, Zukowska Z (2005) NPY and extreme stress: lessons learned from posttraumatic stress disorder. In: Z Zukowska, GZ Feuerstein (eds): *NPY Family of Peptides in Neurobiology, Cardiovascular and Metabolic Disorders: from Genes to Therapeutics*. Birkhäuser Publishing Ltd., Basel, Switzerland

36 Merten N, Beck-Sickinger AG (2005) Molecular ligand-receptor interaction of the NPY/PP peptide family. In: Z Zukowska, GZ Feuerstein (eds): *NPY Family of Peptides in Neurobiology, Cardiovascular and Metabolic Disorders: from Genes to Therapeutics*. Birkhäuser Publishing Ltd., Basel, Switzerland

The NPY family of peptides as growth factors in neurodegenerative diseases and tumor biology

NPY as a pleiotropic growth factor

Joanna Kitlinska, Lydia E. Kuo and Jennifer Pons

Department of Physiology and Biophysics, Georgetown University School of Medicine, 3900 Reservoir Rd, NW, Basic Science Building, Washington DC 20057, USA

Introduction

Neuropeptide Y (NPY) is a pleiotropic peptide with activities ranging from the inhibition of neurotransmitter release and vasoconstriction, to regulation of appetite and anxiety [1]. NPY release is increased during sympathetic hyperactivity, thus its actions are augmented in response to stress [2, 3]. The peptide has been implicated in many clinical disorders, from hypertension to depression [3–5]. Recently, new functions for NPY have emerged with the discovery of its growth-regulatory actions. In CHO-K1 cells transfected with Y1 or Y2 receptors, NPY stimulation leads to the activation of p42/44 mitogen-activated protein kinase (MAPK) [6] and an increase in proliferation [7]. The peptide exerts proliferative effects in a variety of cells, including neuronal precursors [8] and neural crest-derived tumor cells [9], vascular smooth muscle cells (VSMC) [10–13] and endothelial cells (EC) [14], preadipocytes [15], lymphocytes [16–18], retinal glial cells [19], beta cells of pancreatic islets [20] and prostate cells [21]. Its mitogenic actions are tissue and receptor specific. NPY receptor signaling involves activation of PI3K and PKC kinases, leading to phosphorylation of p42/44 and p38 MAPKs [8, 9, 19]. Thus far, the proliferative actions of the peptide have been the most extensively studied in neuronal, vascular and immune cells. Therefore, in this chapter we will focus on NPY growth-regulatory effects in these cells and their clinical implications.

NPY as a vascular growth factor

Mitogenic effect of NPY on vascular smooth muscle cells

The first interest in the role of NPY in the cardiovascular system was associated with its Y1 receptor-mediated vasoconstrictive activity. However, recent discoveries indicated that the peptide can also stimulate the proliferation of VSMCs via its Y1

The NPY Family of Peptides in Immune Disorders, Inflammation, Angiogenesis and Cancer,
edited by Zofia Zukowska and Giora Z. Feuerstein

and Y5 receptors [10–13]. NPY actions in these cells are mediated by $G_{i/o}$-proteins and involve the inhibition of adenylyl cyclase, inositol 1,4,5-triphosphate (IP3)-dependent increase in intracellular Ca^{2+} and the activation of p44/42 MAPK [22–24]. The mitogenic effect of the peptide in VSMCs is observed at sub-picomolar and sub-nanomolar concentrations significantly lower than those necessary for vasoconstriction. Interestingly, NPY stimulates proliferation of VSMCs in bimodal fashion, with two peaks – a high affinity peak at picomolar concentrations and a low affinity peak at nanomolar concentrations [11]. The same bimodal mitogenic effect of NPY has been also observed in other cells, such as neuroblastomas and ECs, which indicates that this phenomenon is a universal feature of the peptide's proliferative actions [9, 25]. The exact mechanism of this bimodal effect has not been fully elucidated yet. However, our recent studies indicate that it may involve interactions of the multiple NPY receptors. In CHO-K1 cells stably transfected with a single Y1 or Y2 receptor, NPY stimulates proliferation only at high concentrations, corresponding to the low affinity peak observed in VSMCs [7, 26]. Thus, the high affinity peak seems to be dependent on the presence of the multiple NPY receptors. This hypothesis is further supported by the recent discovery of the NPY receptor homo- and heterodimerization [27, 28]. The presence of NPY receptor complexes may also be associated with differential coupling to the signaling pathways. We have found that both growth peaks are associated with a bimodal, $G_{i/o}$-protein-dependent decrease in cAMP as well as PKC and MAPK activation. However, the intracellular calcium mobilization is dose-dependent, without bimodality, which differentiates the two mitogenic peaks [29]. Our recent study using proteomics [30] suggests that different isoforms of CAMK may be involved in NPY's low and high affinity mitogenic peaks in VSMCs. However, further studies are necessary to elucidate the mechanisms of NPY's bimodal effect in VSMC and other cells and its physiological significance.

Due to its mitogenic effect in VSMCs, NPY is an important factor contributing to vascular remodeling. In a rat carotid angioplasty model, the vessel injury stimulates proliferation of VSMCs and neointima formation. NPY administrated after angioplasty as a slow release pellet dramatically augments the growth of neointima leading to the formation of vascularized, lipid-filled, atherosclerotic-like plaques and often to the occlusion of the vessel [31] (see chapter by Li et al.). The effects of both angioplasty alone and exogenous NPY are dependent on the Y1 and Y5 receptors, which further support the role of these receptors in VSMC proliferation. Interestingly, chronic stress, known to increase NPY plasma levels, mimics the effect of the exogenous peptide and leads to the complete occlusion of the injured vessel, which is also Y1 receptor-dependent. Therefore, sympathetic nerve-derived NPY is a crucial factor in vascular remodeling (see chapter by Li et al.).

The growth-promoting effect of NPY on VSMCs observed *in vitro* and *in vivo* suggests its important role in the development of atherosclerosis-related diseases, such as heart and peripheral vascular diseases, restenosis following vascular intervention and hypertension. Indeed, a leucine^7 to proline^7 (Leu^{7}Pro^7) polymorphism in NPY signal peptide, known to facilitate release of the peptide, has been associated with increased rate of atherosclerosis in Finnish and Dutch population [32] (see chapter by Pesonnen et al.). Thus, utilizing the Y1 and Y5 receptor antagonists may be an attractive strategy in treatment of these disorders.

Trophic effect of NPY on cardiac myocytes

Like in VSMCs, NPY also exerts a growth-promoting effect on cardiomyocytes, albeit by different mechanisms. The peptide stimulates hypertrophy of cardiac cells, which is associated with an increase in the total cell protein caused by the inhibition of their degradation and augmented *de novo* synthesis [33, 34]. This effect is mediated by $G_{i/o}$-proteins and the PI3K/p70^{s6k} pathway and is accompanied by increased activity of cytosolic creatine kinase [33, 35, 36]. Elevated NPY has been found in patients with cardiac hypertrophy and heart failure (see chapter by Feuerstein). Moreover, in patients with pheochromocytoma, a sympathetic tumor developing in the adrenal gland, the increased levels of plasma NPY correlate with left ventricular hypertrophy [37]. Therefore, chronic exposure to the elevated NPY levels, due to stress or other pathological conditions, may contribute to the development of hypertrophic heart diseases.

Angiogenic activity of NPY

In addition to its growth-promoting effect on VSMCs, NPY has been recently found to be a potent angiogenic factor. Endothelial cells (ECs) express both NPY and its receptors. The peptide stimulates proliferation, migration and capillary tube formation of ECs [14]. Like in VSMCs, the proliferative effect of NPY is bimodal [25], which indicates involvement of multiple receptors. However, as opposed to the Y1/Y5 receptor-mediated proliferation of VSMCs, the angiogenic activities of NPY seem to be Y2/Y5 receptor-dependent. The angiogenic properties of the peptide have been established using a variety of models, such as aortic sprouting, rodent hindlimb ischemia, retinopathy and wound healing [25, 38–41] (see chapters by Björndahl et al.). In all these models, the angiogenic activities of NPY are significantly impaired in Y2 knockout mice or blocked by Y2 receptor antisense oligonucleotides [40–42]. The role of the peptide in physiological angiogenesis has been further supported by reduced angiogenesis in NPY knockout mice, which can be restored by the exogenous peptide [38]. In addition, expression of NPY and its Y2 receptors is stimulated in ischemic tissues, which is also associated with increased plasma levels of the peptide [38]. The angiogenic activities of NPY are mediated by nitric oxide, vascular endothelial growth factor (VEGF) [38] and, according to our recent study, also basic fibroblast growth factor (bFGF). However, further studies are necessary to establish the exact mechanisms of NPY actions and its interactions with other angiogenic factors (see chapter by Lee et al.).

Clinically, the angiogenic activities of NPY may be beneficial for the organism, for instance in revascularization of ischemic tissues and wound healing. However, its elevated levels may also lead to the excessive angiogenesis observed for example in retinopathy. In the rat and mouse models of newborn retinopathy, the progression of angiogenesis is accompanied by an upregulation of NPY and Y2 receptor expression [43], while the vascularization of the retina is blocked in Y2 knockout mice or by Y2 receptor antisense [40, 41]. Moreover, the frequency of the Leu^7Pro^7 polymorphism is increased in patients with type II diabetes associated with retinopathy [40], which further supports the role of NPY in the development of this disease. NPY has been also implicated in other angiogenesis-related disorders. For example, the peptide released from sympathetic nerve-derived tumors, such as pheochromocytoma and neuroblastoma may increase tumor vascularization and in this way facilitate their own growth and spread [9]. Moreover, recent data from our laboratory indicate that the angiogenic activity of NPY, along with its proliferative effect on preadipocytes, may contribute to the development of stress-induced obesity [15]. Thus, depending on the type of the disease, NPY and its Y2 receptors may be targeted to augment or inhibit vascularization. In the future, both Y2 receptor agonists and antagonists may become useful therapeutics in the treatment of angiogenesis-related disorders.

As described above, NPY is a potent growth-promoting factor for both VSMC and ECs. However, the actions of the peptide in these cells are mediated by different receptors – Y1/Y5 in VSMCs and Y2/Y5 in ECs. Additionally, Y1 receptors present on VSMCs are also responsible for NPY's vasoconstrictive effect. These receptor-specific actions of NPY in vascular cells suggest a regulatory role for dipeptidyl peptidase IV (DPPIV). DPPIV is a serine protease abundantly present on ECs [39]. The enzyme cleaves the full length $NPY_{1\text{-}36}$ to $NPY_{3\text{-}36}$ [44]. This shorter form of the peptide is no longer able to bind to the Y1 receptors, but retains its affinity to the Y2 and Y5 receptors. Thus, DPPIV shifts activity of NPY from Y1-mediated actions in VSMCs to Y2-dependent angiogenesis. Therefore, this protease is an integral part of the NPY angiogenic system. DPPIV is upregulated with NPY and Y2 receptors in ischemic tissues [38] and downregulated in aged animals with impaired angiogenesis [25]. Moreover, anti-DPPIV neutralizing antibodies completely block NPY-mediated EC migration and wound closure [39]. The regulatory actions of DPPIV are particularly important in revascularization of ischemic tissues, where the Y1 receptor-mediated vasoconstriction could further decrease blood flow and contribute to the progress of the disease.

Taken together, NPY is a potent growth-promoting factor for vascular cells involved in vascular remodeling and angiogenesis. Its actions are regulated by DPPIV, which shifts activity of the peptide away from its Y1 receptor-mediated effects on VSMCs and enhances its Y2 receptor-dependent angiogenic activities. Thus, NPY receptors and DPPIV may be attractive targets for treatment of atherosclerosis- and angiogenesis-related diseases.

NPY as a neuroproliferative factor

Although NPY can be expressed in a variety of cells, such as platelets and ECs [14, 45], both the peptide and its receptors are the most abundant in neuronal tissues, particularly those of the sympathetic origin [2]. Thus, discovery of growth-promoting activities of NPY in vascular cells raised the question of its effect on neuronal proliferation. The neuroproliferative activities of NPY have been reported for the first time by Hansel et al. in neuronal precursors derived from olfactory epithelium [8]. Postnatal olfactory epithelium is, along with hippocampus, retina and rostral migratory stream, one of the few known locations of stem cells in the central nervous system (CNS) and is used as a model of adult neuronal regeneration [46] (see chapter by Gray and Scharfman). NPY, expressed here by sustentacular cells, stimulates the proliferation of basal cells, which are olfactory neuronal precursors [8]. This effect is mediated by Y1 receptors and involves PKC-dependent activation of p42/44 MAPK. NPY-deficient mice contain significantly fewer proliferating neuronal precursors and olfactory neurons by adulthood, which further supports the crucial role of NPY in neuronal development. Recently, NPY has been also shown to stimulate proliferation of retinal glial (Muller) cells via Y1 receptors and PI3K-mediated MAPK activation [19]. Interestingly, the growth-regulatory effect of the peptide in these cells is biphasic. NPY exerts a mitogenic effect at higher concentrations, whereas at lower concentrations NPY inhibits Muller cell proliferation. Since both Y1 and Y2 receptors are expressed in these cells, the biphasic effect may be associated with interactions of both receptors, as suggested by bimodal mitogenic effect of NPY in VSMCs and ECs. The heterodimerization of the receptors could lead to change of their affinity and possibly signaling pathways. The excessive proliferation of retinal glial cells in proliferative vitreoretinopathy is associated with increased Y1 receptor expression [19]. Thus, it is possible that an increase in Y1 receptors shifts activity of NPY towards the homodimer- or monomer-mediated proliferation.

The discovery of this neuroproliferative effect of NPY, particularly in combination with its angiogenic activities, provides new perspectives on the treatment of degenerative disease and injuries of the nervous system. However, the same actions of the peptide can also contribute to various pathologies, for instance growth of neural crest-derived tumors. NPY is abundantly present in tumors originating from sympathetic neurons, such as neuroblastoma [47–49] (see chapter by Kitlinska). These childhood tumors are often associated with elevated NPY plasma levels, which correlates with poor outcome of the disease [50, 51]. Our results indicate that growth of neuroblastoma is highly dependent on NPY [9]. The neuroblastoma cells express both NPY and its Y2 receptors, which creates a growth-promoting autocrine loop. NPY stimulates proliferation of neuroblastoma cells via activation of p42/44 MAPK, while the Y2 receptor antagonist dramatically inhibits their proliferation. *In vivo*, NPY stimulates growth of neuroblastoma xenografts, which is also associated with increase of tumor vascularization. Thus, NPY released from neuroblastoma tumors facilitates their growth by both direct autocrine effect on neuroblastoma cell proliferation and angiogenesis. Interestingly, both processes are Y2 receptor-dependent, which implicates the Y2 receptor as a new target in therapy of these childhood malignancies.

Although the growth-promoting activities of NPY in neuronal cells are the most extensively studied, the peptide can also exert inhibitory effects, as seen in retinal glial cells [19]. Moreover, we have found that in Ewing's sarcoma tumors, which unlike sympathetic neuroblastoma originate from parasympathetic neurons, NPY stimulates apoptosis and inhibits tumor growth [9] (see chapter by Kitlinska). This effect is mediated by the Y1 and Y5 receptors which are abundantly expressed in Ewing's sarcoma cells. However, despite its growth inhibitory effect, NPY also stimulates angiogenesis, which in turn facilitates tumor growth [9]. Thus, the overall effect of NPY depends on the balance between Y1/Y5-mediated apoptosis and Y2-dependent angiogenesis. These receptor-specific actions of NPY suggest, like in vascular cells, the important role for DPPIV in their regulation. The protease is expressed by both Ewing's sarcoma cells and ECs present in tumor vasculature. By cleaving NPY to its shorter form, which does not activate Y1 receptors but retains affinity to the Y2 receptors, DPPIV prevents NPY-induced apoptosis and enhances the peptide's angiogenic activity. Thus, inhibition of DPPIV may be an effective strategy in Ewing's sarcoma treatment.

Taken together, the effect of NPY on neural cell growth is cell- and receptor-specific and range from stimulation of proliferation to apoptosis. The elucidation of the mechanisms of NPY actions can contribute to the development of new therapies directed against both nerve degeneration and pathological growth of neural cells.

NPY in the immune cells

Apart from its role in the cardiovascular and nervous systems, NPY has been also implicated in the regulation of immune responses. The peptide is constitutively expressed in rat lymphocytes and human lymphoblasts and is inducible in human lymphocytes and monocytes [52, 53]. The elevated plasma NPY levels have been found in children with B-cell precursor leukemia and is associated with a favorable clinical phenotype and better outcome of the disease [54]. Among many immunological functions regulated by NPY, such as cytokine release [55], macrophage chemotaxis and phagocytosis [56], the peptide can also be a lymphoproliferative

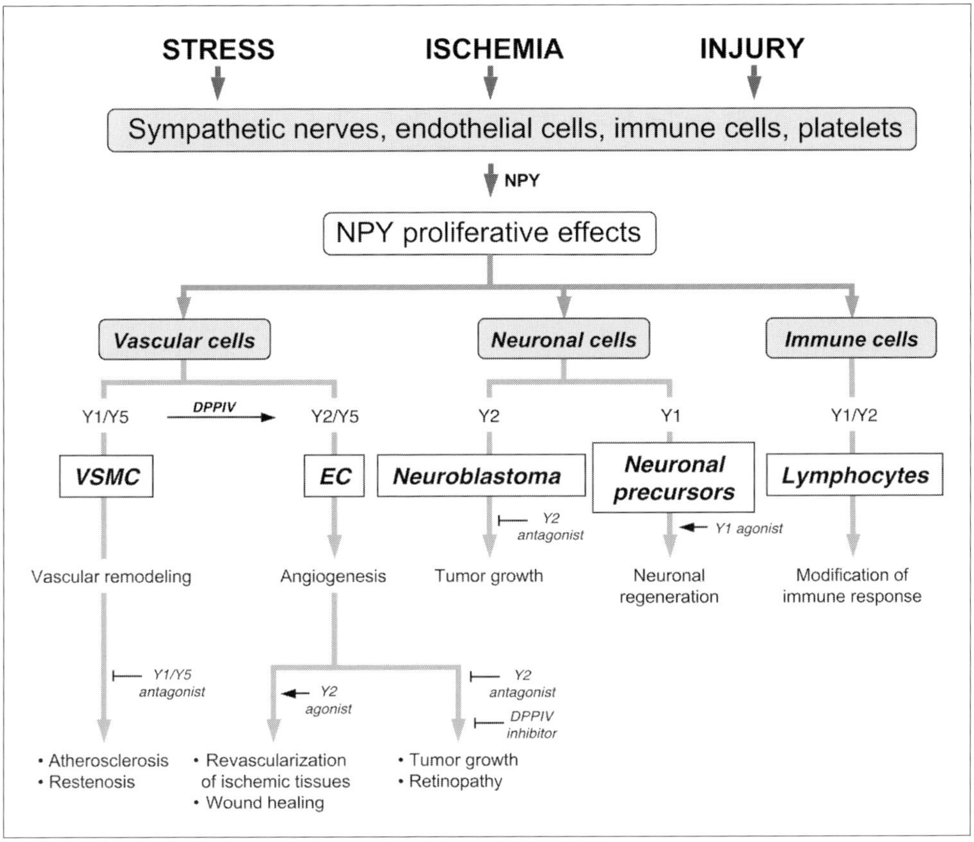

Figure 1
Proliferative actions of NPY and their potential clinical implications.

factor [16–18, 57] (see chapters by von Horsten and De la Fuente). NPY stimulates spontaneous proliferation of lymphocytes derived from different origins, which is mediated by Y1 and possibly Y2 receptors, and is partially dependent on IL1-beta and IL-2 [16, 17, 57]. However, at the same time, the peptide inhibits the lymphoproliferative effect of concanavalin A [17]. Interestingly, these effects are significantly diminished with age [17, 18], which may be associated with decreased expression of NPY receptors in aged animals [25] (see chapter by De la Fuente).

The regulatory effect of NPY on different aspects of immune response is part of the cross talk that occurs between the nervous and immune systems. For example, it is known that stress modulates immune responses of the organism, which, in turn, can lead to the development of other disorders, including cancer [58, 59]. Interestingly, the effects of stress strongly depend upon its intensity and duration. Acute stress enhances immune functions of the organism, while chronic, exhausting stress results in immunosuppression [60]. NPY, as a stress molecule released during sympathetic hyperactivity, can be part of this regulatory mechanism.

Summary

With the discovery of NPY's growth-regulatory effects, emerges a new role for the peptide in regulating cardiovascular, neuronal and immune functions. NPY is involved in angiogenesis, vascular remodeling, proliferation of neuronal cells, and regulation of immune responses. Its actions are cell- and receptor-specific and regulated by the proteolytic activity of DPPIV. Thus, the elucidation of the mechanisms underlying NPY actions can contribute to the development of new therapeutic strategies in the treatment of various disorders, including angiogenesis-related diseases, atherosclerosis, cancer and neuronal degeneration. The growing number of selective receptor agonists and antagonists could potentially become excellent tools to target the NPY system and clinically useful drugs. However, due to the pleiotropic activities of the peptide, the therapy would have to be carefully tailored to the type of the disease, with careful consideration given to the possible side effects.

References

1 Balasubramaniam A (2002) Clinical potentials of neuropeptide Y family of hormones. *Am J Surg* 183: 430–434

2 Zukowska-Grojec Z (1995) Neuropeptide Y. A novel sympathetic stress hormone and more. *Ann NY Acad Sci* 771: 219–233

3 Zukowska-Grojec Z (1993) Origin and actions of neuropeptide Y in the cardiovascular system. In: C Wahlestedt (ed): *The biology of neuropeptide Y and related peptides.* Humana Press, Totowa, NJ 315–388

4 Redrobe JP, Dumont Y, Quirion R (2002) Neuropeptide Y (NPY) and depression: from animal studies to the human condition. *Life Sci* 71: 2921–2937

5 Heilig M (2004) The NPY system in stress, anxiety and depression. *Neuropeptides* 38: 213–224

6 Nie M, Selbie LA (1998) Neuropeptide Y Y1 and Y2 receptor-mediated stimulation of mitogen-activated protein kinase activity. *Regul Pept* 75–76: 207–213

7 Kitlinska J, Pons J, Jacques D, Bkaily G, Sader S, Kurban G, Abdel-Malak N, Perreault C, Zukowska Z (2002) Neuropeptide Y-Y1 and -Y2 receptors: mitogenic activities and intracellular trafficking. *Mol Biol Cell* 13: 233a

8 Hansel DE, Eipper BA, Ronnett GV (2001) Neuropeptide Y functions as a neuroproliferative factor. *Nature* 410: 940–944

9 Kitlinska J, Abe K, Kuo L, Pons J, Yu M, Li L, Tilan J, Everhart L, Lee EW, Zukowska Z et al. (2005) Differential effects of neuropeptide Y on the growth and vascularization of neural crest-derived tumors. *Cancer Res* 65: 1719–1728

10 Erlinge D, Brunkwall J, Edvinsson L (1994) Neuropeptide Y stimulates proliferation of human vascular smooth muscle cells: cooperation with noradrenaline and ATP. *Regul Pept* 50: 259–265

11 Pons J, Kitlinska J, Ji H, Lee EW, Zukowska Z (2003) Mitogenic actions of neuropeptide Y in vascular smooth muscle cells: synergetic interactions with the beta-adrenergic system. *Can J Physiol Pharmacol* 81: 177–185

12 Zukowska-Grojec Z, Pruszczyk P, Colton C, Yao J, Shen GH, Myers AK, Wahlestedt C (1993) Mitogenic effect of neuropeptide Y in rat vascular smooth muscle cells. *Peptides* 14: 263–268

13 Zukowska-Grojec Z, Karwatowska-Prokopczuk E, Fisher TA, Ji H (1998) Mechanisms of vascular growth-promoting effects of neuropeptide Y: role of its inducible receptors. *Regul Pept* 75–76: 231–238

14 Zukowska-Grojec Z, Karwatowska-Prokopczuk E, Rose W, Rone J, Movafagh S, Ji H, Yeh Y, Chen WT, Kleinman HK, Grouzmann E et al (1998) Neuropeptide Y: a novel angiogenic factor from the sympathetic nerves and endothelium. *Circ Res* 83: 187–195

15 Kuo L, Lee EW, Kitlinska J, Johnson MD, Zukowska Z (2005) Neuronal cross-talk with endothelial cells and adipocytes leads to angiogenesis and adipogenesis. *FASEB J* 19: A71

16 Elitsur Y, Luk GD, Colberg M, Gesell MS, Dosescu J, Moshier JA (1994) Neuropeptide Y (NPY) enhances proliferation of human colonic lamina propria lymphocytes. *Neuropeptides* 26: 289–295

17 Medina S, Del Rio M, Hernanz A, De la Fuente M (2000) Age-related changes in the neuropeptide Y effects on murine lymphoproliferation and interleukin-2 production. *Peptides* 21: 1403–1409

18 Medina S, Rio MD, Cuadra BD, Guayerbas N, Fuente MD (1999) Age-related changes in the modulatory action of gastrin-releasing peptide, neuropeptide Y and sulfated cholecystokinin octapeptide in the proliferation of murine lymphocytes. *Neuropeptides* 33: 173–179

19 Milenkovic I, Weick M, Wiedemann P, Reichenbach A, Bringmann A (2004) Neuropeptide Y-evoked proliferation of retinal glial (Muller) cells. *Graefes Arch Clin Exp Ophthalmol* 242: 944–950

20 Cho YR, Kim CW (2004) Neuropeptide Y promotes beta-cell replication via extracellular signal-regulated kinase activation. *Biochem Biophys Res Commun* 314: 773–780

21 Magni P, Motta M (2001) Expression of neuropeptide Y receptors in human prostate cancer cells. *Ann Oncol* 12 (Suppl 2): S27–S29

22 Takahashi E, Berk BC (1998) MAP kinases and vascular smooth muscle function. *Acta Physiol Scand* 164: 611–621

23 Mihara S, Shigeri Y, Fujimoto M (1989) Neuropeptide Y-induced intracellular Ca^{2+} increases in vascular smooth muscle cells. *FEBS Lett* 259: 79–82

24 Mihara S, Shigeri Y, Fujimoto M (1990) Neuropeptide Y receptor in cultured vascular smooth muscle cells: ligand binding and increase in cytosolic free Ca^{2+}. *Biochem Int* 22: 205–212

25 Kitlinska J, Lee EW, Movafagh S, Pons J, Zukowska Z (2002) Neuropeptide Y-induced angiogenesis in aging. *Peptides* 23: 71–77

26 Pons J, Kitlinska J, Jacques D, Bkaily G, Sader S, Kurban G, Abdel-Malak N, Zukowska Z (2002) Bimodal mitogenic action of neuropeptide Y in vascular smooth muscle cells: dimerization of the receptors? *Mol Biol Cell* 13: 80a

27 Berglund MM, Schober DA, Esterman MA, Gehlert DR (2003) Neuropeptide Y Y4 receptor homodimers dissociate upon agonist stimulation. *J Pharmacol Exp Ther* 307: 1120–1126

28 Dinger MC, Bader JE, Kobor AD, Kretzschmar AK, Beck-Sickinger AG (2003) Homodimerization of neuropeptide y receptors investigated by fluorescence resonance energy transfer in living cells. *J Biol Chem* 278: 10562–10571

29 Pons J, Kitlinska J, Jacques D, Bkaily G, Sader S, Kurban G, Abdel-Malak N, Zukowska Z (2003) Interactions between multiple receptors and signaling pathways determine neuropeptide Y (NPY)-mediated proliferation of vascular smooth muscle cells. *Regul Pept* 110: 175

30 Pons J, Kitlinska J, Ji H, Lee EW, Jacques D, Bkaily G, Perreault C, Sader S, Abdel-Malak N, Ciborowski P et al (2004) Neuropeptide Y-mediated vascular cell proliferation: multiple receptors, signaling pathways and interactions with beta-adrenergic receptors. Abstract presented at the 7th International NPY Meeting, Coimbra, Portugal

31 Li L, Lee EW, Ji H, Zukowska Z (2003) Neuropeptide Y-induced acceleration of postangioplasty occlusion of rat carotid artery. *Arterioscler Thromb Vasc Biol* 23: 1204–1210

32 Erkkila AT, Lindi V, Lehto S, Laakso M, Uusitupa MI (2002) Association of leucine 7 to proline 7 polymorphism in the preproneuropeptide Y with serum lipids in patients with coronary heart disease. *Mol Genet Metab* 75: 260–264

33 Millar BC, Schluter KD, Zhou XJ, McDermott BJ, Piper HM (1994) Neuropeptide Y stimulates hypertrophy of adult ventricular cardiomyocytes. *Am J Physiol* 266: C1271–1277

34 Bell D, Allen AR, Kelso EJ, Balasubramaniam A, McDermott BJ (2002) Induction of hypertrophic responsiveness of cardiomyocytes to neuropeptide Y in response to pressure overload. *J Pharmacol Exp Ther* 303: 581–591

35 Zeng C, Zhou Y, Liu G, Sun W (2001) The signal transduction pathway causing the synergistic hypertrophic effects of neuropeptide Y and norepinephrine on primary cardiomyocyte. *Neuropeptides* 35: 211–218

36 Goldberg Y, Taimor G, Piper HM, Schluter KD (1998) Intracellular signaling leads to the hypertrophic effect of neuropeptide Y. *Am J Physiol* 275: C1207–1215

37 Kuch-Wocial A, Slubowska K, Kostrubiec M, Pasierski T, Januszewicz W, Switalska H, Wocial B, Pruszczyk P (2004) Plasma neuropeptide Y immunoreactivity influences left ventricular mass in pheochromocytoma. *Clin Chim Acta* 345: 43–47

38 Lee EW, Michalkiewicz M, Kitlinska J, Kalezic I, Switalska H, Yoo P, Sangkharat A, Ji

H, Li L, Michalkiewicz T et al. (2003) Neuropeptide Y induces ischemic angiogenesis and restores function of ischemic skeletal muscles. *J Clin Invest* 111: 1853–1862
39 Ghersi G, Chen W, Lee EW, Zukowska Z (2001) Critical role of dipeptidyl peptidase IV in neuropeptide Y-mediated endothelial cell migration in response to wounding. *Peptides* 22: 453–458
40 Koulu M, Movafagh S, Tuohimaa J, Jaakkola U, Kallio J, Pesonen U, Geng Y, Karvonen MK, Vainio-Jylha E, Pollonen M et al. (2004) Neuropeptide Y and Y2-receptor are involved in development of diabetic retinopathy and retinal neovascularization. *Ann Med* 36: 232–240
41 Ekstrand AJ, Cao R, Bjorndahl M, Nystrom S, Jonsson-Rylander AC, Hassani H, Hallberg B, Nordlander M, Cao Y (2003) Deletion of neuropeptide Y (NPY) 2 receptor in mice results in blockage of NPY-induced angiogenesis and delayed wound healing. *Proc Natl Acad Sci USA* 100: 6033–6038
42 Lee EW, Grant DS, Movafagh S, Zukowska Z (2003) Impaired angiogenesis in neuropeptide Y (NPY)-Y2 receptor knockout mice. *Peptides* 24: 99–106
43 Yoon HZ, Yan Y, Geng Y, Higgins RD (2002) Neuropeptide Y expression in a mouse model of oxygen-induced retinopathy. *Clin Experiment Ophthalmol* 30: 424–429
44 Mentlein R (1999) Dipeptidyl-peptidase IV (CD26) – role in the inactivation of regulatory peptides. *Regul Pept* 85: 9–24
45 Myers AK, Farhat MY, Vaz CA, Keiser HR, Zukowska-Grojec Z (1988) Release of immunoreactive-neuropeptide by rat platelets. *Biochem Biophys Res Commun* 155: 118–122
46 Hansel DE, Eipper BA, Ronnett GV (2001) Regulation of olfactory neurogenesis by amidated neuropeptides. *J Neurosci Res* 66: 1–7
47 Biedler JL, Roffler-Tarlov S, Schachner M, Freedman LS (1978) Multiple neurotransmitter synthesis by human neuroblastoma cell lines and clones. *Cancer Res* 38: 3751–3757
48 Fruhwald MC, O'Dorisio MS, Cottingham SL, Qualman SJ, O'Dorisio TM (1998) Neuropeptides in developmental tumors of the central and peripheral nervous system. *Ann NY Acad Sci* 865: 420–426
49 O'Hare MM, Schwartz TW (1989) Expression and precursor processing of neuropeptide Y in human and murine neuroblastoma and pheochromocytoma cell lines. *Cancer Res* 49: 7015–7019
50 Dotsch J, Christiansen H, Hanze J, Lampert F, Rascher W (1998) Plasma neuropeptide Y of children with neuroblastoma in relation to stage, age and prognosis, and tissue neuropeptide Y. *Regul Pept* 75–76: 185–190
51 Cohen PS, Cooper MJ, Helman LJ, Thiele CJ, Seeger RC, Israel MA (1990) Neuropeptide Y expression in the developing adrenal gland and in childhood neuroblastoma tumors. *Cancer Res* 50: 6055–6061
52 Schwarz H, Villiger PM, von Kempis J, Lotz M (1994) Neuropeptide Y is an inducible gene in the human immune system. *J Neuroimmunol* 51: 53–61
53 von Horsten S, Exton NG, Exton MS, Helfritz F, Nave H, Ballof J, Stalp M, Pabst R

(1998) Brain NPY Y1 receptors rapidly mediate the behavioral response to novelty and a compartment-specific modulation of granulocyte function in blood and spleen. *Brain Res* 806: 282–286

54 Kogner P, Ericsson A, Barbany G, Persson H, Theodorsson E, Bjork O (1992) Neuropeptide Y (NPY) synthesis in lymphoblasts and increased plasma NPY in pediatric B-cell precursor leukemia. *Blood* 80: 1324–1329

55 Kawamura N, Tamura H, Obana S, Wenner M, Ishikawa T, Nakata A, Yamamoto H (1998) Differential effects of neuropeptides on cytokine production by mouse helper T cell subsets. *Neuroimmunomodulation* 5: 9–15

56 De la Fuente M, Medina S, Del Rio M, Ferrandez MD, Hernanz A (2000) Effect of aging on the modulation of macrophage functions by neuropeptides. *Life Sci* 67: 2125–2135

57 von Horsten S, Ballof J, Helfritz F, Nave H, Meyer D, Schmidt RE, Stalp M, Klemm A, Tschernig T, Pabst R (1998) Modulation of innate immune functions by intracerebroventricularly applied neuropeptide Y: dose and time dependent effects. *Life Sci* 63: 909–922

58 Sampson W (2002) Controversies in cancer and the mind: effects of psychosocial support. *Semin Oncol* 29: 595–600

59 Levy SM, Herberman RB, Maluish AM, Schlien B, Lippman M (1985) Prognostic risk assessment in primary breast cancer by behavioral and immunological parameters. *Health Psychol* 4: 99–113

60 Dhabhar FS, McEwen BS (1997) Acute stress enhances while chronic stress suppresses cell-mediated immunity *in vivo*: a potential role for leukocyte trafficking. *Brain Behav Immun* 11: 286–306

NPY and hippocampal neurogenesis

William P. Gray[1] *and Helen E. Scharfman*[2]

[1]Division of Clinical Neurosciences, University of Southampton, Room 6207, Biomedical Sciences Building, Bassett Crescent East, Biomedical Sciences Building, Southampton SO16 7PX, UK; [2]Helen Hayes Hospital, New York State Department of Health, Columbia University

Introduction

Neuropeptide Y (NPY) is a 36 amino acid polypeptide neurotransmitter widely distributed in the mammalian central nervous system (CNS). It has a very high sequence homology spanning a large evolutionary spectrum from invertebrates to humans [1] suggesting important functions for this peptide. Known physiological functions within the CNS include the regulation of social and feeding behaviour [2–5], circadian rhythm [6, 7] and central cardiovascular function [8]. NPY has a number of cognitive and behavioural effects on alcohol intake [9], learning and memory [10], the anxiety response to stress and the maintenance of normal mood (for review see [11]). A recent addition to the many physiological roles for NPY is as a proliferative factor for neural precursor cells, first demonstrated in the olfactory epithelium of adult mice [12]. Neuropeptides are synthesised as preprohormones, which undergo a number of cleavage steps initially in the endoplasmic reticulum and later in immature large core dense vesicles (LCDV) where they undergo proteolytic cleavage and post-translational modification to the final bioactive neuropeptide. The neuropeptide is released from LCDVs under specific firing conditions which are distinct from those for classical neurotransmitter release [13]. NPY is one of the many neuropeptides that undergoes post-translational amidation, and in mammals, amidated peptides serve as growth factors mediating cell survival and neuronal proliferation [14–16]. Recent studies have begun to identify a role for a number of amidated neuropeptides in neural stem cell biology both in the developing and adult mammalian brain and particularly in the olfactory epithelium (for review see [17]). This review will focus on the emerging role for NPY in the regulation of neural stem cells particularly in the hippocampus beginning with an overview of the development of the dentate gyrus and neurogenesis in the adult hippocampus. This will be followed by a review of relevant experimental studies and finally by a discussion of the effect of NPY on neurogenesis as a possible mechanism for some of the cognitive and behavioural effects of this intriguing neuropeptide molecule.

The NPY Family of Peptides in Immune Disorders, Inflammation, Angiogenesis and Cancer,
edited by Zofia Zukowska and Giora Z. Feuerstein

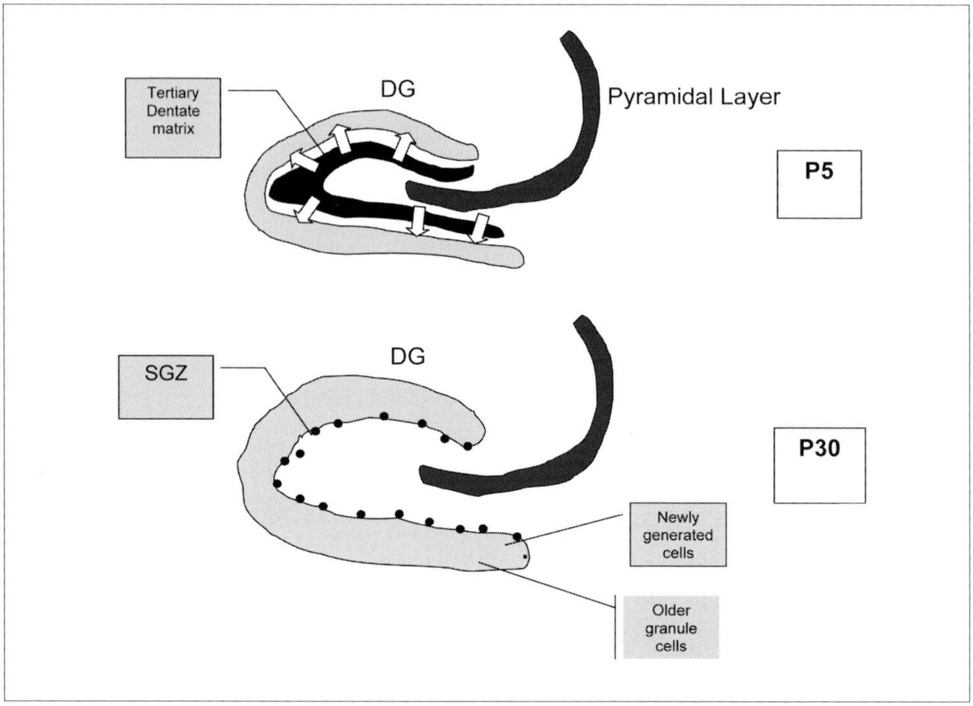

Figure 1
Schematic diagram (modified from [18]) of a cross section through the dentate gyrus (DG) at post natal day 5 (P5) and P30 showing the tertiary matrix in the dentate hilus with arrows indicating the migration of newly born neurons into the developing granule cell layer. The tertiary dentate matrix has disappeared by P30 with migration of precursor cells to the hilar border of the granule cell layer where they reside as a discontinuous layer (black dots), to form the adult neurogenic sub granular zone (SGZ). Because of the hilar position of the tertiary matrix and later SGZ, the granule cell layer is formed in an inside-out pattern with newer cell being successively added from the hilar side.

Development of the granule cell layer of the dentate gyrus

Unlike the rest of the hippocampus, which is generated from the hippocampal anlage at embryonic day 16 (E16), the granule cell layer of the dentate is largely generated postnatally. Cells of the second dentate migration form the tertiary matrix in the hilus of the developing dentate by postnatal day 3 (P3) from where precursor cells migrate to the inner border of the granule cell layer to form the adult neurogenic subgranular zone by P20 (Fig. 1). The granule cell layer therefore develops as an inside-out structure with the newer cells being added successively along the inner margin [18, 19] (see Fig. 1). This neurogenic niche is unique in that it is the site of ongoing neurogenesis throughout adult life in all mammals studied to date including humans [20] and it has been estimated to generate 6% of the total granule cell population every month [21].

Hippocampal neurogenesis

Neurogenesis in the hippocampus is found predominantly in the subgranular zone (SGZ) of the dentate gyrus [22], although there is also evidence of ongoing hippocampal neurogenesis in the cornu ammonis [23]. The production of new neurons in the SGZ exceeds the number that survive and integrate into hippocampal circuitry [24], net production depending on the number and proliferation rate of SGZ precursors as well as the survival of the newly generated neurons [25]. Dentate neurogenesis is highest in the postnatal period and decreases exponentially throughout life in rodents [26, 27] and also probably in humans [20, 28]. The reasons for this age-related decline are unclear, but appear to be related to suppression of precursor proliferation by increasing levels of circulating adrenal corticosteroids with age [29].

The hippocampus has been repeatedly implicated in the modulation of cognition in association with learning and memory, and the demonstration of functional integration of these newly born dentate granule cells into hippocampal circuitry [30, 31] has lead to the hypothesis that neurogenesis may be important in learning and memory formation [32–34]. Hippocampal dependant associative learning tasks increase dentate neurogenesis [34] and pharmacological reduction of newly born neurons in the dentate impairs certain types of hippocampal dependant learning such as trace conditioning [32]. Moreover, prenatal stress produces learning deficits associated with inhibition of dentate neurogenesis [35] and the level of cell proliferation in the SGZ predicts the spatial memory performance of rats in the water maze [36]. Exposure to a novel or enriched environment increases neurogenesis largely by increasing the proportion of cells that survive and integrate and also by increasing the population of proliferating SGZ cells, but in a strain dependant manner [37] (for a critical review see [25]).

In addition, several avenues of research have suggested that decreased hippocampal neurogenesis may be important in the pathogenesis of depression (for review see [38]). Stress, which is often a requisite for depression, reduces hippocampal neurogenesis [39], tricyclic antidepressants increase hippocampal neurogenesis [40] and the antidepressant induced increase in dentate neurogenesis appear to be necessary for their behavioural effects [41]. As newly born neurons functionally integrate into hippocampal circuitry 2–3 weeks after birth [42, 43], this may explain the puzzling 2–3 week delay in mood elevation after commencing antidepressants.

Our knowledge of the control mechanisms underlying hippocampal neurogenesis remains far from complete. The reader is referred to recent reviews for an in-depth analysis of factors controlling adult neurogenesis [44, 45]. Hippocampal neurogenesis is also upregulated in response to brain injury, including seizures [46–48], stroke [49] and traumatic brain injury [50] and there appears to be a proliferative recruitment of radial glia-like precursor cells in the SGZ as part of this response [51].

The dentate gyrus of kainic acid-treated rats shows particularly high NPY-like immunoreactivity [52] and it's expression after status epilepticus parallels the time-course of the proliferative SGZ response [53]. We were therefore motivated to examine the role of NPY as a possible proliferative factor for stem/precursor cells in the hippocampus/dentate gyrus after Hansel et al.'s seminal paper described a proliferative effect of NPY on neuroblasts in the olfactory epithelium [12].

NPY and hippocampal neurogenesis

We began to investigate a possible role for NPY in neurogenesis in the hippocampus using dissociated monolayer and neurosphere cultures harvested from the hippocampi of postnatal rat pups when the level of neurogenesis is at its highest, in order to optimise the number of stem/precursor cells in culture (for detailed methods see [54, 55]). Cell proliferation was measured by the incorporation of the thymidine analogue Bromodeoxyuridine (BrdU) which is incorporated into the DNA of dividing cells in immunohistochemically detectable quantities during the S phase of cell division [21].

Proliferative effects of NPY on hippocampal cell cultures

We followed the progression of cell number, rates of division and phenotypic fate (Class III β-tubulin (neuronal), glial fibrillary acid protein (GFAP) (glial) and nestin (stem/precursor cell)) over 7 days in culture. Time points after 24 h *in vitro* revealed significant increases in cell number, BrdU incorporation, nestin immunopositivity and Class III β-tubulin immunopositivity relative to controls (Fig. 2A–D). Increases in total cell number, BrdU incorporation and nestin immunopositivity all followed

Figure 2
NPY enhances the total number of cells, BrdU positive cells, β-tubulin positive cells and nestin immuno-positive cells with time in culture.
Hippocampal cell cultures were grown for the indicated number of days in standard control conditions, or in the presence of 1 μM NPY. BrdU was added for the final 24 h of each time-

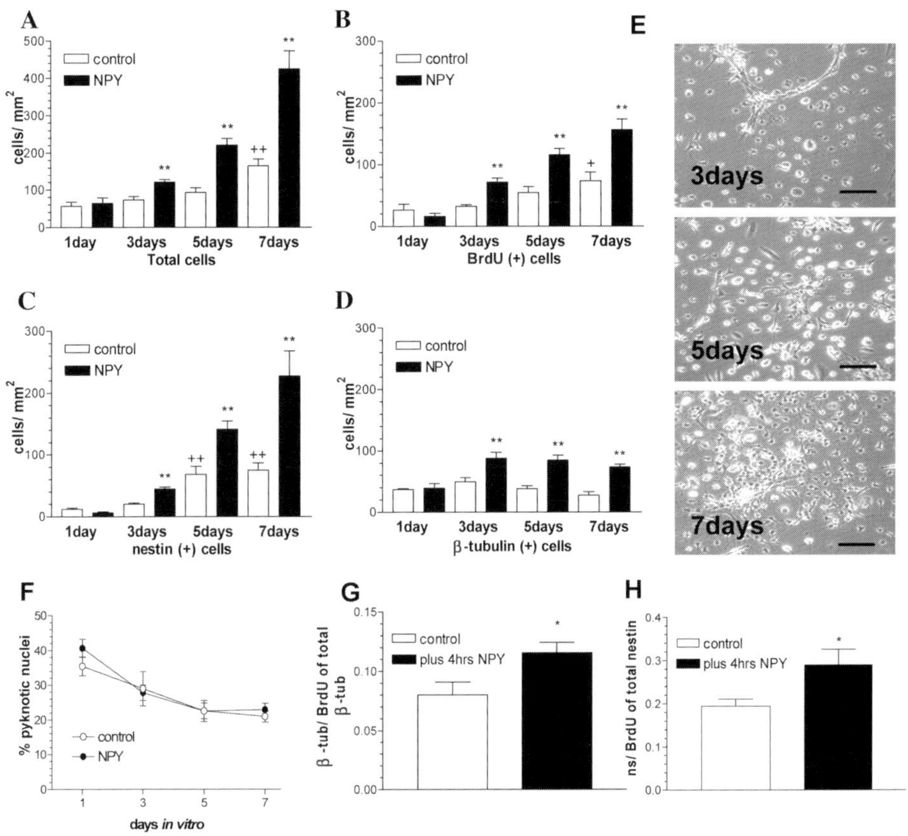

point. (A) Total cell counts increase under standard conditions, and are further significantly increased in the presence of NPY. (B) The number of BrdU incorporating cells is increased between day 1 and 7 in control conditions, and is increased relative to controls at each of days 3, 5, 7 in the presence of NPY. (C) Nestin counts increase with days in vitro, *and are significantly different from controls at days 3, 5 and 7. (D) β-tubulin counts are maximal at day 3, and in control and NPY conditions fall with time in culture. Values represent mean ± SE based on a sample of 10 wells per condition, from three separate experiments. (E) Representative phase contrast images of cells cultured in 1 μM NPY at 3, 5 and 7 div. Scale bar; 100 μm. (F) Cell death was assessed by quantifying chromatin condensation and was similar under control or NPY culturing conditions. NPY increases the proportion of mitotic nestin (G) and β-tubulin positive cells (H). Values represent mean ± SE based on a sample of at least 12 wells per condition from three separate experiments analysed using Student's t test *, $p < 0.05$. BrdU, bromodeoxyuridine; ns, nestin. Reproduced with permission from [55].*

a similar exponential curve over time. The cell type that showed the greatest increase in number and BrdU incorporation were small unipolar or bipolar cells, with phase bright cell bodies (Fig. 2E) and which immunostained for nestin – typical of hippocampal stem/precursor cells previously described by Palmer et al. [56]. The time course of increases relative to controls in β-tubulin expressing cells was different with a peak at 3 and 5 days *in vitro* (DIV) after which the numbers plateaued. The number of GFAP positive cells was not increased at 5 days *in vitro* compared to controls (34.7 ± 2.5 cells/mm^2 in controls *versus* 33.6 ± 2.8 cells/mm^2 in NPY).

NPY exerts a proliferative effect on nestin positive cells and neuroblasts *in vitro*

The NPY induced increase in cells with a neuronal phenotype could be a proliferative and or a neurotrophic effect. To address this question we quantified cell death, identified by nuclear condensation, using DAPI staining in the same cultures. Cell death was maximal on Day 1 *in vitro* but there was no difference in cell death between control cultures and cultures exposed to NPY at any time point examined (Fig. 2F). We confirmed a proliferative effect of NPY on neuroblasts (β-tubulin and BrdU co-expressing cells) and putatively more primitive precursor cells (nestin and BrdU co-expressing cells) by exposing three DIV cultures to both 1 μM NPY and BrdU for 4 h showing a significant increase in the proportion of both β-tubulin cells (Fig. 2G) and nestin cells (Fig. 2H) co-expressing BrdU.

Figure 3
The NPY proliferative effect is mediated via the Y1 receptor.
*(A) The Y1 selective compound, [F7, P34] NPY reproduced the action of 1 μM NPY on total cell numbers, BrdU cell numbers and Class III β-tubulin positive cells. (B) BIBP3226, an NPY Y1 selective antagonist blocks the proliferative and neurogenic effects of NPY. ANOVA with Dunnett's multiple comparison test: *, $p < 0.05$; **, $p < 0.01$; ***, $p < 0.001$. (C) Cultures from Y1$^{-/-}$ mice show no response to NPY in contrast to Y1$^{lox/lox}$ controls. Values represent mean ± SE based on a sample of at 6–12 wells per condition from 2–3 separate experiments. Student's t test *, $p < 0.05$. Y1 staining cells with a neuronal morphology (D) co-localised with staining for the neuronal marker Class III β-tubulin (E and F – arrows). Y1 staining cells with neuronal morphology did not stain for nestin (I – arrow). Y1 staining cells with a round morphology and small short processes (G) stained for the stem cell marker nestin (H and I) and not for Class III β-tubulin (F). Finally small bipolar cells immunostaining with Y1 antibody (J) also co-localised for nestin (K and L). Figures D–F are Z stacked 3D projections. Figures G–L are co-planar 0.3 μm optical sections. Scale bar = 20 μm. Reproduced with permission from [55].*

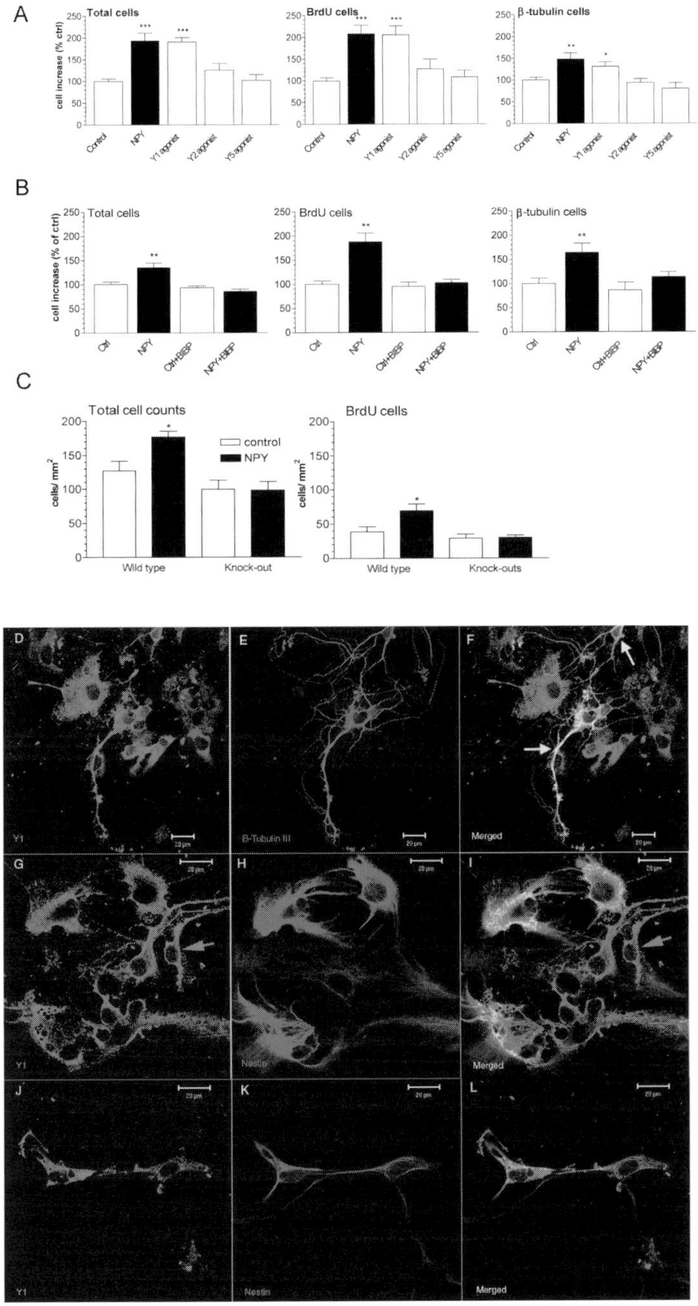
A
Total cells
BrdU cells
β-tubulin cells
cell increase (% ctrl)
Control
NPY
Y1 agonist
Y2 agonist
Y5 agonist
B
Total cells
BrdU cells
β-tubulin cells
cell increase (% of ctrl)
Ctrl
NPY
Ctrl+BIBP
NPY+BIBP
C
Total cell counts
BrdU cells
control
NPY
cells/ mm2
Wild type
Knock-out
Knock-outs
D
E
F
Y1
B-Tubulin III
Merged
G
H
I
Nestin
J
K
L

The proliferative effects of NPY are mediated via the Y1 receptor subtype

To determine if the activity of NPY is mediated via a single receptor subtype or through a mixed population we used Y1, Y2 and Y5 selective peptide agonists in an effort to mimic the effects of NPY (Fig. 3A). The Y1-receptor selective [F7, P34] NPY was uniquely able to mimic the effects of 1 μM NPY. The non-peptide, Y1 selective antagonist, BIBP3226 completely abolished NPY's proliferative activity and increase in Class III β-tubulin positive cells to basal levels (Fig. 3B). Incubation with BIBP3226 alone had no effect. Cultures from NPY-Y1 receptor subunit knockout mice and age and generation matched *lox-lox* controls corroborated the proliferative activity of NPY as a Y1 receptor mediated event (Fig. 3C).

BrdU, β-tubulin and nestin positive cells immunostain for the Y1 receptor

Immunostaining 5 day old cultures grown in 1 μM NPY against the Y1 receptor showed three patterns of Y1 antibody staining. One pattern of staining was clearly neuronal (Fig. 3D) and cells with this morphology almost always stained positively for the early neuronal marker Class III β-tubulin (Lee et al., 1990) (Fig. 3E and F). The second pattern of Y1 antibody staining were flat rounded cells with short simple processes (Fig. 3D and G). These cells did not stain for Class III β-tubulin (Fig. 3F) but did stain for the stem/precursor cell marker nestin (Fig. 3H and I). These cells were typically much smaller that GFAP staining astrocytes and had a different shorter pattern of processes (Fig. 3D and G). Interestingly Y1 immunostaining cells with a neuronal morphology were negative for nestin (Fig. 3G, H and I). The final pattern of Y1 immunostaining was small uni- or bipolar cells which were nestin positive (Fig. 3J–L). 83.5 + 6.6% (n = 400 cells) of nestin immunopositive cells were positive for Y1 receptor antibody staining at three DIV. 100% (n = 400 cells) of BrdU positive cells after a single 4 h pulse were positive for Y1 receptor antibody staining.

The neuroproliferative effect of NPY is confined to neuroblasts from the dentate gyrus

The hippocampus is not uniform in its neurogenic structure and stem/precursor cells reside mainly in the SGZ of the dentate gyrus and the posterior periventricular layer (PPL) of the subventricular zone (SVZ) from where progeny migrate forward in the rostral migratory stream to the olfactory bulb supporting olfactory neurogenesis in the rodent (Fig. 4). In order to ascertain which neurogenic population of precursor cells responds to NPY we generated region-specific microcultures from the dentate gyrus and the hippocampus proper (including the PPL) after Seaburg and van der Kooy [57].

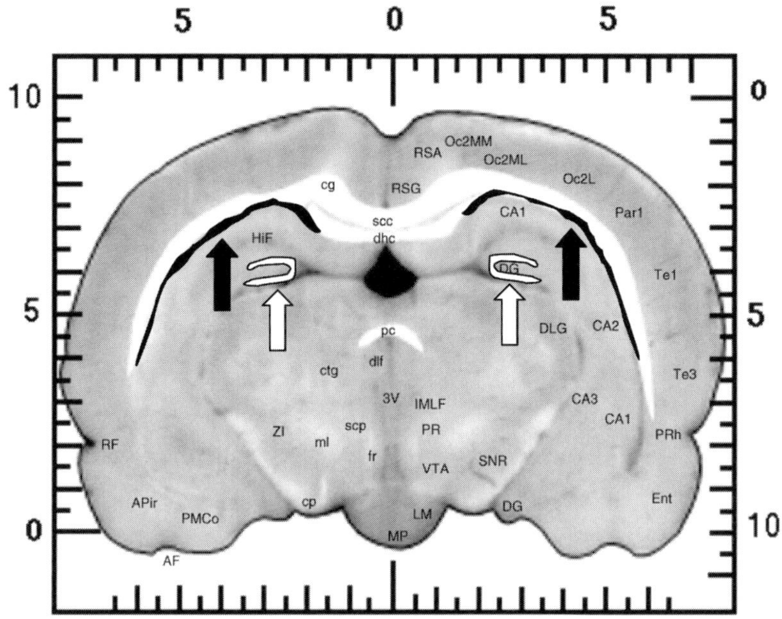

Figure 4
The posterior periventricular layer (black shaded area) lines the subventricular portion of the hippocampus (arrowed) and contains proliferating stem cells that migrate forward in the rostral migratory stream and the olfactory bulb. Cell cultures generated from whole hippocampus preparations will contain these stem cells as well as precursor cells from the subgranular zone of the dentate gyrus (white shaded areas and arrows).

Five days culture in 1 μM NPY significantly increased cell number and cells incorporating BrdU on day 5 in cultures derived from both dentate gyrus and hippocampus-only cultures (Tab. 1). Dentate cultures showed an increase in the number of neurons in response to 1 μM NPY whereas hippocampal cultures did not (Tab. 1). This was not due to a neurotrophic effect as both control and NPY treated dentate cultures had similar proportions of cell death up to day 3 (Fig. 5). 4 h exposure of BrdU and 1 μM NPY to day 3 cultures confirmed a proliferative effect of NPY on class III β-tubulin expressing cells (Fig. 5). NPY had no effect on GFAP cell numbers up to 3 days and no proliferative effect on GFAP-positive cells at 3 days in culture.

*Table 1 - NPY increases the number of neurons only in dentate derived cultures. Data is mean + SE of 8–12 wells per condition from a total of four experiments. Student's t test. **, $p < 0.01$; ***, $p < 0.001$. Reproduced with permission from Howell et al. J Neurochem 2005*

	Dentate cultures (cells/mm²)		**Hippocampal cultures (cells/mm²)**	
Cell marker	Control	1μM NPY	Control	1μM NPY
Total cell count	432 ± 44	719 ± 42**	426 ± 53	696 ± 41**
BrdU	140 ± 23	277 ± 11***	223 ± 22.3	359 ± 34**
β-tubulin	108 ± 10	181 ± 14**	27 ± 4.5	32 ± 4.1

The neuroproliferative effect of NPY on dentate precursors is Y1 receptor mediated and requires ERK 1/2 activation

The proliferative effect of NPY on dentate neuroblasts was abolished by the Y1 antagonist BIBP3226 and was mimicked by the selective agonist [F7, P34] NPY (Fig. 6A–C). BIBP3226 alone had no effect. The ERK kinases (extracellular signal-regulated kinases) are a group of mitogen-activated protein kinases (MAPK) important in control of cellular proliferation [58] including neural progenitors [59]. Inhibition of ERK 1/2 phosphorylation by UO126 abolished the neuroproliferative effect of NPY on dentate neuroblasts (Fig. 6D, E). Confocal imaging confirmed co-localisation of Y1 antibody staining with BrdU and β-tubulin positive cells (Fig. 6F, G).

$Y1^{-/-}$ mice have reduced subgranular zone BrdU incorporation and doublecortin positive cells

As the Y1 receptor was strongly implicated in the mediation of the proliferative effect of NPY on dentate derived neuroblasts we used constitutive Y1 receptor knockout mice and age and generation matched controls to see if NPY was important for dentate neurogenesis *in vivo*. Stereological estimation of subgranular zone cells labelled with BrdU over five consecutive days in 35 day old Y1-/- mice revealed an approximate 40% decrease in the number of BrdU positive cells in comparison to age and generation matched $Y1^{lox/lox}$ controls (6685 + 729 *versus* 4176 + 217 cells per subgranular zone for control and $Y1^{-/-}$ groups, respectively; (Fig. 7a and b). There was no difference in BrdU cell counts from the hilus or areas CA1-3 demonstrating no difference in BrdU bioavailability between the $Y1^{-/-}$ mice and controls. The reduction in SGZ proliferation appeared to translate into a reduction in neurogenesis as staining for the immature neuronal marker doublecortin showed a

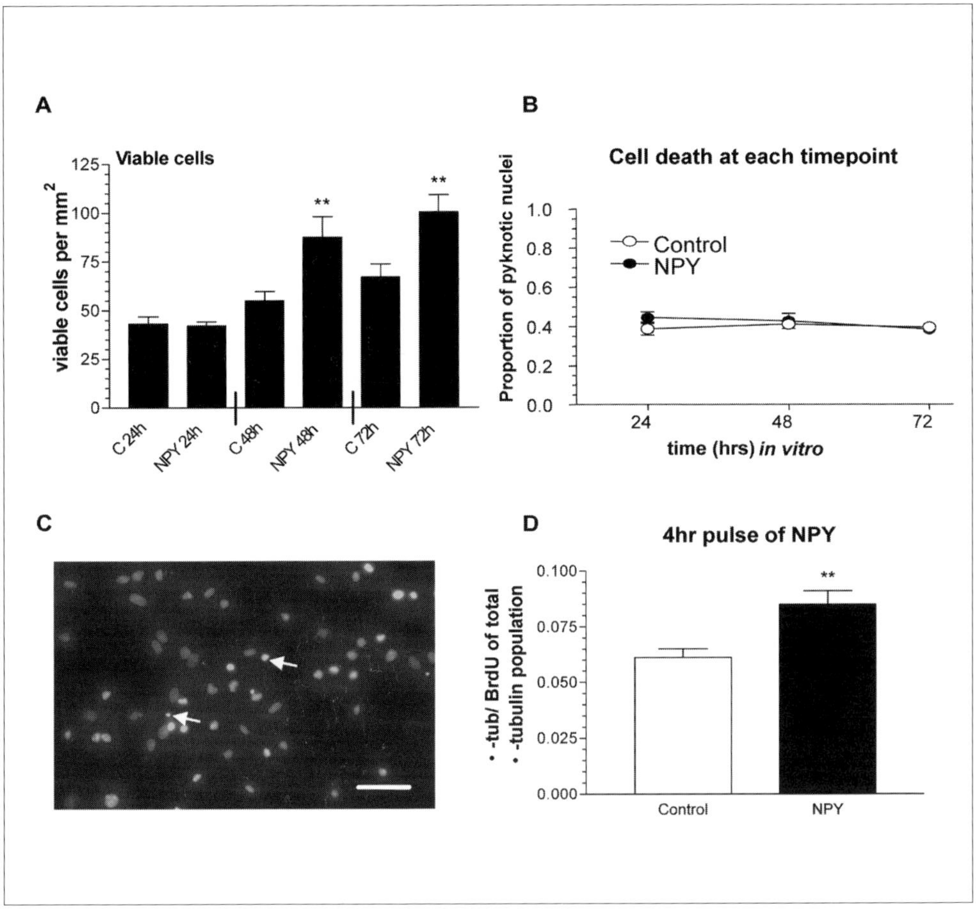

Figure 5

*A short pulse of NPY stimulates neuroblast proliferation while having no effect upon cell survival. (A and B) Cells from dentate tissue grown in 1 µM NPY for 24, 48 and 72 h after plating. Viable cell counts (A) were significantly increased after 48 and 72 h NPY compared to the same time points under control conditions. Cell death was similar between control and NPY groups (B). One-way ANOVA with Bonferroni's post-test (**, $p < 0.01$). (A and B) Values represent mean ± SE based on a sample of eight wells per condition from two separate experiments. (C) A representative field of DAPI stained cells at 72 h in culture. Solid arrowheads indicate nuclei with a condensed morphology. Open arrow indicates a normal healthy nucleus. Scalebar = 50 µm (D) Cells grown for 3 DIV in control conditions were exposed to BrdU and/ or NPY for 4 h. The proportion of β-tubulin cells incorporating BrdU over a 4 h period is increased with a 4 h exposure to NPY implying a proliferative effect. Values are mean ± SE of 20–24 wells per condition from four separate experiments analysed using Student's t test (**, $p<0.01$). Reproduced with permission from [54].*

reduction of similar magnitude (Fig. 7c and d). Interestingly estimates of total number of cells in the granule cell layer of $Y1^{-/-}$ mice (375,800 ± 50,050 cells) did not differ from $Y1^{lox/lox}$ controls (363,200 ± 33,270 cells).

$Y1^{-/-}$ mice have a reduced subgranular zone proliferation rate but an increased survival of newly generated neurons

Adult 90 day old $Y1^{-/-}$ mice showed a lower cell proliferation rate in the subgranular zone (1086 ± 96.06 cells/SGZ) compared to age and generation matched $Y1^{lox/lox}$ controls (1590 ± 141.0 cells/SGZ; $P < 0.001$) as judged by the number of cells incorporating BrdU two hours after a single IP injection (Fig. 7e). However, the number of BrdU positive cells in $Y1^{-/-}$ mice sacrificed 21 days after a single BrdU injection was not significantly different from that in the $Y1^{lox/lox}$ controls (Fig. 7e), indicating a relatively greater survival of newly generated cells in the dentate gyrus of $Y1^{-/-}$ animals. There was no significant difference between the proportion of BrdU positive cells that co-localised for the mature neuronal marker Neu-N at 21 days after a single BrdU injection between $Y1^{-/-}$ mice (98%) and Y1lox/lox controls (96%). This lack of difference in new neuron incorporation by 21 days was confirmed by the absence of any significant difference in total granule cell layer cell counts between 90 day old constitutive $Y1^{-/-}$ mice (346,200 ± 11,690 N=6) and $Y1^{lox/lox}$ controls (342,600 ± 13,700 N=5).

Experimental summary

We have shown that NPY is a proliferative factor for nestin positive stem cells and neuroblast precursor cells from the whole hippocampus and that the neuroproliferative effect of NPY is confined to neuroblasts from the dentate gyrus, both in the early postnatal dentate gyrus *in vitro* and the adult dentate gyrus *in vivo*. We have also shown that the neuroproliferative effect is Y1 receptor mediated and requires ERK1/2 activation (Fig. 6). Given the reduction in proliferation and neurogenesis

Figure 6
*NPY mediates neuroproliferation via the NPY Y1 receptor and requires ERK 1/2 activation. (A) NPY significantly increases the number of cells immunopositive for BrdU following a 4 h pulse of BrdU, which was blocked by the specific Y1 receptor antagonist BIBP3226 and mimicked by the specific Y1 receptor agonist [F7, P34] NPY. This pattern was repeated for β-tubulin positive cells (B) and neuroblasts (BrdU + β-tubulin positive cells) (C). Experiments were performed on 12–16 wells per condition from four separate experiments. ANOVA with Dunnett's post-test, **, $p < 0.01$. (D) ERK1/2 activation is required for NPY neuroprolifera-*

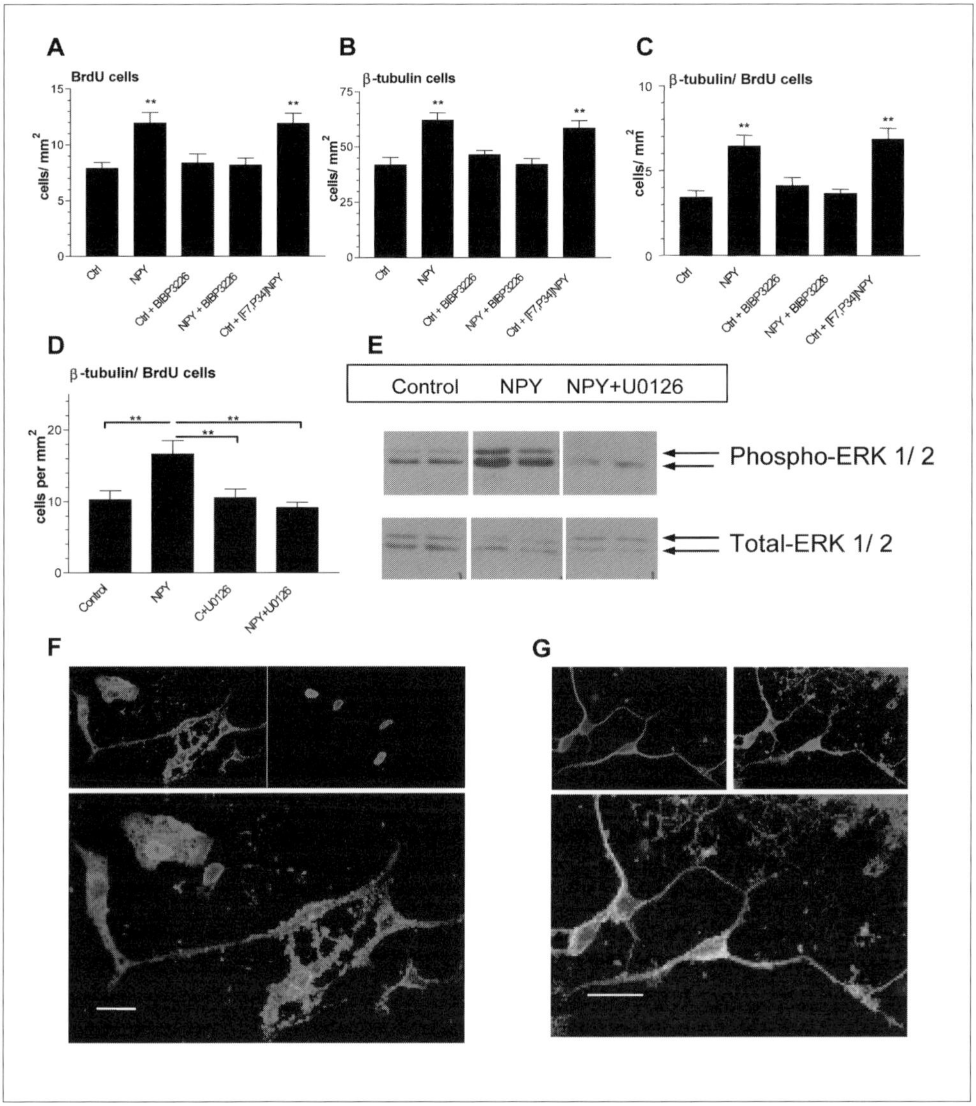

*tion. 1 μM U0126 (ERK kinase inhibitor) abolished the NPY increase in β-tubulin/ BrdU positive cells (ANOVA with Dunnett's post test, **, $p < 0.01$). Values represent mean ± SE from three separate experiments. (E) NPY phosphorylation of ERK1/2 was confirmed by increased levels of activated ERK 1/2 upon the addition of NPY, while NPY plus U0126 had low levels of activated protein. Total ERK levels in cell lysates were comparable. (F and G) Confocal images showing BrdU and Y1 antibody staining (F) and β-tubulin, DAPI and Y1 antibody co-localisation (G) in cells cultured from the dentate gyrus and exposed to BrdU for 4 h prior to fixation. (Scale bar 20 μm). Reproduced with permission from [54].*

seen in naive $Y1^{-/-}$ animals (Fig. 7), it is clear that NPY is important for the maintenance of baseline neurogenesis. It is also clear from the residual level of precursor proliferation as well as the compensatory changes in constitutive $Y1^{-/-}$ mice that many other factors dynamically influence dentate neurogenesis *in vivo*.

NPY control of dentate neurogenesis

The most likely origin of NPY for modulating dentate neurogenesis is from NPY releasing interneurons in the dentate hilus, as granule cell production of NPY is only induced after seizures [60] or perforant path stimulation [61]. The interneurons in the dentate gyrus that express NPY are heterogeneous, receiving diverse inputs and innervating a variety of dentate gyrus cell types, including axon projections to the SGZ, where they could release NPY. The neurogenic SGZ and the molecular layer of the dentate gyrus are rich in Y1 receptors [62], the functional significance of which has heretofore been unclear. NPY interneurons are thus ideally positioned to sample patterns of afferent and efferent activity and thus modulate neurogenesis via NPY release on Y1 receptor positive subgranular zone precursors. Volume transmission by released NPY acting via rapidly internalising Y1 receptors [63] is a well recognised mechanism of action of NPY and is particularly attractive as a means of orchestrating the proliferation of populations of precursors in the dentate SGZ. The other possible origin of NPY that could modulate SGZ precursors is from the sympathetic nervous system, which releases NPY into the circulation [64, 65] that crosses the blood brain barrier [66] to the vascular neurogenic SGZ. Indeed exercise induced systemic NPY release [65] may also play a role in mediating the exercise induced increase in dentate neurogenesis [67].

Figure 7
*Y1 homozygous knockout animals have reduced subgranular zone cell proliferation and lower numbers of immature doublecortin positive neurons. (A) Five $Y1^{lox/lox}$ controls and five $Y1^{-/-}$ 35 day old mice were administered BrdU (50 mg/ kg) intra-peritoneally once per day for 5 days and sacrificed 5 h after the final injection. Immunohistochemistry illustrates cells that incorporated BrdU in the subgranular zone. (B) $Y1^{-/-}$ animals had a significant reduction only in the number of BrdU positive cells counted in the SGZ (ANOVA with Student Newman-Keuls post-test; **, $p < 0.01$). (C) Doublecortin immunoreactivity (brown DAB stain) showing new neurons in the subgranular zone. (D) The number of doublecortin positive cells is significantly reduced in the dentate of $Y1^{-/-}$ animals in comparison to $Y1^{lox/lox}$ controls (n = 5 animals per group). Students t test, **, $p < 0.01$. (E) The number of BrdU positive cells in the SGZ was significantly reduced in $Y1^{-/-}$ mice compared to $Y1^{lox/lox}$ controls 2 h after a single IP injection of BrdU (50 mg/kg) ($P < 0.001$: ANOVA with Bonferronni's post test, n = 4 animals per group and 5 sections sampled from each dentate). However, there was no sig-*

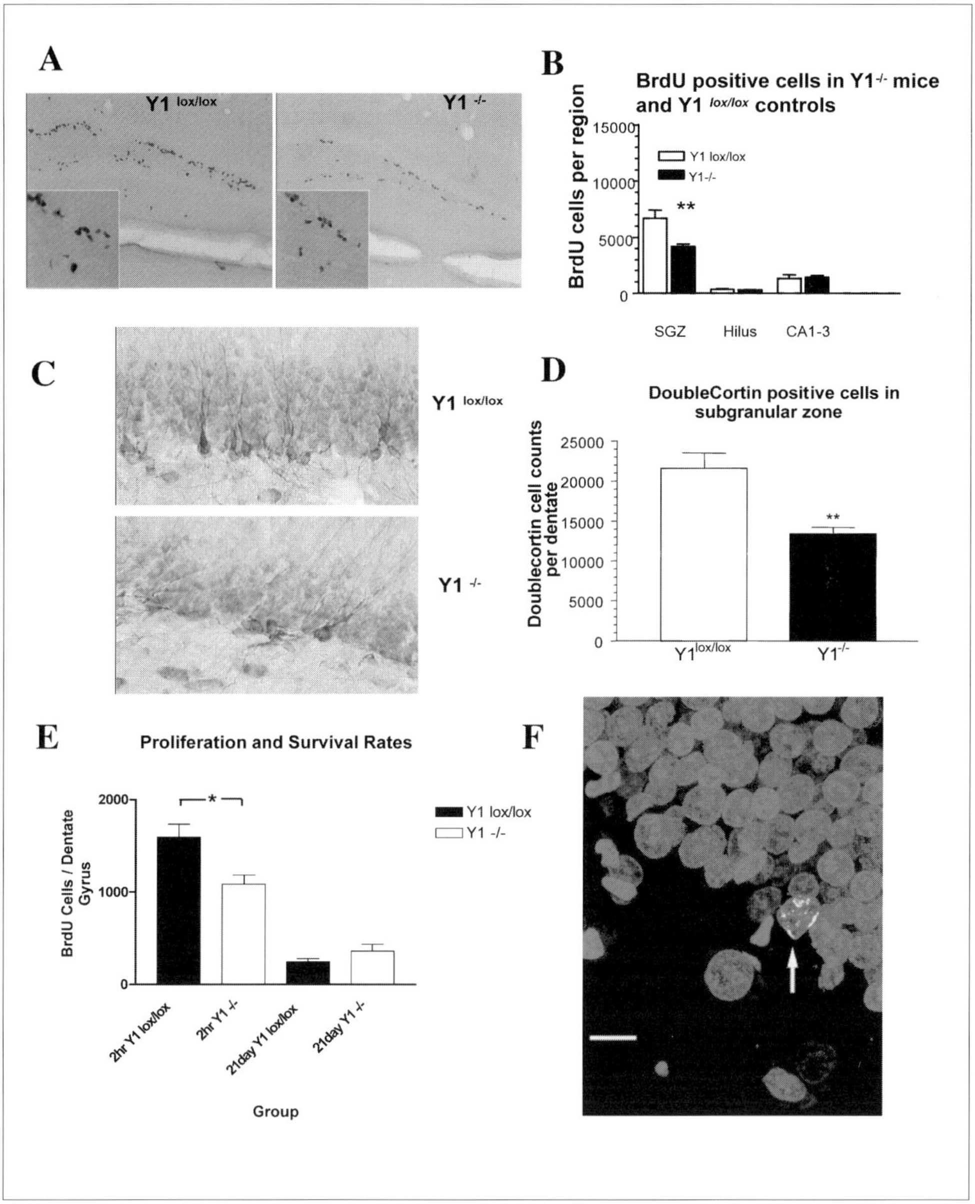

nificant difference in BrdU counts 21 days after BrdU injection (n= 6 animals per group and 5 sections sampled from each dentate). (F) Representative projection image of a 3D confocal Z stack through the granule cell layer and subgranular zone (SGZ) of a Y1$^{-/-}$ KO animal showing Neu-N positive nuclei and a single BrdU positive nucleus co-localising for Neu-N (white arrow) in the SGZ. (Scale bar = 10 μm). Reproduced with permission from [54].

NPY, neurogenesis and cognitive function

Dentate neurogenesis also appears to be necessary for some forms of hippocampal-dependant learning [68]. NPY appears to have a role in learning and memory, albeit less established than in anxiety. Intracerebroventricular injections of NPY fragments have been shown to enhance memory retention, recall and prevent scopolamine induced amnesia [69] especially when injected into the rostral hippocampus [70]. NPY attenuates learning impairments induced by MK-801 [71] and the Y2 receptor also appears to be important in learning and memory processing [72]. Although we found no evidence for a direct Y2 receptor mediated effect on neurogenesis, Redrobe et al.'s Y2 knockout study is consistent with our data supporting a Y1 mediated effect on SGZ proliferation since Y2 receptors are considered to be autoreceptors that negatively feedback to NPY-ergic terminals to reduce NPY release [73] and so Y2 receptor null mice are predicted to have increased endogenous NPY expression. Studies in hippocampal NPY over-expressing rats have shown impaired spatial learning associated with decreased NPY- Y1 binding in young animals [74] but normal spatial learning in old animals [75]. It is interesting to note that NPY levels are reduced in the dentate gyrus of aged animals [76] as is neurogenesis (*vide supra*) and so neurogenesis in aged animals may be resistant to changes in NPY expression.

Although Y1 receptors are found throughout the hippocampus and are not just confined to the neurogenic dentate subgranular zone (Kopp et al., 2002), it is intriguing to speculate that some of the cognitive and behavioural effects of NPY on learning and memory may be mediated via the effect of NPY on dentate neurogenesis, although clearly this hypothesis is currently speculative and will need to be tested by experiment.

NPY, behaviour and mood control

Both laboratory [77] and clinical studies [78, 79] have implicated NPY in the regulation of anxiety-related behaviours and the central anxiolytic effect of NPY appears to be Y1 receptor mediated [80, 81]. Clinical studies have also reported reduced NPY in the cerebrospinal fluid and plasma from depressed patients [82]. Animal models of depression [83, 84] have reduced hippocampal levels of NPY and altered Y1 binding levels, antidepressants increase brain levels of NPY [85] and the antidepressant-like activity of NPY appears to be mediated via the Y1 receptor [86]. NPY has been shown to be involved in the antidepressant actions of lithium, electroconvulsive therapy (ECT) and Citalopram [87], all of which increase neurogenesis [40, 88, 89] and the NPY gene is one of the few genes whose expression is increased only with chronic ECT [90]. For a comprehensive discussion of the role of the NPY system in stress, anxiety and depression, the reader is referred to a recent review by Marcus Heilig [11].

Conclusion

The chronic behavioural and cognitive effects of NPY may be partly mediated by Y1 receptor effects on stem cell proliferation and neurogenesis in the dentate gyrus of the adult hippocampus. Further investigation of this mechanism of NPY function is likely to be helpful is dissociating the acute and chronic effects of NPY on cognition and behaviour and may help generate selective pharmacological strategies for different NPY effects.

Acknowledgements
This work was supported by the Medical Research Council, Hope, The British Council ARC programme, the University of Southampton, and NIH grant 41490.

References

1 Larhammar D (1996) Evolution of neuropeptide Y, peptide YY and pancreatic polypeptide. *Regul Pept* 62: 1–11
2 Wu Q, Wen T, Lee G, Park JH, Cai HN, Shen P (2003) Developmental control of foraging and social behavior by the Drosophila neuropeptide Y-like system. *Neuron* 39: 147–161
3 Sokolowski MB (2003) NPY and the regulation of behavioral development. *Neuron* 39: 6–8
4 de Bono M, Bargmann CI (1998) Natural variation in a neuropeptide Y receptor homolog modifies social behavior and food response in *C. elegans*. *Cell* 94: 679–689
5 Kalra SP, Kalra PS (2004) NPY and cohorts in regulating appetite, obesity and metabolic syndrome: beneficial effects of gene therapy. *Neuropeptides* 38: 201–211
6 Albers HE, Ferris CF, Leeman SE, Goldman BD (1984) Avian pancreatic polypeptide phase shifts hamster circadian rhythms when microinjected into the suprachiasmatic region. *Science* 223: 833–855
7 Albers HE, Ferris CF (1984) Neuropeptide Y: role in light-dark cycle entrainment of hamster circadian rhythms. *Neurosci Lett* 50: 163–168
8 Pedrazzini T, Seydoux J, Kunstner P, Aubert JF, Grouzmann E, Beermann F, Brunner HR (1998) Cardiovascular response, feeding behavior and locomotor activity in mice lacking the NPY Y1 receptor. *Nat Med* 4: 722–726
9 Thiele TE, Sparta DR, Hayes DM, Fee JR (2004) A role for neuropeptide Y in neurobiological responses to ethanol and drugs of abuse. *Neuropeptides* 38: 235–243
10 Redrobe JP, Dumont Y, St-Pierre JA, Quirion R (1999) Multiple receptors for neuropeptide Y in the hippocampus: putative roles in seizures and cognition. *Brain Res* 848: 153–166

11 Heilig M (2004) The NPY system in stress, anxiety and depression. *Neuropeptides* 38: 213–224
12 Hansel DE, Eipper BA, Ronnett GV (2001) Neuropeptide Y functions as a neuroproliferative factor. *Nature* 410: 940–944
13 Hokfelt T, Johansson O, Goldstein M (1984) Chemical anatomy of the brain. *Science* 225: 1326–1334
14 Krsmanovic LZ, Stojilkovic SS, Mertz LM, Tomic M, Catt KJ (1993) Expression of gonadotropin-releasing hormone receptors and autocrine regulation of neuropeptide release in immortalized hypothalamic neurons. *Proc Natl Acad Sci USA* 90: 3908–3912
15 Wettstein JG, Earley B, Junien JL (1995) Central nervous system pharmacology of neuropeptide Y. *Pharmacol Ther* 65: 397–414
16 Waschek JA, Lelievre V, Bravo DT, Nguyen T, Muller JM (1997) Retinoic acid regulation of the VIP and PACAP autocrine ligand and receptor system in human neuroblastoma cell lines. *Peptides* 18: 835–841
17 Hansel DE, Eipper BA, Ronnett GV (2001) Regulation of olfactory neurogenesis by amidated neuropeptides. *J Neurosci Res* 66: 1–7
18 Altman J, Bayer SA (1990) Migration and distribution of two populations of hippocampal granule cell precursors during the perinatal and postnatal periods. *J Comp Neurol* 301: 365–381
19 Schlessinger AR, Cowan WM, Gottlieb DI (1975) An autoradiographic study of the time of origin and the pattern of granule cell migration in the dentate gyrus of the rat. *J Comp Neurol* 159: 149–175
20 Eriksson PS, Perfilieva E, Bjork-Eriksson T, Alborn AM, Nordborg C, Peterson DA, Gage FH (1998) Neurogenesis in the adult human hippocampus. *Nat Med* 4: 1313–1317
21 Cameron HA, McKay RD (2001) Adult neurogenesis produces a large pool of new granule cells in the dentate gyrus. *J Comp Neurol* 435: 406–417
22 Altman J, Das GD (1965) Autoradiographic and histological evidence of postnatal hippocampal neurogenesis in rats. *J Comp Neurol* 124: 319–335
23 Rietze R, Poulin P, Weiss S (2000) Mitotically active cells that generate neurons and astrocytes are present in multiple regions of the adult mouse hippocampus. *J Comp Neurol* 424: 397–408
24 Dayer AG, Ford AA, Cleaver KM, Yassee M, Cameron HA (2003) Short-term and long-term survival of new neurons in the rat dentate gyrus. *J Comp Neurol* 460: 563–572
25 Prickaerts J, Koopmans G, Blokland A, Scheepens A (2004) Learning and adult neurogenesis: survival with or without proliferation? *Neurobiol Learn Mem* 81: 1–11
26 Seki T, Arai Y (1995) Age-related production of new granule cells in the adult dentate gyrus. *Neuroreport* 6: 2479–2482
27 Kuhn HG, Dickinson-Anson H, Gage FH (1996) Neurogenesis in the dentate gyrus of the adult rat: age-related decrease of neuronal progenitor proliferation. *J Neurosci* 16: 2027–2033
28 Blumcke I, Schewe JC, Normann S, Brustle O, Schramm J, Elger CE, Wiestler OD

(2001) Increase of nestin-immunoreactive neural precursor cells in the dentate gyrus of pediatric patients with early-onset temporal lobe epilepsy. *Hippocampus* 11: 311–321
29 Cameron HA, McKay RD (1999) Restoring production of hippocampal neurons in old age. *Nat Neurosci* 2: 894–897
30 Scharfman HE, Goodman JH, Sollas AL (2000) Granule-like neurons at the hilar/CA3 border after status epilepticus and their synchrony with area CA3 pyramidal cells: functional implications of seizure-induced neurogenesis. *J Neurosci* 20: 6144–6158
31 van Praag H, Schinder AF, Christie BR, Toni N, Palmer TD, Gage FH (2002) Functional neurogenesis in the adult hippocampus. *Nature* 415: 1030–1034
32 Shors TJ, Miesegaes G, Beylin A, Zhao M, Rydel T, Gould E (2001) Neurogenesis in the adult is involved in the formation of trace memories. *Nature* 410: 372–376
33 Snyder JS, Kee N, Wojtowicz JM (2001) Effects of adult neurogenesis on synaptic plasticity in the rat dentate gyrus. *J Neurophysiol* 85: 2423–2431
34 Gould E, Beylin A, Tanapat P, Reeves A, Shors TJ (1999) Learning enhances adult neurogenesis in the hippocampal formation. *Nat Neurosci* 2: 260–265
35 Lemaire V, Koehl M, Le Moal M, Abrous DN (2000) Prenatal stress produces learning deficits associated with an inhibition of neurogenesis in the hippocampus. *Proc Natl Acad Sci USA* 97: 11032–11037
36 Drapeau E, Mayo W, Aurousseau C, Le Moal M, Piazza PV, Abrous DN (2003) Spatial memory performances of aged rats in the water maze predict levels of hippocampal neurogenesis. *Proc Natl Acad Sci USA* 100: 14385–14390
37 Kempermann G, Kuhn HG, Gage FH (1997) More hippocampal neurons in adult mice living in an enriched environment. *Nature* 386: 493–495
38 D'Sa C, Duman RS (2002) Antidepressants and neuroplasticity. *Bipolar Disord* 4: 183–194
39 Gould E, McEwen BS, Tanapat P, Galea LA, Fuchs E (1997) Neurogenesis in the dentate gyrus of the adult tree shrew is regulated by psychosocial stress and NMDA receptor activation. *J Neurosci* 17: 2492–2498
40 Malberg JE, Eisch AJ, Nestler EJ, Duman RS (2000) Chronic antidepressant treatment increases neurogenesis in adult rat hippocampus. *J Neurosci* 20: 9104–9410
41 Santarelli L, Saxe M, Gross C, Surget A, Battaglia F, Dulawa S, Weisstaub N, Lee J, Duman R, Arancio O et al (2003) Requirement of hippocampal neurogenesis for the behavioral effects of antidepressants. *Science* 301: 805–809
42 Hastings NB, Gould E (1999) Rapid extension of axons into the CA3 region by adult–generated granule cells. *J Comp Neurol* 413: 146–154
43 Markakis EA, Gage FH (1999) Adult-generated neurons in the dentate gyrus send axonal projections to field CA3 and are surrounded by synaptic vesicles. *J Comp Neurol* 406: 449–460
44 Cameron HA, Hazel TG, McKay RD (1998) Regulation of neurogenesis by growth factors and neurotransmitters. *J Neurobiol* 36: 287–306
45 Mitchell BD, Emsley JG, Magavi SS, Arlotta P, Macklis JD (2004) Constitutive and induced neurogenesis in the adult mammalian brain: manipulation of endogenous precursors toward CNS repair. *Dev Neurosci* 26: 101–117

46 Bengzon J, Kokaia Z, Elmer E, Nanobashvili A, Kokaia M, Lindvall O (1997) Apoptosis and proliferation of dentate gyrus neurons after single and intermittent limbic seizures. *Proc Natl Acad Sci USA* 94: 10432–10437
47 Gray WP, Sundstrom LE (1998) Kainic acid increases the proliferation of granule cell progenitors in the dentate gyrus of the adult rat. *Brain Res* 790: 52–59
48 Parent JM, Yu TW, Leibowitz RT, Geschwind DH, Sloviter RS, Lowenstein DH (1997) Dentate granule cell neurogenesis is increased by seizures and contributes to aberrant network reorganization in the adult rat hippocampus. *J Neurosci* 17: 3727–3738
49 Liu J, Solway K, Messing RO, Sharp FR (1998) Increased neurogenesis in the dentate gyrus after transient global ischemia in gerbils. *J Neurosci* 18: 7768–7778
50 Dash PK, Mach SA, Moore AN (2001) Enhanced neurogenesis in the rodent hippocampus following traumatic brain injury. *J Neurosci Res* 63: 313–319
51 Huttmann K, Sadgrove M, Wallraff A, Hinterkeuser S, Kirchhoff F, Steinhauser C, Gray WP (2003) Seizures preferentially stimulate proliferation of radial glia-like astrocytes in the adult dentate gyrus: functional and immunocytochemical analysis. *Eur J Neurosci* 18: 2769–2778
52 Dumont Y, Martel JC, Fournier A, St-Pierre S, Quirion R (1992) Neuropeptide Y and neuropeptide Y receptor subtypes in brain and peripheral tissues. *Prog Neurobiol* 38: 125–167
53 El Bahh B, Auvergne R, Lere C, Brana C, Le Gal La Salle G, Rougier A (2001) Decreased epileptic susceptibility correlates with neuropeptide Y overexpression in a model of tolerance to excitotoxicity. *Brain Res* 894: 209–217
54 Howell OW, Doyle K, Goodman JH, Scharfman HE, Herzog H, Pringle A, Beck-Sickinger AG, Gray WP (2005) Neuropeptide Y stimulates neuronal precursor proliferation in the post-natal and adult dentate gyrus. *J Neurochem* 93: 560–570
55 Howell OW, Scharfman HE, Herzog H, Sundstrom LE, Beck-Sickinger A, Gray WP (2003) Neuropeptide Y is neuroproliferative for post-natal hippocampal precursor cells. *J Neurochem* 86: 646–659
56 Palmer TD, Takahashi J, Gage FH (1997) The adult rat hippocampus contains primordial neural stem cells. *Mol Cell Neurosci* 8: 389–404
57 Seaberg RM, van der Kooy D (2002) Adult rodent neurogenic regions: the ventricular subependyma contains neural stem cells, but the dentate gyrus contains restricted progenitors. *J Neurosci* 22: 1784–1793
58 Lopez-Ilasaca M (1998) Signaling from G-protein-coupled receptors to mitogen-activated protein (MAP)-kinase cascades. *Biochem Pharmacol* 56: 269–277
59 Li BS, Ma W, Zhang L, Barker JL, Stenger DA, Pant HC (2001) Activation of phosphatidylinositol-3 kinase (PI-3K) and extracellular regulated kinases (Erk1/2) is involved in muscarinic receptor-mediated DNA synthesis in neural progenitor cells. *J Neurosci* 21: 1569–1579
60 Gruber B, Greber S, Rupp E, Sperk G (1994) Differential NPY mRNA expression in granule cells and interneurons of the rat dentate gyrus after kainic acid injection. *Hippocampus* 4: 474–482

61 Causing CG, Makus KD, Ma Y, Miller FD, Colmers WF (1996) Selective upregulation of T alpha 1 alpha-tubulin and neuropeptide Y mRNAs after intermittent excitatory stimulation in adult rat hippocampus *in vivo*. *J Comp Neurol* 367: 132–146

62 Kopp J, Xu ZQ, Zhang X, Pedrazzini T, Herzog H, Kresse A, Wong H, Walsh JH, Hokfelt T (2002) Expression of the neuropeptide Y Y1 receptor in the CNS of rat and of wild-type and Y1 receptor knock-out mice. Focus on immunohistochemical localization(1). *Neuroscience* 111: 443–532

63 Fabry M, Langer M, Rothen-Rutishauser B, Wunderli-Allenspach H, Hocker H, Beck-Sickinger AG (2000) Monitoring of the internalization of neuropeptide Y on neuroblastoma cell line SK-N-MC. *Eur J Biochem* 267: 5631–5637

64 Pernow J, Lundberg JM, Kaijser L, Hjemdahl P, Theodorsson-Norheim E, Martinsson A, Pernow B (1986) Plasma neuropeptide Y-like immunoreactivity and catecholamines during various degrees of sympathetic activation in man. *Clin Physiol* 6: 561–578

65 Lundberg JM, Martinsson A, Hemsen A, Theodorsson-Norheim E, Svedenhag J, Ekblom B, Hjemdahl P (1985) Co-release of neuropeptide Y and catecholamines during physical exercise in man. *Biochem Biophys Res Commun* 133: 30–36

66 Kastin AJ, Akerstrom V (1999) Nonsaturable entry of neuropeptide Y into brain. *Am J Physiol* 276: E479–482

67 van Praag H, Kempermann G, Gage FH (1999) Running increases cell proliferation and neurogenesis in the adult mouse dentate gyrus. *Nat Neurosci* 2: 266–270

68 Shors TJ (2004) Memory traces of trace memories: neurogenesis, synaptogenesis and awareness. *Trends Neurosci* 27: 250–256

69 Flood JF, Hernandez EN, Morley JE (1987) Modulation of memory processing by neuropeptide Y. *Brain Res* 421: 280–290

70 Flood JF, Morley JE (1989) Dissociation of the effects of neuropeptide Y on feeding and memory: evidence for pre- and postsynaptic mediation. *Peptides* 10: 963–966

71 Bouchard P, Maurice T, St-Pierre S, Privat A, Quirion R (1997) Neuropeptide Y and the calcitonin gene-related peptide attenuate learning impairments induced by MK-801 via a sigma receptor-related mechanism. *Eur J Neurosci* 9: 2142–2151

72 Redrobe JP, Dumont Y, Herzog H, Quirion R (2004) Characterization of neuropeptide Y, Y(2) receptor knockout mice in two animal models of learning and memory processing. *J Mol Neurosci* 22: 159–166

73 Colmers WF, Bleakman D (1994) Effects of neuropeptide Y on the electrical properties of neurons. *Trends Neurosci* 17: 373–379

74 Thorsell A, Michalkiewicz M, Dumont Y, Quirion R, Caberlotto L, Rimondini R, Mathe AA, Heilig M (2000) Behavioral insensitivity to restraint stress, absent fear suppression of behavior and impaired spatial learning in transgenic rats with hippocampal neuropeptide Y overexpression. *Proc Natl Acad Sci USA* 97: 12852–12857

75 Carvajal CC, Vercauteren F, Dumont Y, Michalkiewicz M, Quirion R (2004) Aged neuropeptide Y transgenic rats are resistant to acute stress but maintain spatial and non-spatial learning. *Behav Brain Res* 153: 471–480

76 Higuchi H, Yang HY, Costa E (1988) Age-related bidirectional changes in neuropeptide Y peptides in rat adrenal glands, brain, and blood. *J Neurochem* 50: 1879–1886

77 Heilig M, Soderpalm B, Engel JA, Widerlov E (1989) Centrally administered neuropeptide Y (NPY) produces anxiolytic-like effects in animal anxiety models. *Psychopharmacology (Berl)* 98: 524–529

78 Morgan CA 3rd, Wang S, Southwick SM, Rasmusson A, Hazlett G, Hauger RL, Charney DS (2000) Plasma neuropeptide-Y concentrations in humans exposed to military survival training. *Biol Psychiatry* 47: 902–909

79 Morgan CA 3rd, Rasmusson AM, Wang S, Hoyt G, Hauger RL, Hazlett G (2002) Neuropeptide-Y, cortisol, and subjective distress in humans exposed to acute stress: replication and extension of previous report. *Biol Psychiatry* 52: 136–142

80 Sajdyk TJ, Vandergriff MG, Gehlert DR (1999) Amygdalar neuropeptide Y Y1 receptors mediate the anxiolytic-like actions of neuropeptide Y in the social interaction test. *Eur J Pharmacol* 368: 143–147

81 Wahlestedt C, Pich EM, Koob GF, Yee F, Heilig M (1993) Modulation of anxiety and neuropeptide Y-Y1 receptors by antisense oligodeoxynucleotides. *Science* 259: 528–531

82 Westrin A, Ekman R, Traskman-Bendz L (1999) Alterations of corticotropin releasing hormone (CRH) and neuropeptide Y (NPY) plasma levels in mood disorder patients with a recent suicide attempt. *Eur Neuropsychopharmacol* 9: 205–211

83 Husum H, Mathe AA (2002) Early life stress changes concentrations of neuropeptide Y and corticotropin-releasing hormone in adult rat brain. Lithium treatment modifies these changes. *Neuropsychopharmacology* 27: 756–764

84 Caberlotto L, Jimenez P, Overstreet DH, Hurd YL, Mathe AA, Fuxe K (1999) Alterations in neuropeptide Y levels and Y1 binding sites in the Flinders Sensitive Line rats, a genetic animal model of depression. *Neurosci Lett* 265: 191–194

85 Heilig M, Wahlestedt C, Ekman R, Widerlov E (1988) Antidepressant drugs increase the concentration of neuropeptide Y (NPY)-like immunoreactivity in the rat brain. *Eur J Pharmacol* 147: 465–467

86 Redrobe JP, Dumont Y, Fournier A, Quirion R (2002) The neuropeptide Y (NPY) Y1 receptor subtype mediates NPY-induced antidepressant-like activity in the mouse forced swimming test. *Neuropsychopharmacology* 26: 615–624

87 Husum H, Mikkelsen JD, Hogg S, Mathe AA, Mork A (2000) Involvement of hippocampal neuropeptide Y in mediating the chronic actions of lithium, electroconvulsive stimulation and citalopram. *Neuropharmacology* 39: 1463–1473

88 Madsen TM, Treschow A, Bengzon J, Bolwig TG, Lindvall O, Tingstrom A (2000) Increased neurogenesis in a model of electroconvulsive therapy. *Biol Psychiatry* 47: 1043–1049

89 Chen G, Rajkowska G, Du F, Seraji-Bozorgzad N, Manji HK (2000) Enhancement of hippocampal neurogenesis by lithium. *J Neurochem* 75: 1729–1734

90 Altar CA, Laeng P, Jurata LW, Brockman JA, Lemire A, Bullard J, Bukhman YV, Young TA, Charles V, Palfreyman MG (2004) Electroconvulsive seizures regulate gene expression of distinct neurotrophic signaling pathways. *J Neurosci* 24: 2667–2677

NPY family of peptides in neural crest-derived tumors

Joanna B. Kitlinska

Department of Physiology and Biophysics, Georgetown University School of Medicine, Georgetown University School of Medicine, 3900 Reservoir Rd, NW, Basic Science Building, Washington, DC 20057, USA

Introduction

Among many pleiotropic functions of neuropeptide Y (NPY), its activity as a growth factor is one of the most recently discovered. The peptide stimulates proliferation of a variety of cells, such as neuronal precursors and vascular smooth muscle cells [1, 2]. Moreover, due to its stimulatory effect on endothelial cell (EC) proliferation and migration, NPY is also a potent angiogenic factor [3–5]. The fact that both cell proliferation and the formation of new blood vessels are important steps in the growth of solid tumors, suggested the possible role of NPY in their development. NPY, as a sympathetic neurotransmitter, is abundant in neural crest-derived tumors, particularly those originating from the autonomic nervous system. The peptide and its receptors are highly expressed in sympathetic neuroblastomas and pheochromocytomas, as well as in Ewing's sarcoma family of tumors (ESFT) of a putative parasympathetic origin [6–12] (Fig. 1). Expression of NPY is significantly lower in tumors of the central nervous system (CNS). Low levels of the peptide have been found in pituitary adenomas and glioblastomas, whereas it was almost not detectable in astrocytomas and prolactinomas [13]. Moreover, low levels of NPY distinguish central primitive neuroectodermal tumors (cPNET) from their NPY-rich peripheral counterparts (pPNET) belonging to the ESFT family [13]. Thus, in this chapter, we will focus on the effect of NPY on the growth and vascularization of tumors originating from the peripheral autonomic nervous system.

Neuroblastoma

Neuroblastoma is a tumor of the sympathetic origin, which, together with ESFT and rhabdosarcoma, belongs to a family of aggressive, small blue round cell tumors of childhood. It is the most common and, at the same time, the most enigmatic malignant disease of infancy. The tumors develop in the postganglionic sympathetic neu-

The NPY Family of Peptides in Immune Disorders, Inflammation, Angiogenesis and Cancer, edited by Zofia Zukowska and Giora Z. Feuerstein

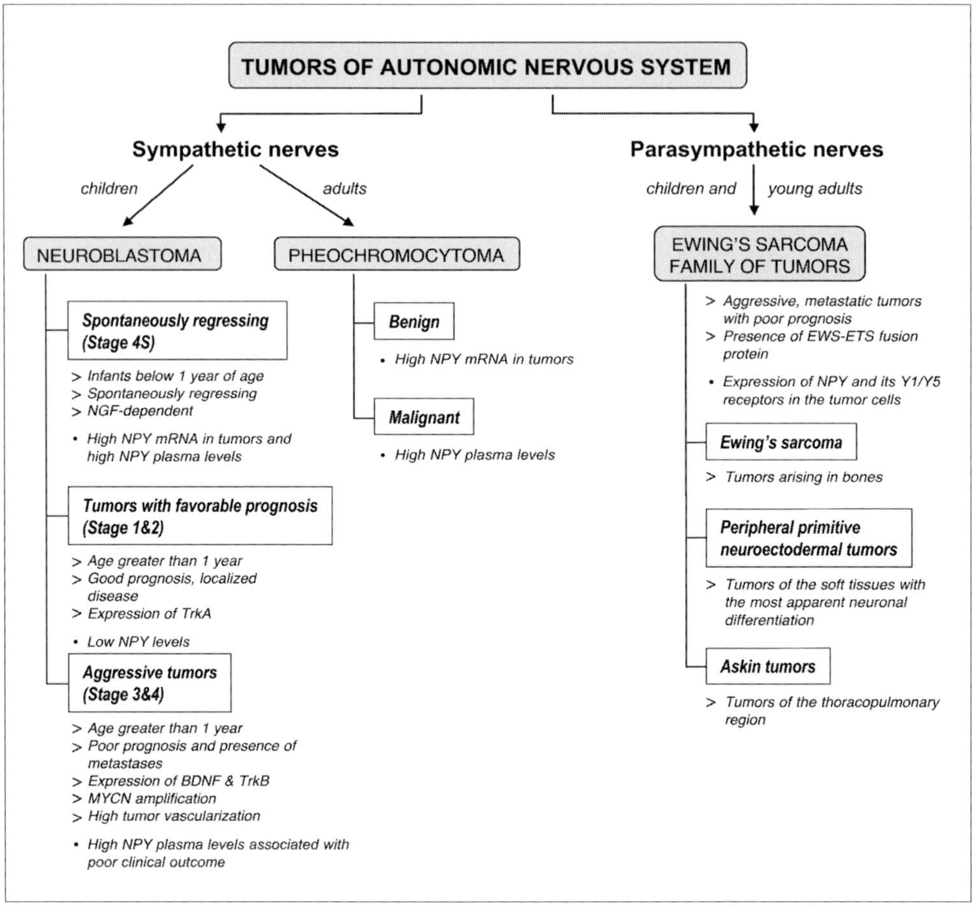

Figure 1
NPY in tumors derived from autonomic nervous system.

rons and are commonly localized to the adrenal gland or sympathetic ganglia [14–16]. The clinical hallmark of neuroblastomas is their diversity. When diagnosed before 1 year of age, they are usually curable and often spontaneously regress or differentiate into benign ganglioneuroma, despite already present metastases. However, when diagnosed in ages greater than 1 year, neuroblastomas are very aggressive, highly vascularized and metastatic, usually leading to death in spite of aggressive chemotherapy [14, 17–20]. Clinically, neuroblastomas are classified into four stages (Fig. 1). At low stages (1 and 2), the disease is localized and has a good prognosis, whereas at high stages (3 and 4), the disease is characterized by the presence of metastases to bones, bone marrow and lymph nodes. Interestingly, no progression from low to high stage neuroblastomas has been observed. Spontaneously regressing neuroblastoma of newborns has been classified as a separate stage, 4S, and often is considered a different type of tumor due to its specific biology and metastatic pattern (mainly liver, skin and bone marrow) [14, 16, 21].

The variety of neuroblastoma phenotypes can be partially explained by differential effects of neurotrophins on their growth. Nerve growth factor (NGF), released in neuroblastomas mostly by Schwann cells and fibroblasts, stimulates differentiation and inhibits proliferation of neuroblastoma cells. Its receptor, TrkA, is preferentially expressed in more favorable stage 1 and 2 neuroblastomas and is considered a good prognostic factor. NGF is also present in 4S neuroblastomas and are implicated in their spontaneous regression [14, 16, 19, 21–24]. On the other hand, brain-derived neurotrophic factor (BDNF), via its TrkB receptor, promotes the survival and invasiveness of neuroblastoma cells. In high stage neuroblastomas, BDNF and TrkB are co-expressed, which forms an autocrine loop and is associated with a poor prognosis and aggressive phenotype of the disease [14, 16, 23–25]. Another important molecular prognostic factor in neuroblastoma is amplification of the MYCN proto-oncogene, which is highly correlated with a more advanced stage, rapid progression and poor outcome of the disease. MYCN amplification is also associated with increased vascularization of these highly angiogenesis-dependent tumors [14, 21, 26–28].

Due to their sympathetic origin, neuroblastomas synthesize and release nerve-specific enzymes and peptides, such as catecholamines and NPY [10–12]. Neuroblastoma is often associated with elevated plasma levels of NPY, which normalizes after treatment [10–12, 29]. However, the initial attempts to use it as a diagnostic factor failed due to the variability of peptide levels between patients. Recently, NPY plasma levels were found to be increased mainly in children over 1 year of age with high stage (3 and 4) neuroblastomas, in association with a poor outcome and MYCN amplification [10, 30, 31] (Fig. 1). Interestingly, high levels of the peptide in plasma, as well as increased mRNA levels in tumors were also found in infants with stage 4S neuroblastoma [10]. Thus, NPY synthesis and release seems to be increased in highly vascularized, advanced tumors, which, unlike low stage neuroblastomas, are under the control of a BDNF/TrkB autocrine loop, as well as in spontaneously regressing neuroblastomas, which are controlled by NGF and its TrkA receptor. Since both NGF and BDNF upregulate NPY expression [32–34], the above associations suggest that NPY may be a mediator of neurotrophin actions.

Neuroblastoma cells express not only NPY, but also high levels of its receptors – mainly Y2, in some cases accompanied by Y5 receptor expression [6, 35]. This simultaneous expression of both ligand and its receptors suggests the possible role for NPY in autocrine regulation of neuroblastoma growth. Indeed, studies performed in our laboratory suggest that the peptide stimulates proliferation of neuroblastoma cells via both Y2 and Y5 receptors and these actions are mediated by activation of ERK1/2 MAPK (Fig. 2). Interestingly, NPY receptor antagonists not only blocked the mitogenic effect of NPY, but also decreased the basal proliferation level by 80% [6]. The latter indicates that NPY released from neuroblastoma cells is an important factor stimulating their growth in an autocrine manner and that the blockade of the NPY pathway leads to significant impairment of neuroblastoma cell growth.

Due to its angiogenic activities, NPY released from the tumors not only exerts direct effects on neuroblastoma cell proliferation, but also stimulates tumor vascularization (Fig. 2). *In vitro*, neuroblastoma conditioned media stimulated proliferation of human microvascular endothelial cells in a dose dependent manner. This effect was completely blocked by NPY receptor antagonists, which indicates that NPY is a crucial factor responsible for angiogenesis in neuroblastoma tumors [6]. *In vivo*, NPY administrated as a slow release pellet significantly increased the growth of neuroblastoma xenografts, which was associated with an increase in tumor vascularization [6] (Fig. 2).

Taken together, NPY released from neuroblastomas enhance their growth by stimulation of both tumor cell proliferation and angiogenesis (Fig. 2). Interestingly, BDNF which is an important growth-stimulatory factor for aggressive neuroblastomas, enhances tumor cell survival, invasiveness and angiogenesis in an autocrine fashion, while also stimulating NPY expression [14, 16, 23–25, 36, 37]. Thus, it is possible that in aggressive neuroblastomas, NPY mediates BDNF's growth-stimulatory actions. Moreover, both NPY-induced angiogenesis and its mitogenic effect in neuroblastoma cells are mediated by the same Y2 receptor, which suggests that blocking the Y2 pathway could be a potential, bi-directional treatment for this disease [6, 38, 39] (Fig. 2). Therapeutic strategies targeting both angiogenesis and tumor cell growth by using low doses of cytostatics combined with angiostatic agents have already been proven to be effective in neuroblastomas [40]. Targeting NPY-mediated angiogenesis in childhood malignancies is further supported by the fact that the peptide does not seem to be involved in developmental angiogenesis, since both Y2 and NPY knockout mice have no defects in organogenesis, despite severely impaired NPY-induced angiogenesis in ischemia and retinopathy [3, 38, 39].

Although the results of the studies described here provided significant insight into understanding the role of NPY in the regulation of neuroblastoma growth, many important questions still remain unanswered: For example, the role of NPY in metastases. Aggressive neuroblastomas are highly vascularized and metastatic [27, 38, 41]. Interestingly, high NPY expression correlates with an unfavorable prognosis and poor outcome of the disease [10, 30]. Also, the concentration of the peptide is significantly higher in bone marrow metastases than in the primary tumor [42]. Do the angiogenic activities of NPY contribute to the metastatic properties of the aggressive neuroblastomas? Does the peptide stimulate migration of neuroblastoma cells in the same way it affected ECs? Also, NPY processing is impaired in neu-

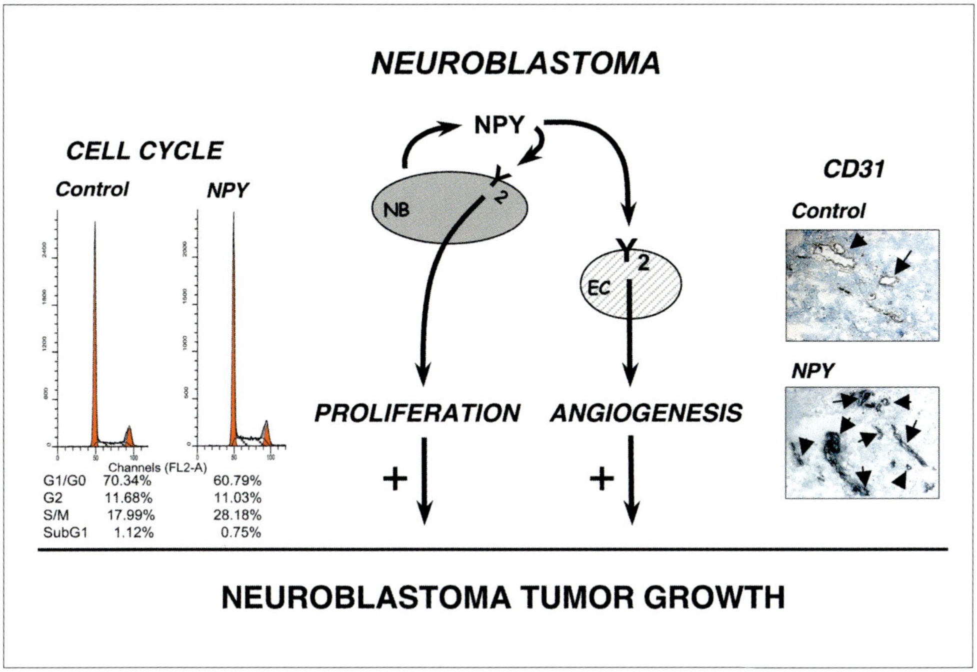

Figure 2
Effect of NPY on neuroblastoma tumor growth.

roblastoma cell lines and increased levels of pro-NPY have been found in advanced neuroblastoma tumors, which was associated with poor prognosis [11, 43]. Is pro-NPY an active form of the peptide? Does it play a role in tumor progression? Does it have specific receptors? Is pro-NPY like pro-NGF, which has recently been seen as an active form of the neurotrophin, binding to p75 receptor with an affinity higher than the mature peptide [44]?

Another interesting aspect of neuroblastoma biology is its spontaneous regression. All known available neuroblastoma cell lines are derived from aggressive tumors. Thus, the data presented here relate only to the advanced disease. What about those enigmatic, spontaneously regressing 4S neuroblastomas? As mentioned before, these tumors, unlike aggressive BDNF-dependent neuroblastomas, are under control of NGF, which stimulates their differentiation and is a favorable prognostic factor [20, 22–24]. NGF is also known to stimulate NPY expression, which correlates with high levels of NPY mRNA in the tumor tissues and its elevated plasma levels in patients with 4S neuroblastomas [10, 30, 32–34]. Does NPY mediate NGF actions in these tumors? Is the effect of NPY on the proliferation of 4S neuroblastomas different than in the aggressive neuroblastomas? Is this perhaps mediated by different receptors? Answering these and other important questions may help to understand the biology of neuroblastoma, the role of NPY in its development and progression and contribute to the new therapeutic strategies in the treatment of this devastating children's disease.

Ewing's sarcoma family of tumors

Ewing's sarcoma family of tumors (ESFT) includes several phenotypic variants, such as: arising in bones Ewing's sarcomas, peripheral primitive neuroectodermal tumors (pPNET) of soft tissues and Askin tumors of the thoracopulmonary region (Fig. 1). The cells exhibit a variable degree of neural differentiation, which is especially apparent in pPNETs. Although their origin is still uncertain, due to their cholinergic properties, ESFT are believed to develop from parasympathetic neurons [15, 45–47]. The tumors are most frequently found in children and young adults between the ages of 10 and 20 years. They are usually aggressive, often leading to death despite intensive chemo- and radiotherapy. The overall survival is 45–60%, however for the patients with metastatic disease present at the time of diagnosis, survival is less then 20% [48, 49].

The characteristic feature of ESFT tumors is a translocation resulting in the fusion of the EWS gene located on chromosome 22 with an ETS transcription factor. The most frequent gene rearrangements include t(11;22)(q24;q12) found in 85% of ESFT, which forms the fusion protein EWS/FLI1 [15, 47, 50, 51]. The EWS-ETS protein acts as an aberrant transcription factor and alters mRNA splicing. The abnormal functions of this chimerical molecule are believed to be the major mechanism leading to the malignant transformation of the cells [47, 52, 53].

ESFT cells, due to their parasympathetic origin, express neuronal markers, including NPY and its receptors [6]. High levels of Y1 receptors, detected by binding assay, have been described by van Valen et al. as a characteristic feature of ESFT cells [54]. Due to the high Y1 receptor expression, the ESFT cell line SK-N-MC has become one of the best known in NPY field, widely used as a model in studies of NPY receptor signaling and trafficking. Since after NPY stimulation Y1 receptors along with their ligand undergo rapid internalization, the Y1 receptor agonists have been used for ligand-mediated delivery of cytotoxic agents to SK-N-MC cells [55]. Interestingly, this cell line has been initially defined as a neuroblastoma [46]. However, cytogenetic and molecular biology techniques allowed detection of the characteristic translocations involving chromosomes 11 and 22, thereby classifying these cells as ESFT [51].

Although Y1 receptors have been extensively studied in ESFT cell lines, sparse and conflicting results have been reported on NPY's role in their growth [54, 56, 57]. A study performed recently in our laboratory strongly supports a growth-

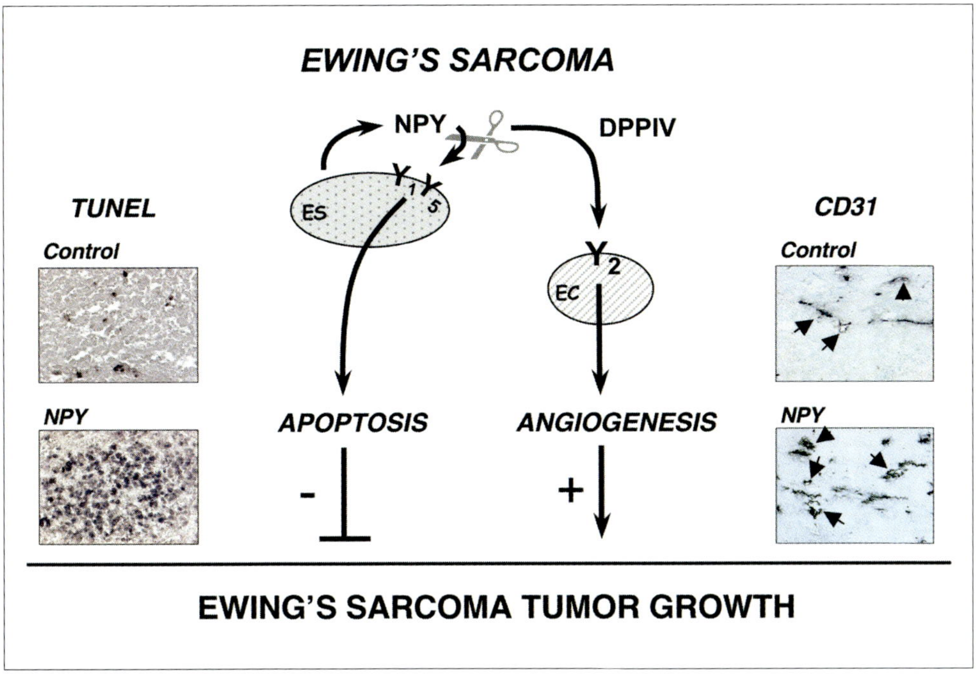

Figure 3
Effect of NPY on Ewing's sarcoma tumor growth.

inhibitory effect of NPY on ESFT cells [6]. We have found that the ESFT cells synthesize both NPY and its receptors. However, in contrast to Y2 receptor-expressing neuroblastomas, ESFT cells express high levels of both Y1 and Y5 receptors. The different receptor patterns are associated with opposite effects of NPY on the growth of these cells. In contrast to the proliferative effect observed in neuroblastomas, in SK-N-MC cells NPY induced apoptosis via activation of both Y1 and Y5 receptors (Fig. 3). Selective Y1 and Y5 antagonists not only blocked apoptotic effect of NPY, but also significantly increased basal cell growth and survival, which strongly indicates the growth-inhibitory role of the endogenous peptide. Additionally, exogenous NPY dramatically increased cell death and reduced growth of SK-N-MC xenografts *in vivo*, which further confirms its apoptotic effect in ESFT cells [6]. Interestingly, despite having this deadly autocrine loop, SK-N-MC cells retain a high proliferation rate. Although ESFT cells may be protected from NPY-induced cell death by several mechanisms, our data implicate DPPIV as a key player in this process. This membrane protease accompanies Y1 receptors in ESFT cell lines [6]. The enzyme cleaves the full length NPY_{1-36} to its shorter form NPY_{3-36}, which is inactive at Y1 receptors, but retains a high affinity to all other NPY receptors [58, 59]. Our data suggests that both Y1 and Y5 receptors need to be activated to inhibit the growth of ESFT cells. Thus, DPPIV, by converting NPY to a non-Y1 agonist, may prevent its apoptotic actions (Fig. 3). Indeed, protease inhibitor augmented NPY-induced caspase 3/7 activation [6].

Although ESFT cells synthesize and release less NPY than sympathetic neuroblastomas, the SK-N-MC conditioned media still stimulated proliferation of ECs and this effect was significantly reduced by NPY receptor antagonists [6]. The above results suggested that NPY plays an important role in vascularization of ESFT tumors. Indeed, NPY-treated SK-N-MC xenografts were better vascularized than the control, despite a dramatic inhibition of tumor growth rate [6] (Fig. 3). Thus, in ESFT tumors NPY exerts two opposite effects – Y1/Y5 receptor-mediated apoptosis of tumor cells and Y2 receptor-dependent angiogenesis (Fig. 3). The overall growth of the tumor depends on balance between these two processes. DPPIV, which converts NPY to the non-Y1 receptor agonist, shifts its activity from Y1 receptor-mediated apoptosis to Y2 receptor-mediated angiogenesis. This may be an important factor regulating the actions of NPY in ESFT tumors and is an attractive target in their treatment (Fig. 3).

Pheochromocytoma

Pheochromocytoma, unlike neuroblastoma and ESFT, is a neoplasm that primarily occurs in adults (Fig. 1). Similarly to neuroblastoma, pheochromocytomas originate from the sympathetic tissue. The tumors arise from chromaffin cells and are most often localized in the adrenal medulla or, sporadically, in extra-adrenal paraganglia. Pheochromocytomas are usually benign; however, in approximately 10% of patients they exhibit an aggressive clinical phenotype. Due to its sympathetic origin, the tumors release high levels of catecholamines and NPY, which contributes to the hypertension often observed in pheochromocytoma patients [7–9, 60–63]. The clinical reports regarding NPY in pheochromocytoma are controversial. The mRNA levels of NPY are higher in adrenal than extra-adrenal tumors [62]. Moreover, NPY mRNA is significantly increased in benign tumors, whereas its plasma levels are elevated in patients with malignant pheochromocytomas [8, 9] (Fig. 1).

The best-known *in vitro* model of pheochromocytoma is the PC12 cell line derived from a rat tumor. The cells express both NPY and its receptors – predominantly the Y1 receptor, accompanied by lower levels of Y2 and Y5 receptors [6, 64, 65]. Interestingly, this receptor expression pattern depends on the degree of cell differentiation. NGF induces neuronal differentiation of PC12 cells, which is associated with an increase in Y2 receptor expression. In contrast, PC12 differentiated toward chromaffin cells by dexamethasone treatment express only Y1 and Y5 receptors [64, 65].

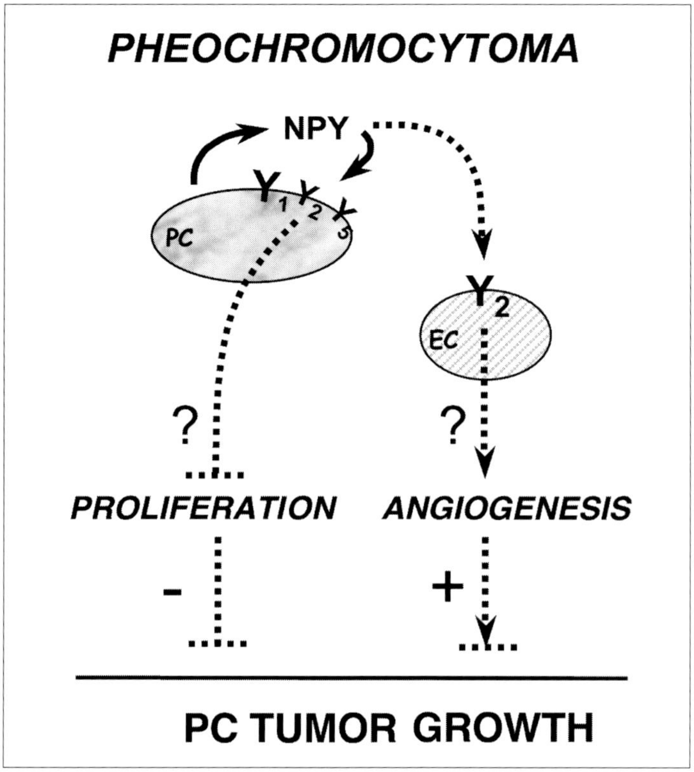

Figure 4
Putative actions of NPY in pheochromocytoma tumors.

Despite high expression of NPY at the mRNA level, PC12 cells do not release the peptide into the culture media. Instead, the synthesized peptide is accumulated on the cell membrane, which can be due to the high levels of Y1 receptors [6]. In spite of elevated receptor expression, the exogenous NPY does not exert a direct effect on PC12 proliferation, which may be explained by the saturation of the receptors with the endogenous peptide. However, the receptor antagonists significantly increase the number of viable cells, suggesting that, like in SK-N-MC cells, NPY may be a growth-inhibitory factor for PC12 cells [6] (Fig. 4).

Since PC12 is the only commonly available pheochromocytoma cell line, *in vitro* studies on the biology of these tumors are very limited. Thus, the role of NPY in the regulation of pheochromocytoma growth requires further investigation. Moreover, the fact that NPY plasma levels are elevated in patients with malignant tumors raises the possibility that, even if NPY does not promote cell proliferation directly, it may facilitate tumor growth and metastases indirectly by stimulating angiogenesis [8] (Fig.4). This hypothesis is supported by the fact that NGF, known to stimulate NPY expression in PC12 cells, increases vascularization of PC12 xenografts via a VEGF pathway, which is also a mediator of NPY-induced angiogenesis [3, 66].

Summary

NPY is an important factor in the regulation of growth of neural crest-derived tumors. The peptide both regulates proliferation of neuroendocrine tumor cells in an autocrine manner and induces tumor vascularization via its Y2 receptor-mediated effect on ECs. The overall effect of NPY is critically dependent on the tumor type and its receptor pattern. Thus, NPY and its receptors may become targets for novel, bidirectional therapies in the treatment of neuroendocrine tumors, directed against both tumor cell proliferation and angiogenesis.

References

1 Hansel DE, Eipper BA, Ronnett GV (2001) Neuropeptide Y functions as a neuroproliferative factor. *Nature* 410: 940–944

2 Pons J, Kitlinska J, Ji H, Lee EW, Zukowska Z (2003) Mitogenic actions of neuropeptide Y in vascular smooth muscle cells: synergetic interactions with the beta-adrenergic system. *Can J Physiol Pharmacol* 81: 177–185

3 Lee EW, Michalkiewicz M, Kitlinska J, Kalezic I, Switalska H, Yoo P, Sangkharat A, Ji H, Li L, Michalkiewicz T et al. (2003) Neuropeptide Y induces ischemic angiogenesis and restores function of ischemic skeletal muscles. *J Clin Invest* 111: 1853–1862

4 Zukowska-Grojec Z, Karwatowska-Prokopczuk E, Rose W, Rone J, Movafagh S, Ji H, Yeh Y, Chen WT, Kleinman HK, Grouzmann E et al (1998) Neuropeptide Y: a novel angiogenic factor from the sympathetic nerves and endothelium. *Circ Res* 83: 187–195

5 Kitlinska J, Lee EW, Movafagh S, Pons J, Zukowska Z (2002) Neuropeptide Y-induced angiogenesis in aging. Peptides 23: 71–77

6 Kitlinska J, Abe K, Kuo L, Pons J, Yu M, Li L, Tilan J, Everhart L, Lee EW, Zukowska Z et al (2005) Differential effects of neuropeptide Y on the growth and vascularization of neural crest-derived tumors. *Cancer Res* 65: 1719–1728

7 Pruszczyk P, Wocial B, Ignatowska-Switalska H, Feltynowski T, Ellafi M, Januszewicz A, Lapinski M, Zukowska-Grojec Z, Januszewicz W (1995) Does plasma neuropeptide-Y immunoreactivity in patients with pheochromocytoma depend on hormonal activity of the tumor? *Clin Chim Acta* 243: 205–212

8 Grouzmann E, Comoy E, Bohuon C (1989) Plasma neuropeptide Y concentrations in patients with neuroendocrine tumors. *J Clin Endocrinol Metab* 68: 808–813

9 Helman LJ, Cohen PS, Averbuch SD, Cooper MJ, Keiser HR, Israel MA (1989) Neu-

ropeptide Y expression distinguishes malignant from benign pheochromocytoma. *J Clin Oncol* 7: 1720–1725

10 Dotsch J, Christiansen H, Hanze J, Lampert F, Rascher W (1998) Plasma neuropeptide Y of children with neuroblastoma in relation to stage, age and prognosis, and tissue neuropeptide Y. *Regul Pept* 75–76: 185–190

11 O'Hare MM, Schwartz TW (1989) Expression and precursor processing of neuropeptide Y in human and murine neuroblastoma and pheochromocytoma cell lines. *Cancer Res* 49: 7015–7019

12 Biedler JL, Roffler-Tarlov S, Schachner M, Freedman LS (1978) Multiple neurotransmitter synthesis by human neuroblastoma cell lines and clones. *Cancer Res* 38: 3751–3757

13 Fruhwald MC, O'Dorisio MS, Cottingham SL, Qualman SJ, O'Dorisio TM (1998) Neuropeptides in developmental tumors of the central and peripheral nervous system. *Ann NY Acad Sci* 865: 420–426

14 Maris JM, Matthay KK (1999) Molecular biology of neuroblastoma. *J Clin Oncol* 17: 2264–2279

15 Beltinger C, Debatin KM (2001) Murine models for experimental therapy of pediatric solid tumors with poor prognosis. *Int J Cancer* 92: 313–318

16 Brodeur GM (2003) Neuroblastoma: biological insights into a clinical enigma. *Nat Rev Cancer* 3: 203–216

17 Shastry P, Basu A, Rajadhyaksha MS (2001) Neuroblastoma cell lines – a versatile *in vitro* model in neurobiology. *Int J Neurosci* 108: 109–126

18 Schwab M (1999) Human neuroblastoma: from basic science to clinical debut of cellular oncogenes. *Naturwissenschaften* 86: 71–78

19 Nakagawara A (1998) Molecular basis of spontaneous regression of neuroblastoma: role of neurotrophic signals and genetic abnormalities. *Hum Cell* 11: 115–124

20 Nakagawara A (1998) The NGF story and neuroblastoma. *Med Pediatr Oncol* 31: 113–115

21 Askin DF (2000) Neonatal cancer: a clinical perspective. *J Obstet Gynecol Neonatal Nurs* 29: 423–431

22 Papac RJ (1998) Spontaneous regression of cancer: possible mechanisms. *In Vivo* 12: 571–578

23 Nakagawara A (2001) Trk receptor tyrosine kinases: a bridge between cancer and neural development. *Cancer Lett* 169: 107–114

24 Nakagawara A, Brodeur GM (1997) Role of neurotrophins and their receptors in human neuroblastomas: a primary culture study. *Eur J Cancer* 33: 2050–2053

25 Brodeur GM, Nakagawara A, Yamashiro DJ, Ikegaki N, Liu XG, Azar CG, Lee CP, Evans AE (1997) Expression of TrkA, TrkB and TrkC in human neuroblastomas. *J Neurooncol* 31: 49–55

26 Westermann F, Schwab M (2002) Genetic parameters of neuroblastomas. *Cancer Lett* 184: 127–147

27 Meitar D, Crawford SE, Rademaker AW, Cohn SL (1996) Tumor angiogenesis correlates

with metastatic disease, N-myc amplification, and poor outcome in human neuroblastoma. *J Clin Oncol* 14: 405–414

28 Erdreich-Epstein A, Shimada H, Groshen S, Liu M, Metelitsa LS, Kim KS, Stins MF, Seeger RC, Durden DL (2000) Integrins alpha(v)beta3 and alpha(v)beta5 are expressed by endothelium of high-risk neuroblastoma and their inhibition is associated with increased endogenous ceramide. *Cancer Res* 60: 712–721

29 Rascher W, Kremens B, Wagner S, Feth F, Hunneman DH, Lang RE (1993) Serial measurements of neuropeptide Y in plasma for monitoring neuroblastoma in children. *J Pediatr* 122: 914–916

30 Cohen PS, Cooper MJ, Helman LJ, Thiele CJ, Seeger RC, Israel MA (1990) Neuropeptide Y expression in the developing adrenal gland and in childhood neuroblastoma tumors. *Cancer Res* 50: 6055–6061

31 Kogner P, Bjork O, Theodorsson E (1994) Plasma neuropeptide Y in healthy children: influence of age, anaesthesia and the establishment of an age-adjusted reference interval. *Acta Paediatr* 83: 423–427

32 Allen JM, Martin JB, Heinrich G (1987) Neuropeptide Y gene expression in PC12 cells and its regulation by nerve growth factor: a model for developmental regulation. *Brain Res* 427: 39–43

33 Li BS, Kramer PR, Zhao W, Ma W, Stenger DA, Zhang L (2000) Molecular cloning, expression, and characterization of rat homolog of human AP-2alpha that stimulates neuropeptide Y transcription activity in response to nerve growth factor. *Mol Endocrinol* 14: 837–847

34 Minth-Worby CA (1994) Transcriptional regulation of the human neuropeptide Y gene by nerve growth factor. *J Biol Chem* 269: 15460–15468

35 Hofliger MM, Castejon GL, Kiess W, Beck Sickinger AG (2003) Novel cell line selectively expressing neuropeptide Y-Y2 receptors. *J Recept Signal Transduct Res* 23: 351–360

36 Barnea A, Roberts J (2001) Induction of functional and morphological expression of neuropeptide Y (NPY) in cortical cultures by brain-derived neurotrophic factor (BDNF): evidence for a requirement for extracellular-regulated kinase (ERK)-dependent and ERK-independent mechanisms. *Brain Res* 919: 57–69

37 Donovan MJ, Lin MI, Wiegn P, Ringstedt T, Kraemer R, Hahn R, Wang S, Ibanez CF, Rafii S, Hempstead BL (2000) Brain derived neurotrophic factor is an endothelial cell survival factor required for intramyocardial vessel stabilization. *Development* 127: 4531–4540

38 Ekstrand AJ, Cao R, Bjorndahl M, Nystrom S, Jonsson-Rylander AC, Hassani H, Hallberg B, Nordlander M, Cao Y (2003) Deletion of neuropeptide Y (NPY) 2 receptor in mice results in blockage of NPY-induced angiogenesis and delayed wound healing. *Proc Natl Acad Sci USA* 100: 6033–6038

39 Lee EW, Grant DS, Movafagh S, Zukowska Z (2003) Impaired angiogenesis in neuropeptide Y (NPY)-Y2 receptor knockout mice. *Peptides* 24: 99–106

40 Klement G, Baruchel S, Rak J, Man S, Clark K, Hicklin DJ, Bohlen P, Kerbel RS (2000)

Continuous low-dose therapy with vinblastine and VEGF receptor-2 antibody induces sustained tumor regression without overt toxicity. *J Clin Invest* 105: R15–R24
41 Chlenski A, Liu S, Cohn SL (2003) The regulation of angiogenesis in neuroblastoma. *Cancer Lett* 197: 47–52
42 Kogner P, Bjork O, Theodorsson E (1993) Neuropeptide Y in neuroblastoma: increased concentration in metastasis, release during surgery, and characterization of plasma and tumor extracts. *Med Pediatr Oncol* 21: 317–322
43 Bjellerup P, Theodorsson E, Jornvall H, Kogner P (2000) Limited neuropeptide Y precursor processing in unfavourable metastatic neuroblastoma tumours. *Br J Cancer* 83: 171–176
44 Srinivasan B, Roque CH, Hempstead BL, Al-Ubaidi MR, Roque RS (2004) Microglia-derived pronerve growth factor promotes photoreceptor cell death via p75 neurotrophin receptor. *J Biol Chem* 279: 41839–41845
45 Rorie CJ, Thomas VD, Chen P, Pierce HH, O'Bryan JP, Weissman BE (2004) The Ews/Fli-1 fusion gene switches the differentiation program of neuroblastomas to Ewing sarcoma/peripheral primitive neuroectodermal tumors. *Cancer Res* 64: 1266–1277
46 Biedler JL, Helson L, Spengler BA (1973) Morphology and growth, tumorigenicity, and cytogenetics of human neuroblastoma cells in continuous culture. *Cancer Res* 33: 2643–2652
47 de Alava E, Gerald WL (2000) Molecular biology of the Ewing's sarcoma/primitive neuroectodermal tumor family. *J Clin Oncol* 18: 204–213
48 Kovar H (2003) Ewing tumor biology: perspectives for innovative treatment approaches. *Adv Exp Med Biol* 532: 27–37
49 Rodriguez-Galindo C, Spunt SL, Pappo AS (2003) Treatment of Ewing sarcoma family of tumors: current status and outlook for the future. *Med Pediatr Oncol* 40: 276–287
50 Burchill SA (2003) Ewing's sarcoma: diagnostic, prognostic, and therapeutic implications of molecular abnormalities. *J Clin Pathol* 56: 96–102
51 Whang-Peng J, Triche TJ, Knutsen T, Miser J, Kao-Shan S, Tsai S, Israel MA (1986) Cytogenetic characterization of selected small round cell tumors of childhood. *Cancer Genet Cytogenet* 21: 185–208
52 Chansky HA, Hu M, Hickstein DD, Yang, L (2001) Oncogenic TLS/ERG and EWS/Fli-1 fusion proteins inhibit RNA splicing mediated by YB-1 protein. *Cancer Res* 61: 3586–3590
53 Lessnick SL, Braun BS, Denny CT, May WA (1995) Multiple domains mediate transformation by the Ewing's sarcoma EWS/FLI-1 fusion gene. *Oncogene* 10: 423–431
54 van Valen F, Winkelmann W, Jurgens H (1992) Expression of functional Y1 receptors for neuropeptide Y in human Ewing's sarcoma cell lines. J *Cancer Res Clin Oncol* 118: 529–536
55 Langer M, Kratz F, Rothen-Rutishauser B, Wunderli-Allenspach H, Beck-Sickinger AG (2001) Novel peptide conjugates for tumor-specific chemotherapy. *J Med Chem* 44: 1341–1348
56 Shorter NA, Pence JC (1997) Retinoic acid-induced regulation of neuropeptide Y recep-

tor expression and function in the neuroepithelioma line SK-N-MC. *J Pediatr Surg* 32: 721–723

57 Reubi JC, Gugger M, Waser B, Schaer JC (2001) Y(1)-mediated effect of neuropeptide Y in cancer: breast carcinomas as targets. *Cancer Res* 61: 4636–4641

58 Mentlein R, Gallwitz B, Schmidt WE (1993) Dipeptidyl-peptidase IV hydrolyses gastric inhibitory polypeptide, glucagon-like peptide-1(7-36)amide, peptide histidine methionine and is responsible for their degradation in human serum. *Eur J Biochem* 214: 829–835

59 Mentlein R (1999) Dipeptidyl-peptidase IV (CD26) – role in the inactivation of regulatory peptides. *Regul Pept* 85: 9–24

60 O'Hare MM, Schwartz TW (1989) Expression and precursor processing of neuropeptide Y in human pheochromocytoma and neuroblastoma tumors. *Cancer Res* 49: 7010–7014

61 Takahashi K, Mouri T, Sone M, Murakami O, Itoi K, Ohneda M, Yoshinaga K (1990) Release of neuropeptide Y from pheochromocytomas. *Endocrinol Jpn* 37: 53–60

62 des Senanayake P, Denker J, Bravo EL, Graham RM (1995) Production, characterization, and expression of neuropeptide Y by human pheochromocytoma. *J Clin Invest* 96: 2503–2509

63 Eurin J, Barthelemy C, Masson F, Maistre G, Soualmia H, Noe E, Sarfati E, Eurin B, Carayon A (2000) Release of neuropeptide Y and hemodynamic changes during surgical removal of human pheochromocytomas. *Regul Pept* 86: 95–102

64 Bournat JC, Allen JM (2001) Regulation of the Y1 neuropeptide Y receptor gene expression in PC12 cells. *Brain Res Mol Brain Res* 90: 149–164

65 DiMaggio DA, Farah JM Jr, Westfall TC (1994) Effects of differentiation on neuropeptide-Y receptors and responses in rat pheochromocytoma cells. *Endocrinology* 134: 719–727

66 Middeke M, Hoffmann S, Hassan I, Wunderlich A, Hofbauer LC, Zielke A (2002) *In vitro* and *in vivo* angiogenesis in PC12 pheochromocytoma cells is mediated by vascular endothelial growth factor. *Exp Clin Endocrinol Diabetes* 110: 386–392

NPY family of peptides in endocrine, breast and prostate tumors

Massimiliano Ruscica, Elena Dozio, Marcella Motta and Paolo Magni

Istituto di Endocrinologia, via G. Balzaretti 9, 20133 Milano, Italy

Introduction

The peptides belonging to the neuropeptide Y (NPY) family (pancreatic polypeptide (PP), peptide YY (PYY) and NPY) have been involved in a wide range of physiological actions, including the control of energy metabolism and food intake, reproduction, neuroendocrine functions, like the secretion of growth hormone, and the regulation of cardiovascular and cognitive functions [1, 2]. Moreover, effects relevant to tumor progression have also been demonstrated for these peptides, specifically on cell proliferation, matrix invasion and metastatization, and angiogenesis. Proliferative or antiproliferative effects of NPY have been shown in different cell systems and appear mediated mainly through the NPY Y1 receptor (Y1-R), which has been involved in the control of the proliferation of vascular smooth muscle and endothelial cells [3, 4] and injured glial cells [5]. NPY has also been shown to stimulate the proliferation of hepatic stellate cells at physiological concentrations [6] and neuroproliferation in rodent olfactory epithelium in a dose dependent manner [7]. NPY released from sympathetic nerves also stimulated endothelial cell adhesion to matrix, migration, proliferation, capillary tube formation on matrigel, and aortic sprouting [8]. Recent evidence has extended the oncological relevance of NPY and related peptides to endocrine-related cancer, and namely to endocrine/neuroendocrine tumors and breast and prostate cancer. To this regard, it should be observed that, although NPY, PYY and PP are structurally closely related [9], most studies focused on the expression and the effects of the NPY system in tumors. For this reason, the information reported in this review mainly refers to NPY and the related receptors (NPY-Rs). The involvement of NPY and NPY-Rs in the progression of endocrine-related cancer may be related to the following conditions: a) expression of NPY-Rs in the context of the tumor, which thus becomes a target of extratumorally-produced NPY, secreted by normal structures (nerve fibers, neuroendocrine cells) proximal to the tumor. The presence of NPY-Rs may be targeted by specific

The NPY Family of Peptides in Immune Disorders, Inflammation, Angiogenesis and Cancer,
edited by Zofia Zukowska and Giora Z. Feuerstein

drugs with either agonist or antagonist activity or by radiolabeled compounds with diagnostic or therapeutical relevance; b) tumoral production of NPY, which may act locally in a paracrine fashion and/or reach the bloodstream and generate systemic effects (the tumor as a source of NPY and related peptides). In this case, the determination of NPY plasma levels may have the value of a diagnostic marker; c) combination of expression of NPY-Rs and production of NPY by the tumor: in this case, NPY may either promote or suppress cancer progression through paracrine mechanisms.

NPY family of peptides in endocrine tumors

The group of endocrine tumors includes both neuroendocrine tumors and epithelial tumors arising from endocrine structures. Neuroendocrine tumors are benign or malignant lesions originating from neuroendocrine structures, like the pituitary and the parafollicular cells of the thyroid, or from the so-called "diffuse neuroendocrine system", and therefore localized in different organs, such as the lung (small cell carcinoma) or the gastrointestinal tract (gastrinoma and other peptide-secreting tumors). They produce several peptides related to their tissue of origin. Regarding the possible role of the NPY family of peptides, it should be recalled that PP, together with chromogranin A (CgA), remains one of the best histopathological markers of gastroenteropancreatic tumors [10, 11]. Epithelial tumors of endocrine structures (carcinomas) may acquire a neuroendocrine phenotype at some point of their progression and thus express NPY-Rs or produce bioactive peptides, including NPY, PYY and PP. The available information on the role of the NPY family of peptides in pituitary tumors, phaeochromocytoma and neuroblastoma, adrenal gland and adrenal cortical lesions, ectopic ACTH-secreting tumors, and thymic carcinoid tumors are reported in the following section.

Pituitary tumors

NPY exerts direct effects in the control of pituitary hormone release, since it potentiates gonadotrophin, growth hormone and prolactin secretion [12]. NPY has also been shown to directly inhibit the secretion of growth hormone *in vitro* by human pituitary adenomas [13]. In a series of 135 cases, including pituitary adenoma, pituitary hyperplasia and normal pituitary, adenomas exhibited a significant increase of NPY immunoreactivity compared to nontumoral tissues (33% *versus* 12%). Interestingly, in this series, five of 23 null cell adenomas exhibited NPY immunoreactivity suggesting that some of these adenomas may produce hormones and may be more differentiated than expected. In this respect, NPY could be regarded as a marker of some null cell adenomas [14].

Pheochromocytoma

Pheochromocytomas are tumors of neuroectodermal origin arising from adrenal chromaffin cells. The tumor is usually unilateral, but about 10% of the tumors occur bilaterally, as part of the multiple endocrine neoplasia IIa syndrome (MENI-Ia). Malignant cases are often diagnosed only after the detection of metastatic lesions [15]. Various peptides stored in chromaffin granules have been identified in the adrenal medulla, including chromogranins, enkephalins and other opioid peptides, neurotensin, somatostatin, substance P (SP), vasoactive intestinal peptide (VIP), adrenomedullin, calcitonin gene-related peptide, vasopressin, and endothelin [15], as well as NPY [16]. In pheochromocytomas, these substances are over-secreted along with catecholamines [15]. The regulation of NPY secretion by these tumors is still not fully clarified. Grouzmann et al. demonstrated the potential role of angiotensin II in mediating noradrenaline (NA) and NPY release in a human pheochromocytoma [17]. NPY is frequently reported to be co-localized and co-released with NA in the symphathetic nervous system [18]. Bradykinin (BK) is well known to increase NA release from the adrenal medulla and it has recently been shown to regulate also NPY release by human pheochromocytoma tissue [19].

The estimation of urinary catecholamines and catecholamine metabolites is currently the principal method used in screening for pheochromocytoma, which is suspected when the total amount of urinary adrenaline and NA exceeds 100 µg or that of metanephrine and normetanephrine, 1 mg. In addition, the measurement of different neuroendocrine marker proteins like CgA, NPY and neuron-specific enolase has recently been introduced as diagnostic tools [20, 21]. Elevated plasma concentrations of CgA have been found in about 80% of patients with pheochromocytoma, whereas NPY levels were found increased in about 43–87% of cases [22], making these parameters useful markers for the diagnosis of pheochromocytoma and paraganglioma, a related tumor arising from parasympathetic ganglia. In parallel with high concentrations of circulating NPY immunoreactivity, in these tumors, especially in the malignant forms, there is also an increase in NPY mRNA levels [23]. The extent of NPY expression, however, does not always correlate with malignancy, and thus it is not a reliable marker for distinguishing malignant pheochromocytomas from benign ones [24]. This last report overturns some previous studies [25, 26], which suggest that, because most pheochromocytoma with strong NPY expression are well-differentiated tumors, the induction of NPY gene expression in pheochromocytoma cells might depend on their differentiation status. Pheochromocytoma is a potentially fatal disease and in most cases it can be cured by surgery. Its manipulation at surgical removal has been shown to produce a remarkable increase in NPY plasma level, which could play an important role especially on the vascular systemic resistance in both isolated pheochromocytomas [27], as well as in bilateral pheochromocytomas [28]. Therefore, NPY, secreted by this tumor into the bloodstream, may be at least in part responsible for the hypertensive symptoms, and its measurement may represent a useful diagnostic marker. Moreover, recent evidence shows that 35% of the 23 pheochromocytomas tested by receptor autoradiography, express Y1-R and Y2-R, also concomitantly with the NPY hormone [29], which might represent a biological cell marker for this tumor.

Neuroblastoma

Neuroblastomas, which more frequently affect small children, are tumors of neuroectodermal origin and usually express significant amounts of NPY. Therefore, similarly to pheochromocytomas, the measurement of plasma NPY has been proposed as a marker for the diagnosis and the follow-up of neuroblastoma in children [30, 31]. Studies conducted *in vitro* using the SH-SY5Y human neuroblastoma cell line, which expresses and secretes NPY, have shown that NPY gene expression and NPY release may be regulated by different agents, including cholinergic agonists [32], retinoic acid (RA) [33], and leptin [34]. RA, which is a well-known differentiating agent, seems to represent an overall downregulator of the NPY system in neuroblastoma, since it is also able to reduce the expression of the Y1-R [35]. In addition, NPY has been shown to inhibit the growth of the Y1-R expressing SK-N-MC neuroblastoma cell line [36]. NPY receptor autoradiography performed on neuroblastoma sections shows, however, that Y2-R is the only subtype detected *in vivo* [29]. Differentiation therapy with retinoids has been proposed for various malignant tumors, including neuroblastomas [37]. RA treatment not only induces cell differentiation, but also reduces the proliferation of various neuroblastoma-derived cell lines, including the SH-SY5Y, with reversion of the malignant phenotype. Neuroblastomas are normally very aggressive, but, in some cases, the tumors show a spontaneous evolution towards more benign forms (stage IVS), which is accompanied by an increased expression of the NPY gene. Taken together, these *in vivo* and *in vitro* observations underline the complexity of the interplay between differentiating agents, differentiation markers and proliferation factors, making difficult to correlate tumor aggressiveness and expression of markers like NPY.

Adrenal cortical lesions

The human adrenal cortex shows numerous nerve structures containing different vasoactive peptides, including NPY. In hyperplastic cortical tissue, where NPY immunoreactivity was observed in the form of bundles and fibres and mainly between or along blood vessels, fewer nerves containing these peptides were seen. A similar pattern of immunopositivity was also found in some adrenal adenomas and carcinomas [38]. The NPY-positive nerve fibres observed between the parenchymal cells may thus directly affect the tumor cell growth by modulating various growth factors [39] and influencing, via vasoconstriction and vasodilation, the blood flow within the tumor [38]. Moreover, it has been shown that, in all cortical adenomas and in two of three carcinomas analyzed, Y1-R is the only receptor subtype detected [29].

Insulinoma

Insulinomas are pancreatic endocrine tumors (PETs), which are also called *islet tumors* or *islet cell tumors*; however, because the cell of origin of most is unknown, the general term PET is preferred. Except for insulinomas, PETs are frequently malignant. NPY has also been detected in one insulinoma, in a patient with multiple endocrine neoplasia I (MENI) and hyperparathyroidism [40].

Medullary thyroid carcinoma

Medullary thyroid carcinoma, a tumor arising from the parafollicular cells of the thyroid, is often associated with other tumors in the context of a MEN IIa syndrome, and has been found to express elevated amounts of NPY [28].

Ectopic ACTH-secreting tumor

Ectopic adreno-corticotrophin-hormone (ACTH) secretion from tumors, resulting in hypersecretion of glucocorticoids and Cushing's syndrome, is the most frequent and well studied of ectopic syndromes. The hormone production from these tumors is important because it is a major cause of morbidity and death in cancer patients, and their therapy can be challenging. Cases reported in the literature [41, 42] show, in a total of four ectopic ACTH-secreting tumors (two bronchial carcinoid and two neuroendocrine tumor of thymus), the presence of some neuropeptides (endothelin, adrenomedullin and corticotropin-releasing hormone) associated with NPY, suggesting a possible pathophysiological role of these molecules in this tumor [42]. In another case with a thymic carcinoid [43] with combined ectopic ACTH and growth-hormone-releasing-hormone (GHRH) syndrome, the analysis of biochemical markers revealed elevated levels of plasma CgA and total daily excretion of urinary 5-hydroxyindolacetic acid (5-HIAA) as well as plasma NPY. After surgery, these markers were further suppressed, and CgA and daily urinary 5-HIAA, but not plasma NPY, were normalized. Moreover, immunocytochemical analysis of consecutive sections of the tumor tissue indicated that most cells (80–90%) produced GHRH, ACTH and NPY and that these peptides were co-expressed. In conclusion, the possible autocrine and/or paracrine relationships of these responses to ectopic hormone secretion remain to be investigated [43].

NPY family of peptides in breast tumor

Breast carcinoma is one of the most frequent and harmful cancers. NPY-Rs expression has been found in as high as 85% of primary breast cancer in a series of 95 cases, and in 100% of lymph node metastases of receptor-positive primaries (n = 27) [36]. In these cases, the Y1-R predominated and was often expressed in high density and great homogeneity, whereas Y2-R was present only in 24% of the cases. In normal breast tissue, however, Y1-R was only found in a minority of the cases and concomitantly with Y2-R, which seemed to be predominant in non-neoplastic breast. The neoplastic condition of breast tissue may thus induce a switch of expression from receptor Y2-R to Y1-R. Therefore, both ductal and lobular breast cancers, of the *in situ* and invasive types, as well as all lymph node metastases, may express NPY receptors [36]. Breast tumors expressing gastrin-releasing peptide (GRP) receptor or Y1-R, or both, were found in 93% of the cases by *in vitro* receptor autoradiography and lymph node metastases showed a similar receptor profile to the corresponding primary breast tumor [44]. On the basis of these findings, it is suggested that the combination of radiolabeled GRP and NPY Y1-R analogs should allow targeting of breast carcinomas and their lymph node metastases for *in vivo* peptide receptor scintigraphy and radiotherapy [45]. In addition to breast cancer, the presence of NPY-Rs was also evaluated in ovarian tumors, and found to be present in all granulosa cell tumors studied, but only in 32% of a series of 22 ovarian adenocarcinomas. These receptors corresponded to either Y1-R or Y2-R or both [46]. Taken together, these findings indicate that NPY may play a role in the progression of breast cancer and suggest that targeting NPY-Rs with radiolabeled analogs in this tumor may have a diagnostic and therapeutic value.

NPY family of peptides in prostate tumor

Prostate cancer (PCa) represents one of the most common malignant diseases among men in the Western world, where there is a 10% chance of developing PCa and a 3/4% chance of dying of causes directly related to it [47]. PCa is initially androgen dependent, and it may later progress to the androgen independence stage, which is associated with a lack of efficacy of hormonal therapy, and it appears to be promoted, at least, in part by several growth factors and neurohormones. This neuroendocrine differentiation of PCa has been associated to a worse prognosis. The secretory products of neuroendocrine cells, which are present in 50% of PCa cases, contribute to the pathophysiology of PCa and to its progression to androgen independence [47, 48]. An additional component to these mechanisms is given by the specific neuroendocrine differentiation of cancer cells, which occurs with greater frequency in PCa than in other malignancies [49]. In the human prostate, particularly in the smooth muscle layer, NPY is mainly localized in nerve fibers and neuroendocrine cells [50, 51]. Moreover, immunopositivity for NPY has been shown in 75% of the PCa specimens obtained from a series of patients [52], suggesting a possible participation of this factor in PCa growth and progression. Within the prostate, NPY could be released by neuroendocrine cells or by nerve terminals to reach the NPY-Rs present on nearby epithelial cells of tumoral origin. In a recent study from our laboratory (unpublished observation), it was found that the genes coding for the NPY receptor isoforms (Y1-R, Y2-R, Y4-R and Y5-R) were differently expressed in three human PCa cell lines (the androgen dependent LNCaP and the androgen independent DU145 and PC3). Specifically, Y1-R was expressed in all cell lines, whereas NPY and Y5-R in none. The Y1-R protein was expressed in all PCa cell lines, as well as in PCa specimens obtained from surgery. Treatment with NPY at different concentrations resulted in a reduced proliferation of LNCaP and DU145 cells and in an increased proliferation of PC3 cells [53]. These effects were abolished by incubating the cells with the specific Y1-R antagonist BIBP3226. In DU145 and PC3 cells, such effects of NPY were associated with an increased phosphorylation of the mitogen activated protein kinase/extracellular signal-regulated protein kinases 1/2. The final effects of exogenous NPY on PCa cells, either proliferative or antiproliferative, may be the result of different factors, like the clone-specific pattern of expression of NPY-Rs, the coupling of these NPY-Rs to different intracellular pathways and the responsivity to androgens, all contributing to drive the NPY-related proliferative potential of each cell line in one or in the other direction. Further studies seem necessary to evaluate the role of the activation of NPY-Rs, particularly in PCa patients at different stages of the disease. Along with information on the pathophysiology of advanced PCa, these studies might also give indications about novel future lines for the treatment of this frequent disease.

Conclusions and perspectives

As mentioned in the present review, in the last several years, it was shown that the receptors of regulatory peptides, and specifically of NPY-related peptides, are not only present in normal human tissue, but are also overexpressed in a large variety of human neoplasms [54], highlighting the diagnostic and therapeutical relevance of their evaluation. Neuropeptide receptors are potential targets for imaging and radiotherapy by radiolabeled peptide hormone analogues [55]. For example, the high level of expression of somatostatin receptors on various tumor cells has already provided the molecular basis for successful use of radiolabeled somatostatin analogues as tumor tracers in nuclear medicine [56]. Similarly, the evaluation of the

NPY system with a similar aim may result of clinical utility [36, 44, 57]. In addition to radiolabeled analogues for peptides such as somatostatin, cholecystokinin (CCK), gastrin, bombesin, substance P, and vasoactive intestinal peptide (VIP), analogues for NPY-related peptides have also been introduced [58, 59]. NPY analogues for the Y1-Rs, which are overexpressed in prostate and breast cancer, are currently under development and in different phases of (pre) clinical investigation. Moreover, multireceptor tumor targeting using the combination of bombesin and NPY Y1-R analogues is promising for scintigraphy and peptide receptor radionuclide therapy of breast carcinomas and their lymph node metastases [60].

In conclusion, the current evidence indicates that the NPY-family of peptides may play an important role in the progression of neuroendocrine tumors, and breast and prostate cancer. Future studies will better clarify some aspects of the involvement of these neuropeptides in tumor biology, as well as indicate novel diagnostic markers and therapeutical approaches.

References

1 Pedrazzini T, Pralong F, Grouzmann E (2003) Neuropeptide Y: the universal soldier. *Cell Mol Life Sci* 60: 350–377

2 Magni P (2003) Hormonal control of the neuropeptide Y system. *Curr Protein Pept Sci* 4: 45–57

3 Erlinge D, Brunkwall J, Edvinsson L (1994) Neuropeptide Y stimulates proliferation of human vascular smooth muscle cells: cooperation with noradrenaline and ATP. *Regul Pept* 50: 259–265

4 Zukowska-Grojec Z, Karwatowska-Prokopczuk E, Rose W, Rone J, Movafagh S, Ji H, Yeh Y, Chen WT, Kleinman HK, Grouzmann E et al (1998) Neuropeptide Y: a novel angiogenic factor from the sympathetic nerves and endothelium. *Circ Res* 83: 187–195

5 Canto Soler MV, Gallo JE, Dodds RA, Hokfelt T, Villar MJ, Suburo AM (2002) Y1 receptor of neuropeptide Y as a glial marker in proliferative vitreoretinopathy and diseased human retina. *Glia* 39: 320–324

6 Oben JA, Yang S, Lin H, Ono M, Diehl AM (2003) Norepinephrine and neuropeptide Y promote proliferation and collagen gene expression of hepatic myofibroblastic stellate cells. *Biochem Biophys Res Commun* 302: 685–690

7 Hansel DE, Eipper BA, Ronnett GV (2001) Neuropeptide Y functions as a neuroproliferative factor. *Nature* 410: 940–944

8 Zukowska Z, Grant DS, Lee EW (2003) Neuropeptide Y: a novel mechanism for ischemic angiogenesis. *Trends Cardiovasc Med* 13: 86–92

9 Tatemoto K, Carlquist M, Mutt V (1982) Neuropeptide Y-a novel brain peptide with structural similarities to peptide YY and pancreatic polypeptide. *Nature* 313: 404–406

10 Yazawa K, Kuroda T, Watanabe H, Shimozawa N, Nimura Y, Nakata S, Fujimori Y,

Koide N, Koike S, Kajikawa S et al (1998) Multiple carcinoids of the duodenum accompanied by type I familial multiple endocrine neoplasia. *Surg Today* 28: 636–639

11 Gurevich L, Kazantseva I, Isakov VA, Korsakova N, Egorov A, Kubishkin V, Bulgakov G (2003) The analysis of immunophenotype of gastrin-producing tumors of the pancreas and gastrointestinal tract. *Cancer* 98: 1967–1976

12 Chabot JG, Enjalbert A, Pelletier G, Dubois PM, Morel G (1988) Evidence for a direct action of neuropeptide Y in the rat pituitary gland. *Neuroendocrinology* 47: 511–517

13 Adams EF, Venetikou MS, Woods CA, Lacoumenta S, Burrin JM (1987) Neuropeptide Y directly inhibits growth hormone secretion by human pituitary somatotropic tumours. *Acta Endocrinol (Copenh)* 115: 149–154

14 Grouzmann E, Deruaz JP, Gomez F, Waeber G (1998) Immunolocalization of neuropeptide Y in human pituitary tumours. *Regul Pept* 75–76: 89–92

15 Nakao K, Itoh H, Takaya K (2000) Current topics in pheochromocytoma. *Biomed Pharmacother* 54 (Suppl 1): 124–128

16 deS Senanayake, Denker J, Bravo EL, Graham RM (1995) Production, characterization, and expression of neuropeptide Y by human pheochromocytoma. *J Clin Invest* 96: 2503–2509

17 Grouzmann E, Werffeli-George P, Fathi M, Burnier M, Waeber B, Waeber G (1994) Angiotensin-II mediates norepinephrine and neuropeptide-Y secretion in a human pheochromocytoma. *J Clin Endocrinol Metab* 79: 1852–1856

18 Lundberg JM (1996) Pharmacology of cotransmission in the autonomic nervous system: integrative aspects on amines, neuropeptides, adenosine triphosphate, amino acids and nitric oxide. *Pharmacol Rev* 48: 113–178

19 Eurin J, Barthelemy C, Masson F, Soualmia H, Sarfati E, Carayon A (2002) Bradykinin-induced neuropeptide Y release by human pheochromocytoma tissue. *Neuropeptides* 36: 257–262

20 O'Connor DT, Deftos LJ (1986) Secretion of chromogranin A by peptide-producing endocrine neoplasms. *N Engl J Med* 314: 1145–1151

21 Boomsma F, Bhaggoe UM, Man in 't Veld AJ, Schalekamp MA (1995) Sensitivity and specificity of a new ELISA method for determination of chromogranin A in the diagnosis of pheochromocytoma and neuroblastoma. *Clin Chim Acta* 239: 57–63

22 Tabarin A, Minot AP, Dallochio M, Roger P, Ducassou D (1992) Plasma concentration of neuropeptide Y in patients with adrenal hypertension. *Regul Pept* 42: 51–61

23 Senanayake P, Denker J, Bravo EL, Graham RM (1995) Production, characterization, and expression of neuropeptide Y by human pheochromocytoma. *J Clin Invest* 96: 2503–2509

24 Liu J, Kahri AI, Heikkila P, Voutilainen R (1999) Regulation of neuropeptide Y mRNA expression in cultured human pheochromocytoma cells. *Eur J Endocrinol* 141: 431–435

25 Osamura RY, Tsutsumi Y, Yanaihara N, Imura H, Watanabe K (1987) Immunohistochemical studies for multiple peptide-immunoreactivities and co-localization of Met-enkephalin-Arg6-Gly7-Leu8, neuropeptide Y and somatostatin in human adrenal medulla and pheochromocytomas. *Peptides* 8: 77–87

26 Helman LJ, Cohen PS, Averbuch SD, Cooper MJ, Keiser HR, Israel MA (1989) Neuropeptide Y expression distinguishes malignant from benign pheochromocytoma. *J Clin Oncol* 7: 1720–1725
27 Eurin J, Barthelemy C, Masson F, Maistre G, Soualmia H, Noe E, Sarfati E, Eurin B, Carayon A (2000) Release of neuropeptide Y and hemodynamic changes during surgical removal of human pheochromocytomas. *Regul Pept* 86: 95–102
28 Connell JM, Corder R, Asbury J, Macpherson S, Inglis GC, Lowry P, Burt AD, Semple PF (1987) Neuropeptide Y in multiple endocrine neoplasia: release during surgery for phaeochromocytoma. *Clin Endocrinol (Oxf)* 26: 75–84
29 Korner M, Waser B, Reubi JC (2004) High expression of neuropeptide y receptors in tumors of the human adrenal gland and extra-adrenal paraganglia. *Clin Cancer Res* 10: 8426–8433
30 Kogner P, Bjork O, Theodorsson E (1993) Neuropeptide Y in neuroblastoma: increased concentration in metastasis, release during surgery, and characterization of plasma and tumor extracts. *Med Pediatr Oncol* 21: 317–322
31 O'Dorisio MS, Hauger M, O'Dorisio TM (2002) Age-dependent levels of plasma neuropeptides in normal children. *Regul Pept* 109: 189–192
32 Magni P, Maggi R, Pimpinelli F, Motta M (1998) Cholinergic muscarinic mechanisms regulate neuropeptide Y gene expression via protein kinase C in human neuroblastoma cells. *Brain Res* 798: 75–82
33 Magni P, Beretta E, Scaccianoce E, Motta M (2000) Retinoic acid negatively regulates neuropeptide Y expression in human neuroblastoma cells. *Neuropharmacology* 39: 1627–1635
34 Magni P, Vettor R, Pagano C, Calcagno A, Martini L, Motta M (2001) Control of the expression of human neuropeptide Y by leptin: *in vitro* studies. *Peptides* 22: 415–420
35 Mannon PJ, Kaiser LM (1997) Retinoic acid is a negative regulator of the neuropeptide Y/peptide YY Y1 receptor gene in SK-N-MC cells. *J Neurochem* 68: 20–25
36 Reubi JC, Gugger M, Waser B, Schaer JC (2001) Y1-Mediated effect of neuropeptide Y in cancer: Breast carcinomas as targets. *Cancer Res* 61: 4636–4641
37 Finklestein JZ, Krailo MD, Lenarsky C, Ladisch S, Blair J, Reynolds CP, Sitarz AL, Denmann-Hammon G (1992) 13-cis retinoic acid (NSC 122758) in the treatment of children with metastatic neuroblastoma unresponsive to conventional chemotherapy: report from Children's Cancer Study Group. *Med Pediatr Onc* 20: 307–311
38 Li Q, Johansson H, Grimelius L (1999) Innervation of human adrenal gland and adrenal cortical lesions. *Virchows Arch* 435: 580–589
39 Voutilainen R (1998) Adrenocortical cells are the site of secretion and action of insulin-like growth factors and TNF-alpha. *Horm Metab Res* 30: 432–435
40 Waeber G, Gomez F, Bishof-Delaloye A, Chaubert P, Francke ML, Haefliger JA, Scherrer U, Centeno G, Temler E, Nicod P (1997) Insulinoma associated with a case of multiple endocrine neoplasia type I: Functional somatostatin receptors and abnormal glucose-induced insulin secretion. *Horm Res* 48: 76–82
41 Murakami O, Takahashi K, Sone M, Totsune K, Ohneda M, Itoi K, Yoshinaga K, Mouri

T (1993) An ACTH-secreting bronchial carcinoid: presence of corticotropin-releasing hormone, neuropeptide Y and endothelin-1 in the tumor tissue. *Acta Endocrinol (Copenh)* 128: 192–196

42 Murakami O, Takahashi K, Satoh F, Totsune K, Sone M, Arihara Z, Andoh N, Mouri T (1998) Expression of adrenomedullin and adrenomedullin mRNA in ectopic ACTH-secreting tumors. *Eur J Endocrinol* 138: 436–439

43 Jansson JO, Svensson J, Bangtsson BA, Frohman LA, Ahlman H, Wangberg B, Nilsson O, Nilsson M (1998) Acromegaly and Cushing's syndrome due to ectopic production of GHRH and ACTH by a thymic carcinoid tumour: *In vitro* responses to GHRH and GHRP-6. *Clin Endocrinol* 48: 243–250

44 Reubi JC, Gugger M, Waser B (2002) Co-expressed peptide receptors in breast cancer as a molecular basis for *in vivo* multireceptor tumour targeting. *Eur J Nucl Med* 29: 855–862

45 de Jong M, Kwekkeboom D, Valkema R, Krenning EP (2003) Radiolabelled peptides for tumour therapy: current status and future directions. Plenary lecture at the EANM 2002. *Eur J Nucl Med Mol Imaging* 30: 463–469

46 Korner M, Waser B, Reubi JC (2004) Neuropeptide Y receptor expression in human primary ovarian neoplasms. *Lab Invest* 84: 71–80

47 Abrahamsson PA (1999) Neuroendocrine cells in tumor growth of the prostate. *Endocr Relat Cancer* 6: 503–519

48 Cohen RJ, Glezerson G, Haffejee Z, Afrika D (1990) Prostatic carcinoma: histological and immunohistological factors affecting prognosis. *Br J Urol* 66: 405–410

49 Nagakawa O, Ogasawara M, Murata J, Fuse H, Saiki I (2001) Effect of prostatic neuropeptides on migration of prostate cancer cell lines. *Int J Urol* 8: 65–70

50 Tainio H (1995) Peptidergic innervation of the human prostate, seminal vesicle and vas deferens. *Acta Histochem* 97: 113–119

51 Ventura S, Pennefather JN, Mitchelson F (2002) Cholinergic innervation and function in the prostate gland. *Pharmacol Ther* 94: 93–112

52 Mack D, Hacker GW, Hauser-Kronberger C, Frick J, Dietze O (1997) Vasoactive intestinal polypeptide (VIP) and neuropeptide tyrosine (NPY) in prostate carcinoma. *Eur J Cancer* 33: 317–318

53 Magni P, Motta M (2001) Expression of neuropeptide Y receptors in human prostate cancer cells. *Ann Oncol* 12: S27–S29

54 Reubi JC (1997) Regulatory peptide receptors as molecular targets for cancer diagnosis and therapy. *Q J Nucl Med* 41: 63–70

55 Kwekkeboom DJ, Bakker WH, Kam BL, Teunissen JJ, Kooij PP, de Herder WW, Feelders RA, van Eijck CH, de Jong M, Srinivasan A et al (2003) Treatment of patients with gastro-entero-pancreatic (GEP) tumours with the novel radiolabelled somatostatin analogue [177Lu-DOTA(0),Tyr3]octreotate. *Eur J Nucl Med Mol Imaging* 30: 417–422

56 Blum J, Handmaker H, Lister-James J, Rinne N (2000) A multicenter trial with a somatostatin analog (99m)Tc depreotide in the evaluation of solitary pulmonary nodules. *Chest* 117: 1232–1238

57 Reubi JC (2004) Somatostatin and other Peptide receptors as tools for tumor diagnosis and treatment. *Neuroendocrinology* 80 (Suppl 1): 51–56

58 Virgolini I, Kurtaran A, Raderer M, Leimer M, Angelberger P, Havlik E, Li S, Scheithauer W, Niederle B, Valent P et al (1995) Vasoactive intestinal peptide receptor scintigraphy. *J Nucl Med* 36: 1732–1739

59 Weiner RE, Thakur ML (2002) Radiolabeled peptides in the diagnosis and therapy of oncological diseases. *Appl Radiat Isot* 57: 749–763

60 Krenning EP, Kwekkeboom DJ, Valkema R, Pauwels S, Kvols LK, De Jong M (2004) Peptide receptor radionuclide therapy. *Ann NY Acad Sci* 1014: 234–245

Summary

Future directions and therapeutic perspectives for NPY/PYY-based anti-inflammatory and tumor suppressor drugs

Zofia Zukowska[1] *and Giora Z. Feuerstein*[2]

[1]Department of Physiology and Biophysics Georgetown University Medical Center, 3900 Reservoir Rd., NW, Washington, DC 20057, USA; [2]Wyeth Research, Building A3153, 500 Arcola Rd, Collegeville, PA 19426, USA

The scientific areas that are presented and analyzed in this book lay the foundation for understanding the roles of this family of peptides in host-defense and adaptation to stress in its broadest sense – regulation of immune and other basic cellular functions leading to tissue repair and regeneration, in physiology and pathology. Stress conditions necessitate adjustments at multiple levels, from cells to tissues to a whole organism's integrated response. These, in turn, require communication between the neural, endocrine and immune systems, to achieve proper changes in functions of specific cells, supply of nutrients (blood vessels) and tissue's performance. The neurogenic NPY system appears to be a major stress and immune mediator positioned up-stream of many of these signaling cascades, and complemented by its endocrine/exocrine arm of gut-derived peptides, PYY and PP.

As discussed in chapters by Bedoui and von Hoersten, and de la Fuente and Puerto, NPY/PYY peptides are potent activators of immune functions, such as lymphocyte proliferation, monocyte migration and phagocytic cell activation. These activities often occur at concentrations well below those that elicit the known actions in neurotransmission and vasoconstriction, which suggests that the primary function of these peptides in physiology may be that of a cytokine and a growth factor rather than that of a neurotransmitter. The above mentioned studies indicate that NPY and its Y1 receptors may be pathogenic factors in immune aberrations occurring in autoimmune diseases such as systemic lupus erythematosus (SLE). In these conditions, as well as in inflammatory vascular remodeling occurring following vascular interventions and in atherosclerosis, as suggested by research from Zukowska's laboratory the 'stress factor' is prominent and mediated by NPY's Y1 receptors. Hence antagonists to Y1 receptors may become potent new drugs for treatment of these conditions.

Conversely, Y1 agonists offer promise as antinociceptive therapy, particularly in chronic inflammation but also in neuropathic pain, as discussed by Taylor. Interestingly, as presented in the chapter by Gray and Scharfman, the same compounds, when acting centrally in the hippocampus, may be beneficial in depression and

The NPY Family of Peptides in Immune Disorders, Inflammation, Angiogenesis and Cancer,
edited by Zofia Zukowska and Giora Z. Feuerstein

memory losses, by stimulating neurogenesis of neural stem cells. These studies open a completely new chapter of NPY's role in stem cell biology and neural cell plasticity, and may have far reaching implications for understanding brain development, function and its restoration after injury.

The other NPY receptors, of the Y2 type, also appear to have important pathophysiological functions. As discussed by Lee et al., and Bjorndahl et al., the Y2 receptor is the main type which mediates NPY's potent angiogenic and arteriogenic activity. The role of NPY in control of new vessel formation, as in the immune functions, appears to be upstream of many other mediators. This may be the reason why the peptide is able to elicit a coordinated activation of multiple cells and factors, leading to formation of more normal vessels than the ones formed by activation of a single angiogenic factor, e.g., vascular endothelial growth factor. An important auxiliary component of NPY's angiogenic activity is activation of Y5 receptors as well as dipeptidyl peptidase IV (DPPIV), an enzyme that forms NPY's Y2/Y5-selective agonist. Thus, Y2/Y5 agonists may become useful in revascularization therapies in ischemic heart and peripheral vascular diseases. The same type of drugs, as pointed out by studies of Lacroix et al., are potentially beneficial in the treatment of allergic rhinitis and asthma, due to the ability of Y2 agonists to reduce cholinergic vasodilation, produce a long-lasting vasoconstriction, and decrease mucosal secretion and edema.

In contrast, anti-angiogenic therapy with Y2/Y5-antagonists may be useful in treatment of retinopathy and tumors, particularly those derived from the neural crest. Kitlinska et al. have shown that in neuroblastomas, endogenous NPY stimulates tumor cell proliferation and angiogenesis, both via Y2/Y5 receptors. Thus, blocking the Y2/Y5 pathway may become promising bidirectional therapy in treatment of these childhood malignancies.

Finally, as studies by Grouzmann and Mentlein suggest, the NPY/PYY system can be manipulated not only by agonists or antagonists to specific receptors but also via drugs altering the activity of their processing enzymes. Of particular interest here are DPPIV inhibitors, which prevent conversion of these peptides into Y2/Y5-selective fragments and augment peptides' Y1-mediated activities such as vasoconstriction and vascular hypertrophy. These drugs are currently in phase III testing for the treatment of diabetes due to their effects on another peptide, GLP-1, but the consequences of their actions on NPY/PYY's activities may need to be considered too. While they might be beneficial in conditions where NPY-Y1 activity is desired, e.g. SLE and inflammatory pain, they could be detrimental in situations where either Y1 actions are pathogenic or increased Y2 receptor activation is required e.g. revascularization of ischemic tissues.

In spite of the fascinating research and promising potentials for drug development, the role of the NPY family of peptides and their receptors in the fields of immune/inflammatory and tissue remodeling/proliferative disorders has not received adequate attention. Novel growth factor and immunoregulatory functions of NPY, which have been recently discovered, created many new possibilities of therapies with selective receptor agonists and antagonists. However, the diversity of the peptide functions can also contribute to a variety of side effects, which need to be taken into consideration when designing NPY-based treatments. Also, large gaps of knowledge exist about the specific roles of other members of the NPY family, peptide YY and pancreatic polypeptide. These family members share many features with NPY, but also may possess unique biological and pathophysiological dimensions, which need to be researched in depth. Paramount in such future research is the role of the NPY family members and the various NPY receptors in stress, where the peptides and/or their receptors are manipulated to augment or diminish their function in a cell/tissue selective fashion. Redundancies of functions between the NPY peptides and their receptors in reference to their ability to launch effective adaptation responses are still largely unknown. Research of these unexplored frontiers is likely to be critical to efforts in drug development since the key player(s) in each disease process has yet to be identified.

Index

The PIR-Series
Progress in Inflammation Research

Homepage: http://www.birkhauser.ch

Up-to-date information on the latest developments in the pathology, mechanisms and therapy of inflammatory disease are provided in this monograph series. Areas covered include vascular responses, skin inflammation, pain, neuroinflammation, arthritis cartilage and bone, airways inflammation and asthma, allergy, cytokines and inflammatory mediators, cell signalling, and recent advances in drug therapy. Each volume is edited by acknowledged experts providing succinct overviews on specific topics intended to inform and explain. The series is of interest to academic and industrial biomedical researchers, drug development personnel and rheumatologists, allergists, pathologists, dermatologists and other clinicians requiring regular scientific updates.

Available volumes:

T Cells in Arthritis, P. Miossec, W. van den Berg, G. Firestein (Editors), 1998
Chemokines and Skin, E. Kownatzki, J. Norgauer (Editors), 1998
Medicinal Fatty Acids, J. Kremer (Editor), 1998
Inducible Enzymes in the Inflammatory Response,
D.A. Willoughby, A. Tomlinson (Editors), 1999
Cytokines in Severe Sepsis and Septic Shock, H. Redl, G. Schlag (Editors), 1999
Fatty Acids and Inflammatory Skin Diseases, J.-M. Schröder (Editor), 1999
Immunomodulatory Agents from Plants, H. Wagner (Editor), 1999
Cytokines and Pain, L. Watkins, S. Maier (Editors), 1999
In Vivo *Models of Inflammation*, D. Morgan, L. Marshall (Editors), 1999
Pain and Neurogenic Inflammation, S.D. Brain, P. Moore (Editors), 1999
Anti-Inflammatory Drugs in Asthma, A.P. Sampson, M.K. Church (Editors), 1999
Novel Inhibitors of Leukotrienes, G. Folco, B. Samuelsson, R.C. Murphy (Editors), 1999
Vascular Adhesion Molecules and Inflammation, J.D. Pearson (Editor), 1999
Metalloproteinases as Targets for Anti-Inflammatory Drugs,
K.M.K. Bottomley, D. Bradshaw, J.S. Nixon (Editors), 1999
Free Radicals and Inflammation, P.G. Winyard, D.R. Blake, C.H. Evans (Editors), 1999
Gene Therapy in Inflammatory Diseases, C.H. Evans, P. Robbins (Editors), 2000
New Cytokines as Potential Drugs, S. K. Narula, R. Coffmann (Editors), 2000
High Throughput Screening for Novel Anti-inflammatories, M. Kahn (Editor), 2000
Immunology and Drug Therapy of Atopic Skin Diseases,
C.A.F. Bruijnzeel-Komen, E.F. Knol (Editors), 2000
Novel Cytokine Inhibitors, G.A. Higgs, B. Henderson (Editors), 2000
Inflammatory Processes. Molecular Mechanisms and Therapeutic Opportunities,
L.G. Letts, D.W. Morgan (Editors), 2000

Cellular Mechanisms in Airways Inflammation, C. Page, K. Banner, D. Spina (Editors), 2000
Inflammatory and Infectious Basis of Atherosclerosis, J.L. Mehta (Editor), 2001
Muscarinic Receptors in Airways Diseases, J. Zaagsma, H. Meurs, A.F. Roffel (Editors), 2001
TGF-β and Related Cytokines in Inflammation, S.N. Breit, S. Wahl (Editors), 2001
Nitric Oxide and Inflammation, D. Salvemini, T.R. Billiar, Y. Vodovotz (Editors), 2001
Neuroinflammatory Mechanisms in Alzheimer's Disease. Basic and Clinical Research, J. Rogers (Editor), 2001
Disease-modifying Therapy in Vasculitides, C.G.M. Kallenberg, J.W. Cohen Tervaert (Editors), 2001
Inflammation and Stroke, G.Z. Feuerstein (Editor), 2001
NMDA Antagonists as Potential Analgesic Drugs, D.J.S. Sirinathsinghji, R.G. Hill (Editors), 2002
Migraine: A Neuroinflammatory Disease? E.L.H. Spierings, M. Sanchez del Rio (Editors), 2002
Mechanisms and Mediators of Neuropathic pain, A.B. Malmberg, S.R. Chaplan (Editors), 2002
Bone Morphogenetic Proteins. From Laboratory to Clinical Practice, S. Vukicevic, K.T. Sampath (Editors), 2002
The Hereditary Basis of Allergic Diseases, J. Holloway, S. Holgate (Editors), 2002
Inflammation and Cardiac Diseases, G.Z. Feuerstein, P. Libby, D.L. Mann (Editors), 2003
Mind over Matter – Regulation of Peripheral Inflammation by the CNS, M. Schäfer, C. Stein (Editors), 2003
Heat Shock Proteins and Inflammation, W. van Eden (Editor), 2003
Pharmacotherapy of Gastrointestinal Inflammation, A. Guglietta (Editor), 2004
Arachidonate Remodeling and Inflammation, A.N. Fonteh, R.L. Wykle (Editors), 2004
Recent Advances in Pathophysiology of COPD, P.J. Barnes, T.T. Hansel (Editors), 2004
Cytokines and Joint Injury, W.B. van den Berg, P. Miossec (Editors), 2004
Cancer and Inflammation, D.W. Morgan, U. Forssmann, M.T. Nakada (Editors), 2004
Bone Morphogenetic Proteins: Bone Regeneration and Beyond, S. Vukicevic, K.T. Sampath (Editors), 2004
Antibiotics as Anti-Inflammatory and Immunomodulatory Agents, B.K. Rubin, J. Tamaoki (Editors), 2005
Antirheumatic Therapy: Actions and Outcomes, R.O. Day, D.E. Furst, P.L.C.M. van Riel, B. Bresnihan (Editors), 2005
Regulatory T-Cells in Inflammation, L. Taams, A.N. Akbar, M.H.M Wauben (Editors), 2005
Sodium Channels, Pain, and Analgesia, K. Coward, M. Baker (Editors), 2005
Turning up the Heat on Pain: TRPV1 Receptors in Pain and Inflammation, A.B Malmberg, K.R. Bley (Editors), 2005